IDENTIFICATION AND CONTROL IN SYSTEMS GOVERNED BY PARTIAL DIFFERENTIAL EQUATIONS

SIAM PROCEEDINGS SERIES LIST

Neustadt, L.W., Proceedings of the First International Congress on Programming and Control (1966)

Hull, T.E., Studies in Optimization (1970)

Day, R.H. & Robinson, S.M., Mathematical Topics in Economic Theory and Computation (1972)

Proschan, F. & Serfling, R.J., Reliability and Biometry: Statistical Analysis of Lifelength (1974)

Barlow, R.E., Reliability & Fault Tree Analysis: Theoretical & Applied Aspects of System Reliability & Safety Assessment (1975)

Fussell, J.B. & Burdick, G.R., Nuclear Systems Reliability Engineering and Risk Assessment (1977)

Duff, I.S. & Stewart, G.W., Sparse Matrix Proceedings 1978 (1979)

Holmes, P.J., New Approaches to Nonlinear Problems in Dynamics (1980)

Erisman, A.M., Neves, K.W. & Dwarakanath, M.H., Electric Power Problems: The Mathematical Challenge (1981)

Bednar, J.B., Redner, R., Robinson, E. & Weglein, A., Conference on Inverse Scattering: Theory and Application (1983)

Voigt, R.G., Gottlieb, D. & Hussaini, M. Yousuff, Spectral Methods for Partial Differential Equations (1984)

Chandra, Jagdish, Chaos in Nonlinear Dynamical Systems (1984)

Santosa, Fadil, Symes, William W., Pao, Yih-Hsing & Holland, Charles, Inverse Problems of Acoustic and Elastic Waves (1984)

Gross, Kenneth I., Mathematical Methods in Energy Research (1984)

Babuska, I., Chandra, J. & Flaherty, J., Adaptive Computational Methods for Partial Differential Equations (1984)

Boggs, Paul T., Byrd, Richard H. & Schnabel, Robert B., Numerical Optimization 1984 (1985)

Angrand, F., Dervieux, A., Desideri, J.A. & Glowinski, R., Numerical Methods for Euler Equations of Fluid Dynamics (1985)

Wouk, Arthur, New Computing Environments: Parallel, Vector and Systolic (1986)

Fitzgibbon, William E., Mathematical and Computational Methods in Seismic Exploration and Reservoir Modeling (1986)

Drew, Donald A. & Flaherty, Joseph E., Mathematics Applied to Fluid Mechanics and Stability: Proceedings of a Conference Dedicated to R.C. DiPrima (1986)

Heath, Michael T., Hypercube Multiprocessors 1986 (1986)

Papanicolaou, George, Advances in Multiphase Flow and Related Problems (1987)

Wouk, Arthur, New Computing Environments: Microcomputers in Large-Scale Computing (1987)

Chandra, Jagdish & Srivastav, Ram, Constitutive Models of Deformation (1987)

Heath, Michael T., Hypercube Multiprocessors 1987 (1987)

Glowinski, R., Golub, G.H., Meurant, G.A. & Periaux, J., First International Conference on Domain Decomposition Methods for Partial Differential Equations (1988)

Salam, Fathi M.A. & Levi, Mark L., Dynamical Systems Approaches to Nonlinear Problems in Systems and Circuits (1988)

Datta, B., Johnson, C., Kaashoek, M., Plemmons, R. & Sontag, E., Linear Algebra in Signals, Systems and Control (1988)

Ringeisen, Richard D. & Roberts, Fred S., Applications of Discrete Mathematics (1988)

McKenna, James & Temam, Roger, ICIAM '87: Proceedings of the First International Conference on Industrial and Applied Mathematics (1988)

Rodrigue, Garry, Parallel Processing for Scientific Computing (1989)

Chan, Tony F., Meurant, Gérard, Périaux, Jacques & Widlund, Olof B., Domain Decomposition Methods (1989)

Caflish, Russel E., Mathematical Aspects of Vortex Dynamics (1989)

Wouk, Arthur, Parallel Processing and Medium-Scale Multiprocessors (1989)

Flaherty, Joseph E., Paslow, Pamela J., Shephard, Mark S. & Vasilakis, John D., Adaptive Methods for Partial Differential Equations (1989)

Kohn, Robert V. & Milton, Graeme W., Random Media and Composites (1989)

Mandel, Jan, McCormick, S.F., Dendy, J.E., Jr., Farhat, Charbel, Lonsdale, Guy, Parter, Seymour V., Ruge, John W. & Stüben, Klaus, Proceedings of the Fourth Copper Mountain Conference on Multigrid Methods (1989)

Colton, David, Ewing, Richard & Rundell, William, Inverse Problems in Partial Differential Equations (1990)

Chan, Tony F., Glowinski, Roland, Periaux, Jacques & Widlund, Olof B., Third International Symposium on Domain Decomposition Methods for Partial Differential Equations (1990)

Dongarra, Jack, Messina, Paul, Sorensen, Danny C. & Voigt, Robert G., Proceedings of the Fourth SIAM Conference on Parallel Processing for Scientific Computing (1990)

Glowinski, Roland & Lichnewsky, Alain, Computing Methods in Applied Sciences and Engineering (1990)

Coleman, Thomas F. & Li, Yuying, Large-Scale Numerical Optimization (1990)

Aggarwal, Alok, Borodin, Allan, Gabow, Harold, N., Galil, Zvi, Karp, Richard M., Kleitman, Daniel J., Odlyzko, Andrew M., Pulleyblank, William R., Tardos, Éva & Vishkin, Uzi, Proceedings of the Second Annual ACM-SIAM Symposium on Discrete Algorithms (1990)

Cohen, Gary, Halpern, Laurence & Joly, Patrick, Mathematical and Numerical Aspects of Wave Propagation Phenomena (1991)

Gómez, S., Hennart, J. P., & Tapia, R. A., Advances in Numerical Partial Differential Equations and Optimization: Proceedings of the Fifth Mexico-United States Workshop (1991)

Glowinski, Roland, Kuznetsov, Yuri A., Meurant, Gérard, Périaux, Jacques & Widlund, Olof B., Fourth International Symposium on Domain Decomposition Methods for Partial Differential Equations (1991)

Alavi, Y., Chung, F. R. K., Graham, R. L. & Hsu, D. F., Graph Theory, Combinatorics, Algorithms, and Applications (1991)

Wu, Julian J., Ting, T. C. T. & Barnett, David M., Modern Theory of Anisotropic Elasticity and Applications (1991)

Shearer, Michael, Viscous Profiles and Numerical Methods for Shock Waves (1991)

Griewank, Andreas & Corliss, George F., Automatic Differentiation of Algorithms: Theory, Implementation, and Application (1991)

Frederickson, Greg, Graham, Ron, Hochbaum, Dorit S., Johnson, Ellis, Kosaraju, S. Rao, Luby, Michael, Megiddo, Nimrod, Schieber, Baruch, Vaidya, Pravin, & Yao, Frances, Proceedings of the Third Annual ACM-SIAM Symposium on Discrete Algorithms (1992)

Field, David A. & Komkov, Vadim, Theoretical Aspects of Industrial Design (1992)

Field, David A. & Komkov, Vadim, Geometric Aspects of Industrial Design (1992)

Bednar, J. Bee, Lines, L. R., Stolt, R. H. & Weglein, A. B., Geophysical Inversion (1992)

O'Malley, Robert E. Jr., Proceedings of the Second International Conference on Industrial and Applied Mathematics (1992)

Keyes, David E., Chan, Tony F., Meurant, Gérard, Scroggs, Jeffrey S., & Voigt, Robert G., Fifth International Symposium on Domain Decomposition Methods for Partial Differential Equations (1992)

Dongarra, Jack, Messina, Paul, Kennedy, Ken, Sorensen, Danny C., & Voigt, Robert G., Proceedings of the Fifth SIAM Conference on Parallel Processing for Scientific Computing (1992)

Corones, James P., Kristensson, Gerhard, Nelson, Paul, & Seth, Daniel L., Invariant Imbedding and Inverse Problems

Sincovec, Richard F., Keyes, David E., Leuze, Michael R., Petzold, Linda R., & Reed, Daniel A., Proceedings of the Sixth SIAM Conference on Parallel Processing for Scientific Computing (1993)

Kleinman, Ralph, Angell, Thomas, Colton, David, Santosa, Fadil, & Stakgold, Ivar, Second International Conference on Mathematical and Numerical Aspects of Wave Propagation (1993)

Banks, H. T., Fabiano, R. H., and Ito, K., Identification and Control in Systems Governed by Partial Differential Equations (1993)

IDENTIFICATION AND CONTROL IN SYSTEMS GOVERNED BY PARTIAL DIFFERENTIAL EQUATIONS

Edited by H. T. Banks
North Carolina State University
R. H. Fabiano
Texas A & M University
K. Ito
North Carolina State University

siam.

Philadelphia

Society for Industrial and Applied Mathematics

IDENTIFICATION AND CONTROL IN SYSTEMS GOVERNED BY PARTIAL DIFFERENTIAL EQUATIONS

Proceedings of the conference on Control and Identification of Partial Differential Equations, part of the Summer Research Conferences in the Mathematical Sciences. Mount Holyoke College, South Hadley, Massachusetts, July 11-16, 1992.

This conference was sponsored by the American Mathematical Society, the Society for Industrial and Applied Mathematics, and the Institute of Mathematical Statistics and was supported in part by a grant from the National Science Foundation.

All rights reserved. Printed in the United States of America. No part of this book may be reproduced, stored, or transmitted without the permission of the Publisher. For information, write the Society for Industrial and Applied Mathematics, 3600 University City Science Center, Philadelphia, PA 19104-2688.

Copyright © 1993 by the Society for Industrial and Applied Mathematics.

siam. is a registered trademark.

PREFACE

The 1992 AMS-IMS-SIAM Joint Summer Research Conference on Control and Identification of Partial Differential Equations was held at Mount Holyoke College, July 11-16. It was an active and stimulating conference, with about 40 participants from the United States as well as Canada, France, and Austria. The objectives of this conference were to provide up-to-date information on recent developments and results in control, identification, and mathematical modeling of partial differential equations and to stimulate further research in the area of control and identification for distributed parameter systems. This volume, which provides a record of some of the presentations and discussions that took place during this meeting, reflects current trends in control and system identification for partial differential equations. In the following we summarize the contributions contained in this volume. The summaries are grouped according to research topics.

Mathematical Modeling

Mathematical models for piezoceramic actuators in smart material structures are presented by H. T. Banks and R. C. Smith. The models are developed for actuation via bending moments and in-plane forces induced by imbedded piezoceramic patches. Use of the control concepts is illustrated by feedback synthesis for an example from structural acoustics. H. T. Tran, J. F. Scroggs, and K. J. Bachmann consider mathematical models for flow dynamics in a vertical reactor for high pressure vapor transport of materials for compound semiconductors. Numerical simulations are performed to address design issues for the Scholz geometry of the reactor. In addition, various control problems related to the optimal reactor design are outlined.

Parameter Estimation

G. Crosta considers the problem of estimating spatially varying conductivity in the one dimensional heat equation. Injectivity and stability of the inverse of parameter-to-solution mappings are studied under various Cauchy data on the parameter. A statistical analysis of error distribution in the least squares parameter estimation problem is studied by B. G. Fitzpatrick and G. Yin. A bootstrap method for inference, including confidence intervals and tests of normality, is developed. M. Kroller and K. Kunisch develop a software based on MATLAB for the estimation of parameters in two point boundary value problems. This can be used to investigate how different discretizations and regularization terms affect the behavior of solutions to an ill-posed inverse problem. The problem of estimating flexural rigidity in a von Kármán plate equation using dynamical point-observations of deformations is discussed by L. White. The problem is formulated as a norm-constrained minimization problem in Hilbert spaces. A numerical approximation based on Galerkin approximations is developed, analyzed, and illustrated with numerical test examples.

Optimal Control

A. Bensoussan and P. Bernhard consider robust stabilization of infinite dimensional systems. The problem is formulated as a standard problem of H_∞-optimal control. The optimal feedback synthesis based on Riccati equations is derived. A shape optimization problem that arises in design of a forebody simulator for a free-jet engine test facility is studied by J. Borggard, J. Burns, E. Cliff, and M. Gunzberger. Sensitivity equations for the sate equation with respect to the design parameters are derived and used for accurate gradient calculations. A linear quadratic regulator problem for systems governed by second order elliptic equations with Neumann boundary control is studied by L. Ji and G. Chen. The cost functional is defined by the square of norm of tracking error at the sensored points on the boundary. Regularity results for the optimal state and numerical results based on the boundary element method are presented. J. A. Reneke considers a control algorithm based on the reproducing kernel Hilbert space method for hereditary systems where the system is described by the input-output covariance function. Numerical examples are given to illustrate the approach.

Feedback Stabilization

The problem of global exponential stabilization of a von Kármán plate by boundary velocity feedback is discussed by M. E. Bradley and I. Lasiecka. The energy multiplier method and microlocal analysis are used to obtain global exponential stability of the controlled dynamics. C. I. Byrnes, D. S. Gilliam, and V. I. Shubov consider a boundary control for Burgers' equation with flux control at one end. It is shown that the uncontrolled dynamics are not asymptotically stable. Global existence, stability, and compactness of solutions to controlled dynamics with velocity feedback are established. Well-posedness of feedback control systems described by the triple (A, B, C) on Hilbert spaces is discussed by K. A. Morris. It is shown that well-posed systems remain stable under bounded feedback control. H. Ozbay and J. Turi consider a class of control systems described by singular integro-differential equations. Existence of a finite dimensional stablizing compensator (output feedback control) is shown employing modern frequency domain techniques. Feedback control of a one-link flexible robot arm is discussed by T. J. Tarn, A. K. Bejczy, and C. Guo. The problem is formulated as a nonlinear distributed parameter control problem. A sampled output feedback control with periodic gain is proposed and analyzed for non-colocated sensor/actuator system dynamics.

The success of the conference was greatly enhanced by active participation of attendees, and the organizers/editors are most grateful to them for their contributions. Special thanks and appreciation are due Carole Kohanski, the conference coordinator for the American Mathematical Society, and the AMS staff for their help in all organization, preparation, and administrative matters. The conference was sponsored by the National Science Foundation through a grant to the AMS.

CONTENTS

1   Modeling of Flow Dynamics and Its Impact on the Optimal Reactor Design Problem
Hien T. Tran, Jeffrey S. Scroggs, and Klaus J. Bachmann

14   Sensitivity Calculations for a 2D, Inviscid, Supersonic Forebody Problem
Jeff Borggaard, John Burns, Eugene Cliff, and Max Gunzberger

26   Models for Control in Smart Material Structures
H.T. Banks and R.C. Smith

45   Bootstrap Methods for Inference in Least Squares Identification Problems
B.G. Fitzpatrick and G. Yin

59   MATLAB-Software for Parameter Estimation in Two-Point Boundary Value Problems
M. Kroller and K. Kunisch

69   Some Stability Estimates for the Identification of Conductivity in the One-Dimensional Heat Equation
Giovanni Crosta

87   Estimation of Material Parameters in a Dynamic Nonlinear Plate Model with Norm Constraints
L. W. White

101   Global Stabilization of a von Kármán Plate without Geometric Conditions
M. E. Bradley and I. Lasiecka

117   On the Standard Problem of H_∞ Optimal Control for Infinite Dimensional Systems
A. Bensoussan and P. Bernhard

141   Perturbation of Well-Posed Systems by State Feedback
K. A. Morris

155   Point Observation in Linear-Quadratic Elliptic Distributed Control Systems
Link Ji and Goong Chen

171   Boundary Control for a Viscous Burgers' Equation
Christopher I. Byrnes, David S. Gilliam, and Victor I. Shubov

186   Feedback Control of Singular Integro-Differential Systems: An Input/Output Approach
Hitay Özbay and Janos Turi

203   Stable and Unstable Zero Dynamics of Infinite Dimensional Systems
Tzyh-Jong Tarn, Antal K. Bejczy, and Chuanfan Guo

223   Covariance Based Control of Linear Hereditary Systems
James A. Reneke

Chapter 1
Modeling of Flow Dynamics and Its Impact on the Optimal Reactor Design Problem*

Hien T. Tran [†]   Jeffrey S. Scroggs [‡]   Klaus J. Bachmann [§]

Abstract

The flow dynamics of a homogeneous gas inside a vertical reactor for high pressure vapor transport (HPVT) of compound semiconductors is modeled. The modeling is for the growth of selected III-V, II-IV-V_2 and II-VI compounds, and addresses the flow of dense phosphorus gas with pressures in the range $1 \leq P \leq 30$ atm. The mathematical model for transport processes in chemical vapor deposition are described by conservation of mass, momentum, energy, and mass transfer equations. In addition, buoyancy effects are included in the model through the gravitational term in the momentum equation. The coupled set of nonlinear equations are solved numerically for a range of parameters which include: reactor geometry and operating pressure, gravitational vector, thermal conductivities, heat capacities, viscosities, and densities which are computed from the ideal gas law. Results of a 2-D, steady, axi-symmetrical flow are presented using a finite element method with non-uniform, quadratic triangular meshes. The numerical simulations were performed to study the flow dynamics and temperature distribution inside the reactor chamber and to illustrate the feasibility of an optimal reactor design study. Work is in progress to refine the model by incorporating multiple species, thermal diffusion effects, gas-phase and surface chemical reactions, and to extend the simulations to three dimensions.

1 Introduction

Modern computers and communication systems are based on semiconductor technologies demanding precisely controlled electrical, optical and mechanical properties that are only obtained in nearly perfect single crystals and epitaxial structures, grown on such crystals by chemical and physical vapor transport techniques. *Physical* vapor transport proceeds via evaporation at the polycrystalline source and condensation at the surface of the single crystal, which must be held at a lower temperature than the source. *Chemical* vapor transport relies on the temperature dependence of the equilibrium constant of a chemical reaction. In this case, transport may proceed from a hot source to a cooler or hotter crystal, depending on the sign of the enthalpy associated with the transport reaction. Generally low pressure conditions, either in the viscous flow or molecular beam range, are preferred because of the interference of uncontrolled convective flows with the uniformity

---

*This work was supported by National Aeronautics and Space Administration grant #NAGW-2865, and by computer time provided by the North Carolina Supercomputer Center. The second author was also supported by National Science Foundation grant #DMS-9201252

[†]Center for Research in Scientific Computation, Department of Mathematics, North Carolina State University, Raleigh, North Carolina 27695

[‡]Center for Research in Scientific Computation, Department of. Mathematics, North Carolina State University, Raleigh, North Carolina 27695

[§]Department of Materials Science and Engineering, North Carolina State University, Raleigh, North Carolina 27695

of the growth process. However, such low pressure methods are only applicable for the growth of very thin crystal layers, i.e. film thickness $\leq 1\mu m$. If thick films or even bulk crystals are required for the construction of devices that cannot be made in ultrathin films, higher density of the nutrient phase is essential for achieving a fast enough growth rate to make manufacturing feasible. In this paper, the focus is on physical vapor transport of compound semiconductors at high vapor density which, in view of the necessarily non-uniform temperature distribution in the growth ampoule, implies convective mixing at normal gravity.

Crystal growth by vapor transport is important because it frequently permits the fabrication of high quality crystals of materials that, for principal reasons, cannot be grown by directional solidification, e.g. incongruently melting compounds. It is also helpful in the growth of crystals of materials that undergo solid state phase transformations in between their melting temperatures and room temperature. In such cases, the growth by vapor transport below the transformation temperature prevents the cracking of the crystals during cooling. Furthermore vapor transport is convenient for the growth of crystals having both high melting temperatures and high decomposition pressures, e.g. II-VI compounds such as CdS, ZnSe and CdTe [9]. The high perfection of vapor grown crystals makes vapor transport also a favorable method of crystal growth at low temperatures which has been applied to mercury compounds, e.g. HgI_2 which is an excellent choice for the fabrication of nuclear and X-ray detectors [5].

Preliminary studies in our laboratory have also shown that crystals grown by HPVT of certain pnictides, e.g. $ZnGeP_2$, exhibit superior properties as compared to crystals of the same material grown from the melt and even low temperature epitaxial growth techniques [15]. There is interest in high quality crystals of this material because of its utility for non-linear optical applications, e.g. frequency mixing [4] and harmonic generation [2], [1]. The partial pressure of phosphorus over a stoichiometric melt of $ZnGeP_2$ is ≈ 3.5 atm [4]. Another pnictide compound considered by us is indium phosphide which is a well established substrate material for the fabrication of light sources and detectors used in optical communications and holds promise as a material for the fabrication of fast microelectronic circuits. Indium phosphide melts at 1062°C and exhibits an equilibrium phosphorus pressure of 27.5 atm at this temperature [3]. Since in the case of both InP and $ZnGeP_2$ the rate of crystal growth is controlled by the transport of the cation forming elements, which exhibit considerably lower partial pressures than the phosphorus partial pressure [8],the source temperature must be close to the melting temperatures. Therefore, our modeling efforts focus on the phosphorus pressure range $1 \leq P \leq 30$ atm at high temperatures. In the formulation of the growth kinetics, in addition to convective mass transport, diffusion of precursors to crystal growth and of waste products in the nutrient vapor phase, as well as surface diffusion and the kinetics of attachment of atoms to the crystal surface must be considered. Both diffusion, driven by a concentration gradient across the boundary layer at the crystal/vapor interface, and Soret diffusion, driven by the temperature gradient that tends to settle the heavier vapor components in the colder region, contribute to the diffusive flux. In this paper, the goal is the simulation of flow rather than a full description of the crystal growth kinetics which will be addressed after preliminary experimentation that validates the calculations of flow and optimization of the growth ampoule geometry and temperature distribution. The ultimate goal of our research is the complete integration of modeling and crystal growth, using the results of simulations to control the growth kinetics of crystal and thereby the uniformity of its physical properties and morphology.

The mathematical process of designing an optimal ampoule geometry requires the formulation of optimization problems for the flow in these reactors. This, in turn, calls for the derivation of a suitable cost functional. The reactor chamber design has the ultimate goal of uniformity of the epitaxial layer; however, measuring the uniformity, even in the computational setting, may not be possible due to simplifications required to formulate the model of the deposition process. Thus, minimizing the cost functional for optimal reactor design can exploit the possibility of modifying the following parameters:

1. shape of the top of the reactor chamber (e.g. convex up, down, or a combination of both),

2. shape of the bottom,

3. width of chamber,

4. aspect ratio (ratio of height to width),

5. orientation of the reactor chamber with respect to the gravity vector,

6. total pressure

7. and temperature variations.

The goal is minimization of a cost functional that is a mathematical description of the following quantities:

1. distance reactants travel from their source to the substrate–a minimal distance is expected to increase the growth rate and purity,

2. variations in the relative fluxes of reactants in ternary reactions (three reactants)–this determines the stoichiometry which must be controlled to within 1%,

3. and the variation in the temperature of the substrate (especially at the surface)–this should provide increased uniformity and purity.

For example, the cost functional corresponding to matching the velocity field inside the reactor with a desired field given by $\vec{u}^d$ is

$$\int_0^T \int_\Omega \|\nabla(\vec{u} - \vec{u}^d)\|^2 \, dx \, dt,$$

where $\Omega \subset R^n$, $n = 2$ or 3, denotes the geometry of the reactor, and $\vec{u}$ is the velocity field of the flow.

The purpose of this paper is to show the feasibility of the above formulation of optimal control problems by studying the flow dynamics and temperature distribution of a homogeneous gas in a vertical reactor. In particular, the effects of the following parameters on the flow dynamics and temperature distribution are studied: aspect ratio, thickness of the quartz window, gravitational vector orientation, and operating pressure. Namely, these numerical experiments are for a simplified two-dimensional, steady state, axi-symmetrical mathematical model for the flow process. This simplified model is also an intermediate step for the development of a more complete three-dimensional model which will include thermal diffusion, multiple species, and gas-phase and surface chemical reactions. Thus in §2 the design of the reactor is described, followed by a discussion of the mathematical model for the transport process in §3. The results of our numerical simulations are presented §4, and §5 contains our concluding remarks.

2 Physical Experiment

The initial geometry was introduced in the early 1970s by Scholz et al. [11], [12] and is shown schematically in Figure 1. The source material and the growing crystal are hermetically sealed in a fused silica ampoule that is located in a furnace equipped with an isothermal furnace liner that assures uniform heating at its outer cylindrical surface. A ring-shaped booster heater below the polycrystalline source permits its heating to a temperature above the temperature of the outer cylindrical surface. Fused to the center portion of the ampoule is a fused silica tube, terminating in a flat fused silica window W, from which heat can be extracted in a controlled manner by a jet of helium gas directed onto its outer surface. Due to the lower temperature of the inner surface of this window, a condition of supersaturation is established that drives the nucleation and growth of the single crystal. See [11] and [12] for various schemes of assuring seed selection in the initial phase of crystal growth. The top dome surface of the ampoule can be either heated as the side faces by use of a closed furnace liner or allowed to leak heat which may be necessary to retain supersaturated conditions at the upper surface of the crystal after prolonged growth. This is particularly important for bulk crystal growth of materials exhibiting low thermal conductivities which prevent efficient heat transfer through the crystal to W. The purpose of the initial modeling of the gas flow and temperature distribution inside the ampoule, for given outside heating and heat extraction conditions, is to help in the design of the optimum ampoule geometry which optimizes the flow and the uniformity of the supercooled region. The flow and the temperature distribution at the top surface of the crystal determines the boundary conditions for diffusive transport and the supersaturation, driving its growth.

3 Transport Equations

The mathematical model for the transport phenomena involved in the HPVT process contains the Navier-Stokes equations coupled with system of conservation equations for mass, energy and species, and the ideal gas law. In order to provide a basis for our discussions in subsequent sessions, these equations in their general forms are now presented. Our goal is the modeling of a binary vapor mixture containing a carrier gas $\dot{a}$ and reactant gas b (the complete models will contain a larger number of species). With our binary mixture, the gas properties are determined by the carrier gas [7]. Thus, in Cartesian coordinates, the transport equations are:

Conservation of mass

$$\frac{\partial(\rho)}{\partial t} + \frac{\partial(\rho u)}{\partial x} + \frac{\partial(\rho v)}{\partial y} + \frac{\partial(\rho w)}{\partial z} = 0, \tag{1}$$

conservation of x-momentum

$$\frac{\partial(\rho u)}{\partial t} + \frac{\partial(\rho u u)}{\partial x} + \frac{\partial(\rho u v)}{\partial y} + \frac{\partial(\rho u w)}{\partial z} = \frac{\partial}{\partial x}\left[\frac{4}{3}\mu\frac{\partial u}{\partial x} - \frac{2}{3}\mu\frac{\partial v}{\partial y} - \frac{2}{3}\mu\frac{\partial w}{\partial z}\right]$$
$$+ \frac{\partial}{\partial y}\left[\mu\frac{\partial u}{\partial y} + \mu\frac{\partial v}{\partial x}\right] + \frac{\partial}{\partial z}\left[\mu\frac{\partial u}{\partial z} + \mu\frac{\partial w}{\partial x}\right] - \frac{\partial P}{\partial x} + \rho g_x, \tag{2}$$

conservation of y-momentum

$$\frac{\partial(\rho v)}{\partial t} + \frac{\partial(\rho v u)}{\partial x} + \frac{\partial(\rho v v)}{\partial y} + \frac{\partial(\rho v w)}{\partial z} = \frac{\partial}{\partial y}\left[\frac{4}{3}\mu\frac{\partial v}{\partial y} - \frac{2}{3}\mu\frac{\partial u}{\partial x} - \frac{2}{3}\mu\frac{\partial w}{\partial z}\right]$$
$$+ \frac{\partial}{\partial x}\left[\mu\frac{\partial v}{\partial x} + \mu\frac{\partial u}{\partial y}\right] + \frac{\partial}{\partial z}\left[\mu\frac{\partial v}{\partial z} + \mu\frac{\partial w}{\partial y}\right] - \frac{\partial P}{\partial y} + \rho g_y, \tag{3}$$

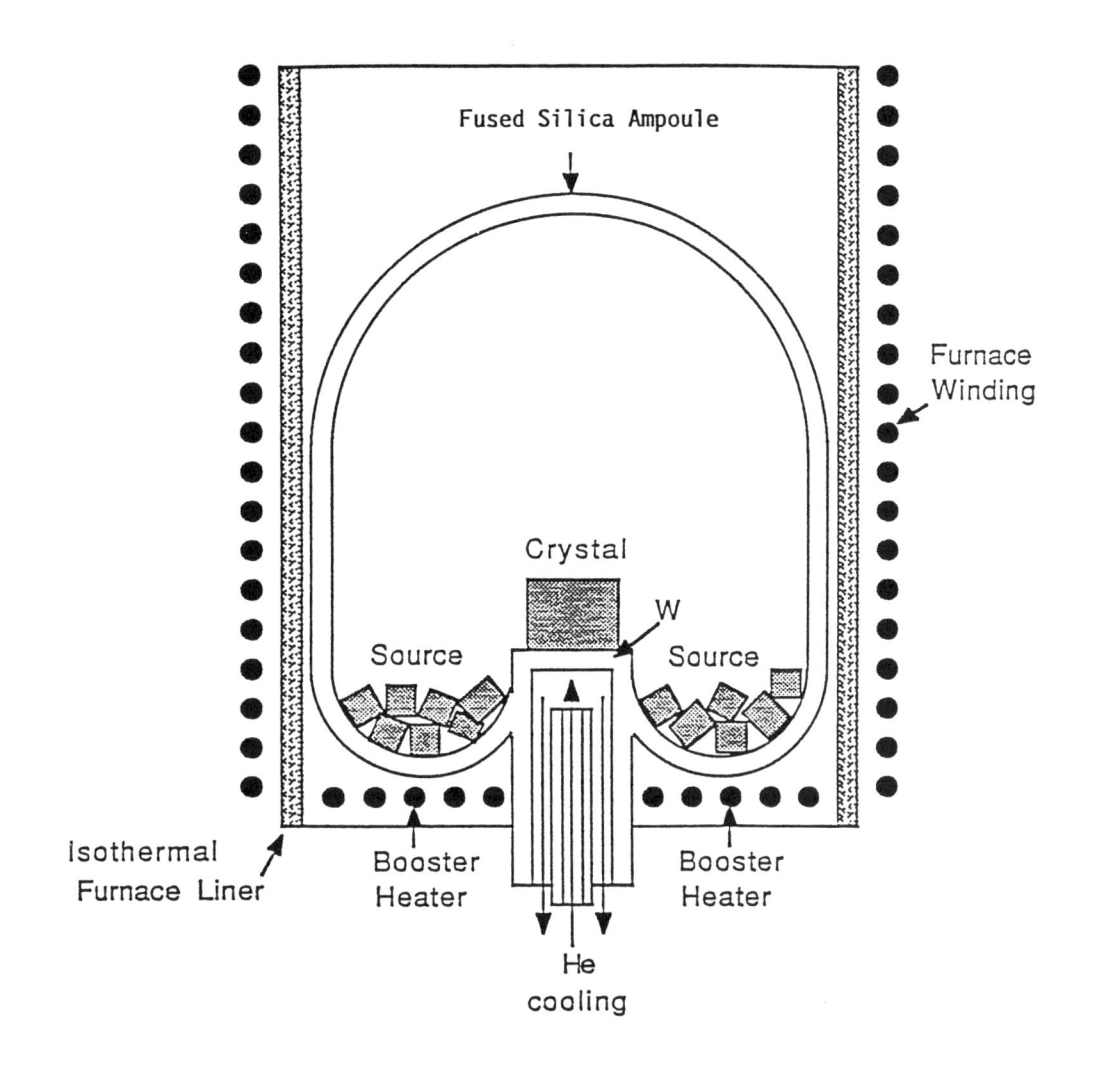

FIG. 1. *Schematic diagram of the Scholz geometry*

conservation of z-momentum

$$\text{(4)} \quad \frac{\partial(\rho w)}{\partial t}+\frac{\partial(\rho w u)}{\partial x}+\frac{\partial(\rho w v)}{\partial y}+\frac{\partial(\rho w w)}{\partial z} = \frac{\partial}{\partial z}\left[\frac{4}{3}\mu\frac{\partial w}{\partial z}-\frac{2}{3}\mu\frac{\partial u}{\partial x}-\frac{2}{3}\mu\frac{\partial v}{\partial y}\right] \\ +\frac{\partial}{\partial x}\left[\mu\frac{\partial w}{\partial x}+\mu\frac{\partial u}{\partial z}\right]+\frac{\partial}{\partial y}\left[\mu\frac{\partial v}{\partial z}+\mu\frac{\partial w}{\partial y}\right]-\frac{\partial P}{\partial z}+\rho g_z,$$

conservation of energy

$$\text{(5)} \quad \frac{\partial(\rho H)}{\partial t}+\frac{\partial(\rho u H)}{\partial x}+\frac{\partial(\rho v H)}{\partial y}+\frac{\partial(\rho w H)}{\partial z} = \frac{\partial}{\partial x}\left[\frac{k}{c_p}\frac{\partial H}{\partial x}\right]+\frac{\partial}{\partial y}\left[\frac{k}{c_p}\frac{\partial H}{\partial y}\right] \\ +\frac{\partial}{\partial z}\left[\frac{k}{c_p}\frac{\partial H}{\partial z}\right]$$

species (diffusion) equation

$$\text{(6)} \quad \frac{\partial(\rho W_b)}{\partial t}+\frac{\partial(\rho u W_b)}{\partial x}+\frac{\partial(\rho v W_b)}{\partial y}+\frac{\partial(\rho w W_b)}{\partial z} = \\ \frac{\partial}{\partial x}\left[\rho D_{ab}\frac{\partial W_b}{\partial x}\right]+\frac{\partial}{\partial y}\left[\rho D_{ab}\frac{\partial W_b}{\partial y}\right]+\frac{\partial}{\partial z}\left[\rho D_{ab}\frac{\partial W_b}{\partial z}\right] \\ +\frac{\partial}{\partial x}\left[\rho D_{ab}\alpha_T W_a W_b\frac{\partial \log T}{\partial x}\right]+\frac{\partial}{\partial y}\left[\rho D_{ab}\alpha_T W_a W_b\frac{\partial \log T}{\partial y}\right] \\ +\frac{\partial}{\partial z}\left[\rho D_{ab}\alpha_T W_a W_b\frac{\partial \log T}{\partial z}\right]$$

and the mixture density ρ is calculated from the perfect gas law

$$\text{(7)} \quad P = \frac{\rho R T}{M}.$$

In the above equations, μ is the dynamic viscosity of carrier gas, g_x, g_y and g_z are the x-, y-, and z-component of the gravity vector $\vec{g}$, H is the enthalpy, k is the thermal conductivity, c_p is the mixture specific heat, W_a is the mass fraction of carrier gas, W_b is the mass fraction of reactant, D_{ab} is the binary diffusion constant of reactant in carrier gas, α_T is the thermal diffusion factor, R is the universal gas law constant, T is the temperature, and M is the molecular weight of the mixture. Implicit in this formulation is that the flow is dominated by the carrier gas. Using these equations for the gas dynamics, the deposition process is modeled through special boundary conditions at the substrate. This is a topic of current study.

The momentum equations (2-4) have utilized Stokes hypothesis [10]. Also, body force enters into the transport model through the momentum equation, and thermal diffusion is modeled in the species equation by the terms

$$\frac{\partial}{\partial \xi}\left[\rho D_{ab}\alpha_T W_a W_b\frac{\partial \log T}{\partial \xi}\right]$$

where ξ stands for x, y, or z. This is the separation of species of different mass or size due to thermal gradients.

4 Numerical Experiments

In this section, numerical results for a simplified two-dimensional model of a HPVT process are presented. Here an axi-symmetric steady flow of a homogeneous gas with the properties of P_2 at 1 and 10 atm is considered. That is, the species equation (6) is not part of our transport equations. In addition, instead of the ideal gas law (7), the state equation is assumed to be the so-called Boussinesq approximation [6]. This implies that the density ρ is constant in all equations except that in the presence of a gravitational field a buoyancy force exists due to density variations. Thus, the body force term $\rho \vec{g}$ now has the representation

$$(8) \qquad \rho[1 - \beta_T(T - T_{\text{ref}})]\vec{g},$$

where β_T is the volume expansion coefficient and T_{ref} is the reference temperature.

In all simulations, the temperature under the substrate is fixed at 1300°K, and 1350°K otherwise (see Figure 1). This specification of the temperature on the boundary results in a discontinuous temperature profile at the corner of the substrate and curvature side walls. Also, from the practical point of view, the temperature is not maintained at 1350°K near these corners. Therefore, the temperature varies linearly from 1350°K to 1300°K near these corners. No-slip boundary conditions are applied everywhere except along the axis of symmetry where $u_r = 0$ and u_z is free. Here u_r and u_z are respectively the r- and z-component in the cylindrical coordinates of the velocity field $\vec{u}$. At the substrate, there is a fused quartz with thickness of 0.2 cm and length of 1.0 cm. The large bowl is a semi-circle with radius 2.5 cm while the smaller bowls are semi-circles with radius 1.0 cm. The height of reactor is 7.0 cm and is the dimension which will be changed in the study involving the aspect ratio of the chamber. The transport properties used in the computations were taken from data in references [13] and [14].

We now report several simulation studies which were designed to provide insight into the bulk flow as a function of several parameters. The computations were performed on the Cray Y-MP at the North Carolina Supercomputer Center using a commercially available finite element code FIDAP. The CPU times for a typical simulation were 3-5 minutes. In all figures displaying temperature profiles, the isotherm nearest to the substrate has value 1301°K where as the one nearest to the side wall has value 1349°K. The increment between those extreme isotherms is between 2°K to 3°K.

4.1 Case 1: reactor orientation

For this case, flow and temperature distributions were studied to determine the the effect on the growth of the orientation of the ampoule. Two configurations for the reactor were considered: one has the substrate at the bottom (stalagmitic orientation) and the other has the substrate at the top (stalactitic orientation). The velocity fields and isotherms at 1 atm are shown in Figures 2 and 3. The curves clearly indicate a larger convective roll for the stalactitic geometry. Fluid dynamic considerations would suggest a decrease in the growth rate due to longer travel paths by reactants. In this case, however, the velocity field is also faster which compensates for the large vortex. On the other hand, temperature contour plots reveal that the isotherms are more uniformed near the substrate for the stalagmitic configuration and the flow appears to be convective dominated near the sink. Since physical vapor transport relies on the condensation at the cooler temperature surface (substrate), a uniform temperature profile will probably yield a uniform growth process. Thus, the best configuration of these two is the stalagmitic.

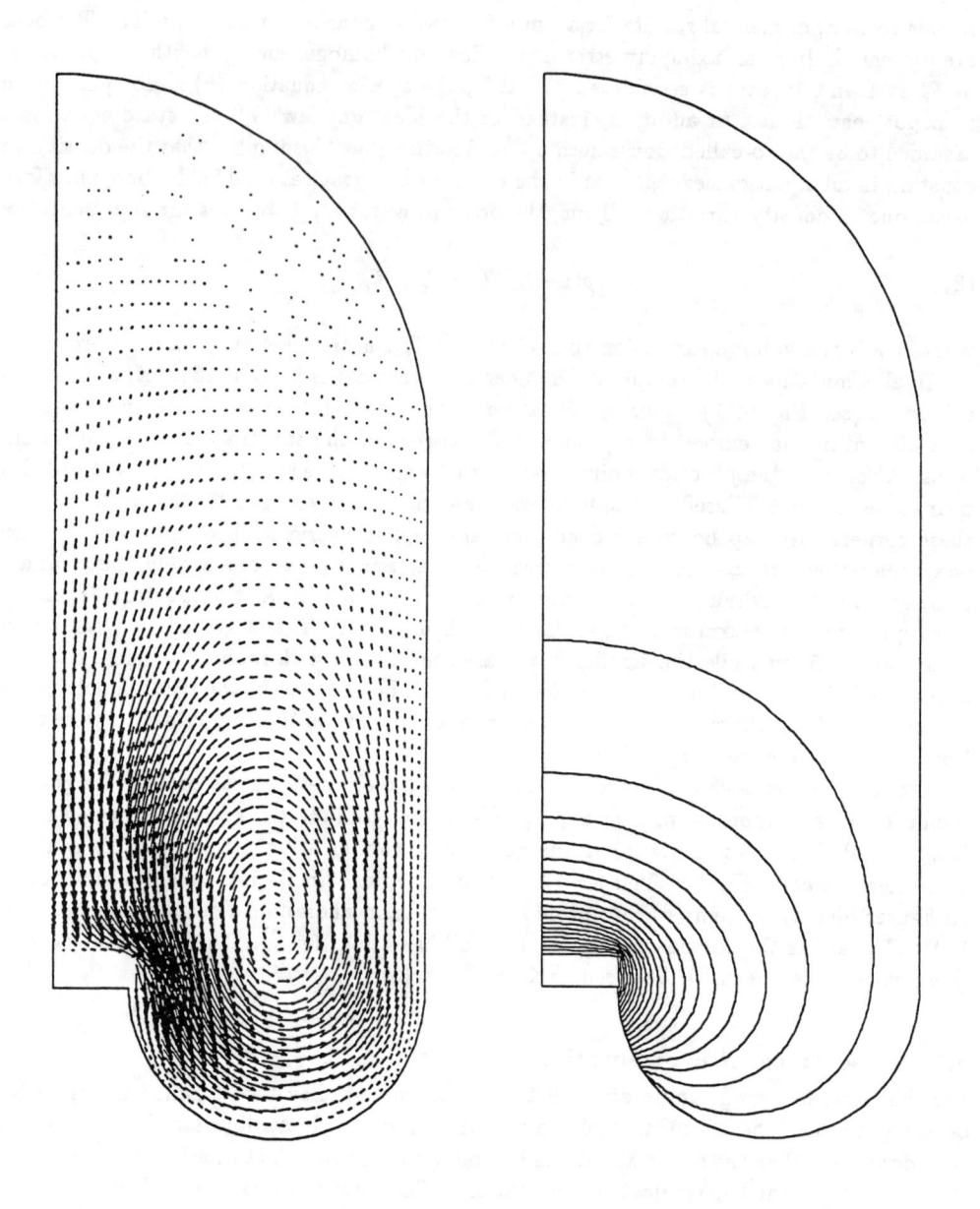

FIG. 2. *Case 1: Velocity fields and temperature contours for the stalagmitic orientation. The velocity has maximum value of 0.015 m/sec. (Whole ampoule is 5cm × 7cm).*

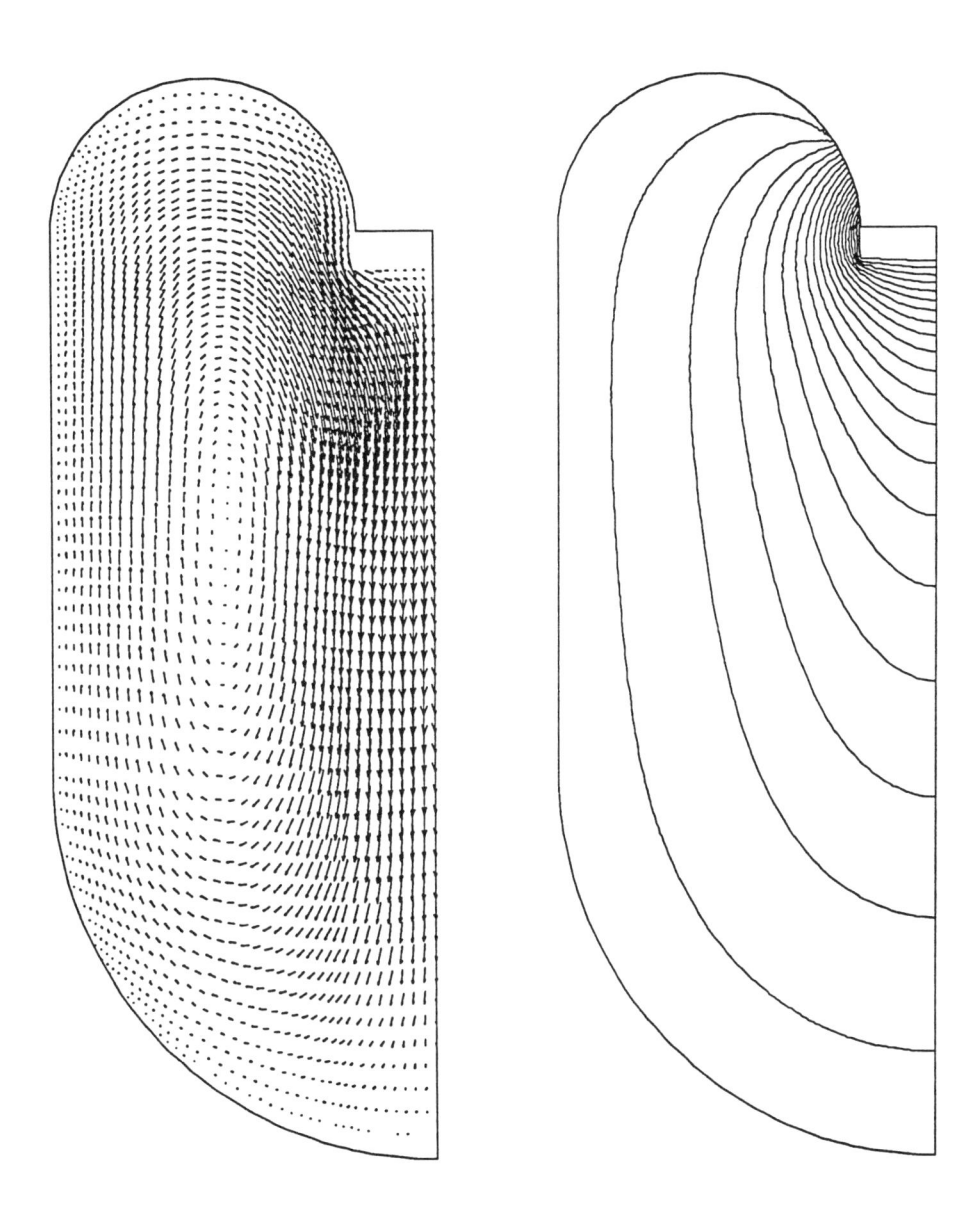

FIG. 3. *Case 1: Velocity fields and temperature contours for the stalactitic orientation with modified gravity. The velocity has maximum value of 0.035 m/sec. (Whole ampoule is 5cm × 7cm).*

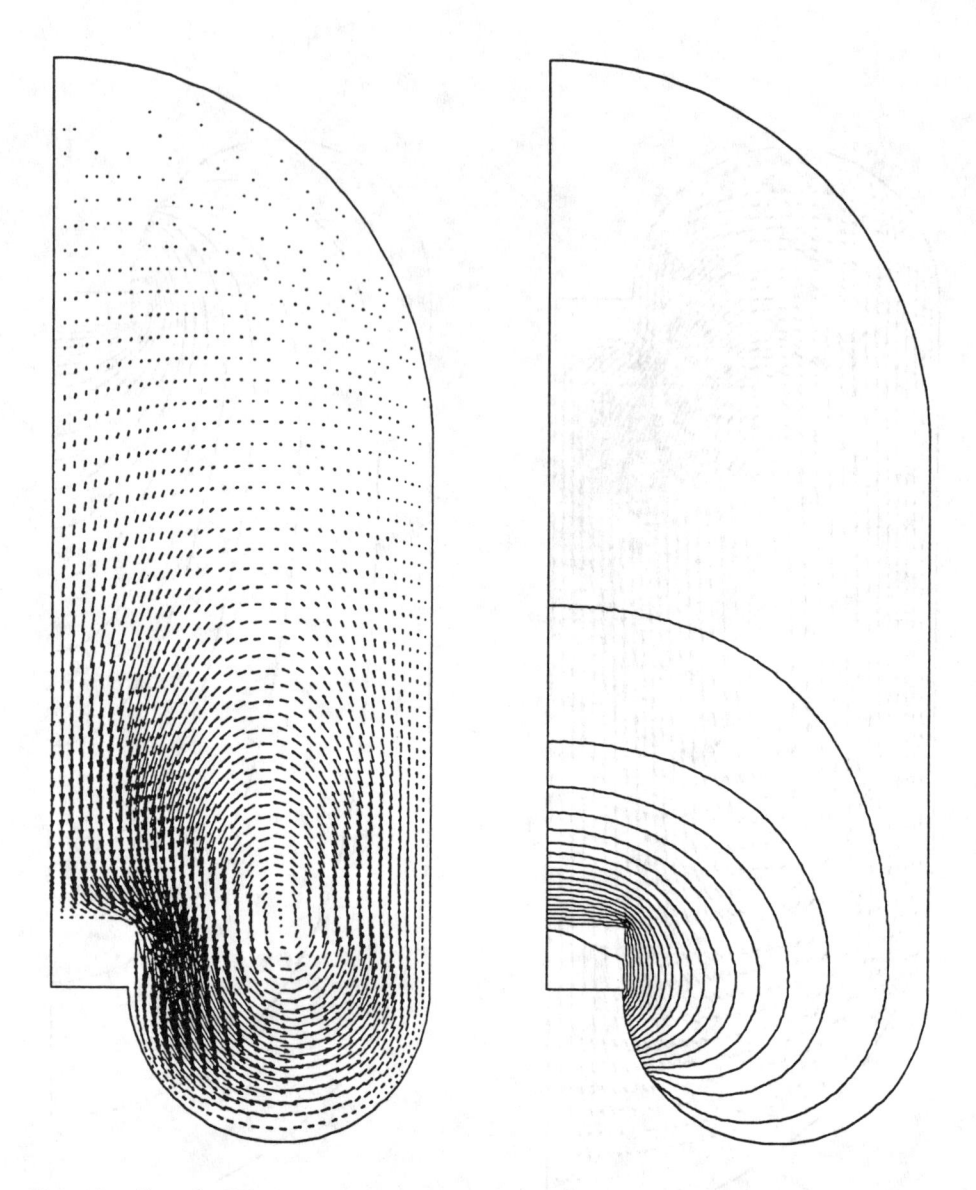

FIG. 4. *Case 2: Velocity fields and temperature contours for the stalagmitic orientation with increased window thickness. The velocity has maximum value of 0.016 m/sec. (Whole ampoule is 5cm × 7cm).*

4.2 Case 2: increase the thickness of the quartz window

In this case, the thickness of the quartz window is increased from 0.2 cm to 0.5 cm. The numerically computed velocity and temperature plots at 1 atm are shown in Figure 4. The plot shows, as expected, an increase in the temperature value (1304°K) near the substrate. The plot, however, indicates little effect on velocity fields.

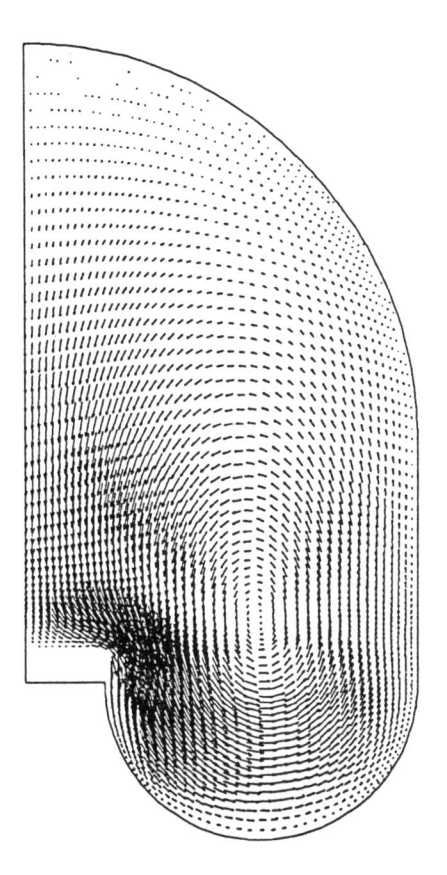

FIG. 5. *Case 3: Velocity fields for the stalagmitic orientation with modified aspect ratio. The velocity has maximum value of 0.015 m/sec. (Whole ampoule is 5cm × 5cm).*

4.3 Case 3: reduce the aspect ratio

For these numerical experiments, the height of the ampoule is reduced from 7.0 cm to 5.0 cm while keeping the width fixed. The velocity fields at 1 atm are shown in Figure 5. The curves clearly indicate no dependence of the flow on the aspect ratio for the stalagmitic orientation. The isotherms are also not affected. On the other hand, the calculations with the stalactitic geometry show that a shorter ampoule moves the advection-transport-dominated region closer to the bottom of the reactor (source).

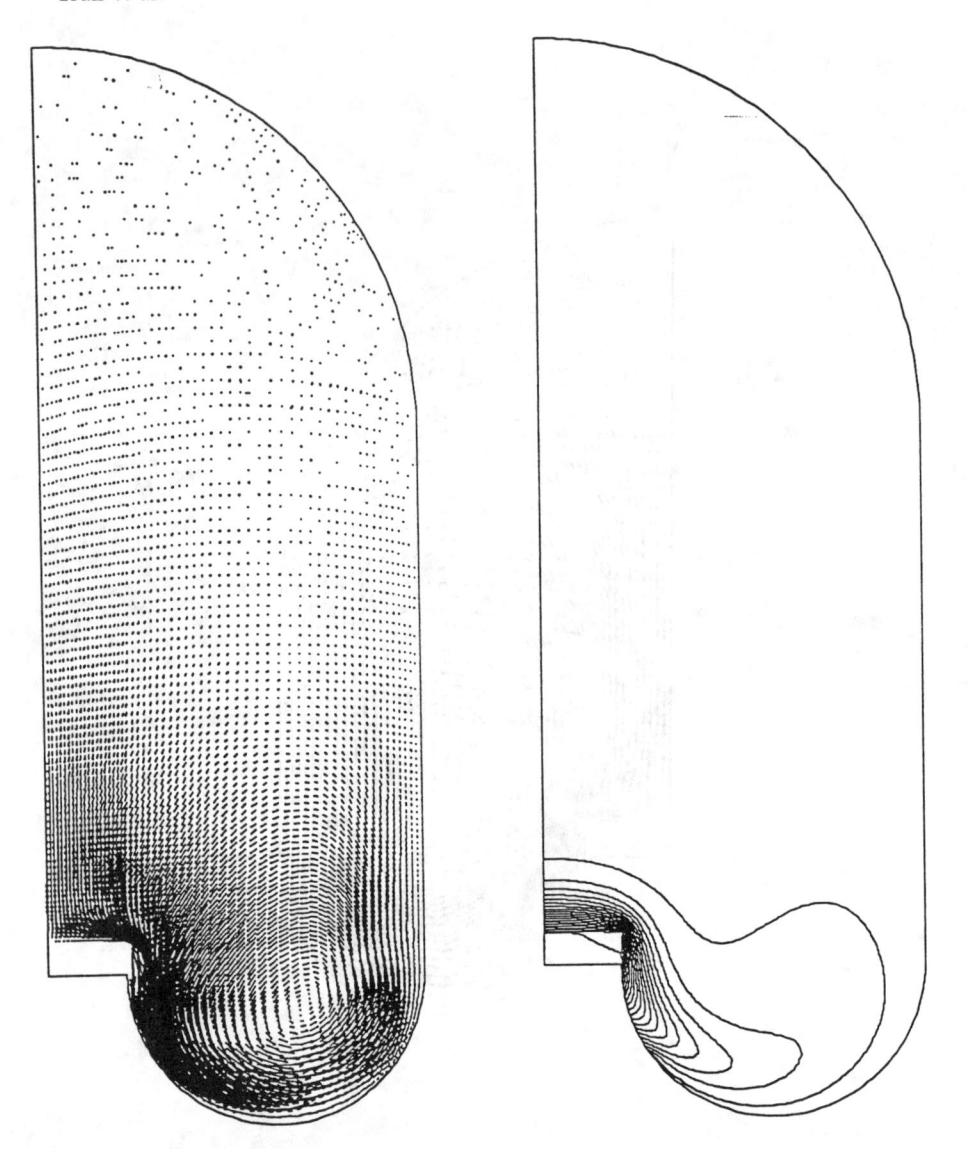

FIG. 6. *Case 4: Velocity fields and isotherms for the stalagmitic orientation at high pressure. The velocity has maximum value of 0.022 m/sec. (Whole ampoule is 5cm × 7cm).*

4.4 Case 4: increase the operating pressure

In this case, the pressure is increased to 10 atm. The velocity profile and temperature contours are shown in Figure 6. Figure 6 shows an increase in the flow rate with increase in the pressure (or density). The curves also reveal a smaller advection-transport region which moves closer to the bottom crucible. This clearly indicates where the source material should be located so that it is closest to the largest velocities. Note also the closer spacing of the isotherms as compared to case 1 which results in warmer temperature near the substrate. Until physical experiments are performed, the effects on the deposition rate and on the uniformity of the growth process will not be known.

5 Concluding Remarks

Numerical simulations have been carried out which examine the effects of varying the reactor height, orientation of the reactor geometry, quartz window thickness, and the pressure. The results indicate that the best configuration of the Scholz geometry has the substrate on the bottom; the flow rate increases with increase in pressure (or density); change in aspect ratio (wider reactors) has no influence on the velocity and temperature profiles for the stalagmitic orientation; and the temperature increases with increase in the quartz window thickness. The simplified two-dimensional model presented in this paper needs further enhancements for modeling growth at high pressure. The refinements currently being made in the model include gas-phase chemical reactions, multiple species, thermal diffusion, compressibility, and three-dimensional modeling.

6 Acknowledgments

The first two authors would like to thank Michael Andreas for many helpful discussions. The encouragement and scientific leadership of H. T. Banks is also gratefully acknowledged.

References

[1] Y. M. Andreev, V. Y. Baranov, V. G. Voevodin, P. Geiko, A. I. Gribenyukov, S. V. Izyumov, S. M. Kozochkin, V. D. Pis'mennyi, Y. A. Satov, and B. Strel'tsov, *Efficient frequency doubling of CO_2 laser radiation*, Sov. J. Quantum Electron., 17 (1987), pp. 1435–1436.

[2] Y. M. Andreev, A. D. Belykh, V. G. Voevodin, P. Geiko, A. I. Gribenyukov, V. A. Gurashvili, and S. M. Izyumov, *Frequency doubling with 3% efficiency of Carbon Monoxide laser radiation*, Sov. J. Quantum Electron., 17 (1987), p. 490.

[3] K. Bachmann and E. Buehler, *Phase equilibria and vapor pressures of pure Phosphorus and of the Indium/Phosphorus system and their implications regarding crystal growth of InP*, j. Electrochem. Soc.: Solid-State Science and Technology, (1974), pp. 835–846.

[4] G. Boyd, E. Buehler, and F. G. Storz, *Linear and non-linear optical properties of Zinc Germanium Diphosphide and Cadmium Selenide*, Appl. Phys. Lett., 18 (1971), p. 301.

[5] A. J. Danbrowski, W. M. Szmyszyk, J. S. Iwanszyk, J. H. Kumiss, W. Drummond, and L. Ames, *Progress in energy resolution of Mercury Iodide (HgI_2) X-ray spectrometer*, Nucl. Instr. and Methods, 213 (1983), p. 89.

[6] D. D. Gray and A. Giorgini, *The validity of the Boussinesq approximation for liquids and gases*, Int. J. Heat Mass Transfer, (1976), pp. 545–551.

[7] R. Mahajan and C. Wei, *Buoyancy, Soret, Dufour, and variable property effects in silicon epitaxy*, Transactions of the ASME, 113 (1991), pp. 688–695.

[8] M. B. Panish and J. B. Arthur, *Phase equilibria and vapor pressures of the system In-P*, J. Chem. Thermodyn., 2 (1970), p. 299.

[9] W. W. Piper and S. J. Polich, *Vapor-phase growth of single crystals of II-VI compounds*, J. Appl. Phys., 32 (1961), p. 1278.

[10] H. Schlichting, *Boundary Layer Theory*, McGraw-Hill, New York, 1955.

[11] H. Scholz, *Crystal growth by temperature alternating methods*, Phillips Technical Reviews, 28 (1967), p. 316.

[12] ——, *On crystallization by temperature gradient reversal*, Acta Electronica, 17 (1974), p. 69.

[13] R. A. Svehla, *Estimated viscosities and thermal conductivites of gases at high temperatures*, Technical Report NASA TR R-132, NASA, Washington, D.C., 1962.

[14] R. C. Weast, *CRC Handbook of Chemistry and Physics*, CRC Press, Inc., Ohio, 1977.

[15] G. C. Xing, K. J. Bachmann, J. B. Posthill, G. S. Solomon, and M. L. Timmons, in Heteroepitaxial Approaches in Semiconductors: Lattice Mismatch and its Consequences, A. T. Macrander and T. J. Drummond, eds. The Electrochemical Society, Pennington, NJ, (1989), p. 132.

Chapter 2
Sensitivity Calculations for a 2D, Inviscid, Supersonic Forebody Problem*

Jeff Borggaard *[1]   John Burns *[2]   Eugene Cliff *[3]   Max Gunzburger *[4]

ABSTRACT

In this paper, we discuss the use of a sensitivity equation method to compute derivatives for optimization based design algorithms. The problem of designing an optimal forebody simulator is used to motivate the algorithm and to illustrate the basic ideas. Finally, we indicate how an existing CFD code can be modified to compute sensitivities and a numerical example is presented.

1 Introduction

A large number of identification, control and design problems may be formulated as infinite dimensional optimization problems. These problems arise in almost all fields of science and engineering and range in scope from inverse problems in seismology, to LQR and H^∞ control, to shape optimization in fluid/structure dynamics. See [4,7] for typical applications. Although there are numerous approaches to solving these problems, each approach requires that some type of approximation be introduced at some point in the design process. Moreover, it is often the case that some iterative scheme is needed to solve the state equations (in black-box methods [3,8]) and the adjoint equations (in adjoint and "one-shot" methods [9]). Also, the optimization algorithm may itself be iterative. In any case, the development of computational methods for optimal design and control can produce several levels of approximation and the convergence properties of the overall algorithm are very much dependent on the approximations. In this paper we concentrate on the problem of computing accurate sensitivities for gradient based optimization algorithms. In order to keep the paper short and, at the same time illustrate the basic idea, we concentrate on a particular application. We give a brief description of the problem and use this problem to motivate the algorithm presented below.

---

*Interdisciplinary Center for Applied Mathematics, Virginia Polytechnic Institute and State University, Blacksburg, VA 24061.

[1] Supported in part by the Air Force Office of Scientific Research under Grant F-49620-92-J-0078.

[2] Supported in part by the Air Force Office of Scientific Research under Grant F-49620-92-J-0078, the National Science Foundation under Grant INT-89-22490 and by the National Aeronautics and Space Administration under Contract Nos. NAS1-18605 and NAS1-19480 while the author was a visiting scientist at the Institute for Computer Applications in Science and Engineering (ICASE), NASA Langley Research Center, Hampton, VA 23681-0001.

[3] Supported in part by the Air Force Office of Scientific Research under Grant F-49620-92-J-0078.

[4] Supported in part by the Air Force Office of Scientific Research under Grant AFOSR-90-0179.

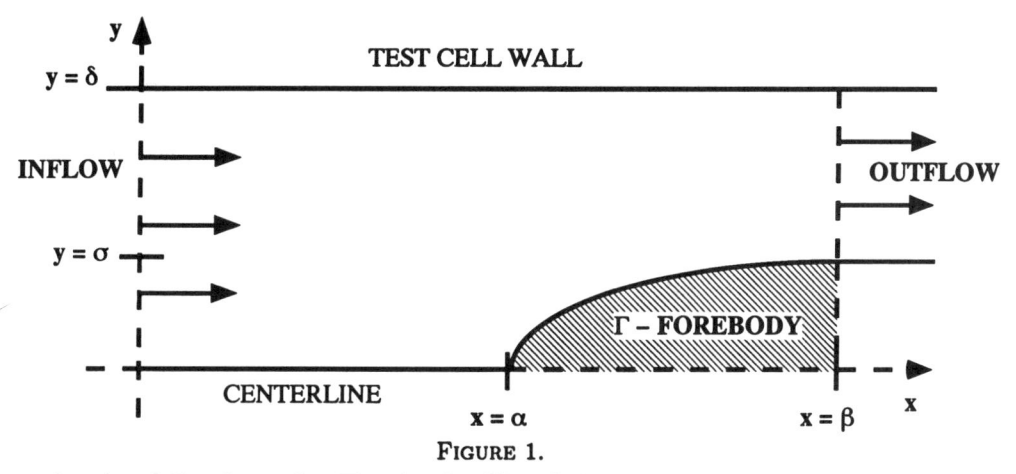

```
                    y ▲
      y = δ ─────────┼──────── TEST CELL WALL ─────────────
                     │                                    │
      INFLOW  ──▶    │                                    │   OUTFLOW
                     │                                    │ ──▶
                     │ ──▶                                │
      y = σ ─┼──     │                                    │
                     │ ──▶                                │
                     │                ░░░░░░              │
                     │             ░░Γ - FOREBODY░░       │
              ─ ─ ─ ─┼────────────░░░░░░░░░░░░░░░─ ─ ─ ─ ─▶
                     │  CENTERLINE                        │   x
                     │              x = α        x = β
```

FIGURE 1.

2 Optimal Design of a Forebody Simulator

This problem is a 2D version of the problem described in [1,2,6]. The Arnold Engineering Development Center (AEDC) is developing a free-jet test facility for full-scale testing of engines in various free flight conditions. Although the test cells are large enough to house the jet engines, they are too small to contain the full airplane forebody and engine. Thus, the effect of the forward fuselage on the engine inlet flow conditions must be "simulated." One approach to solving this problem is to replace the actual forebody by a smaller object, called a "forebody simulator" (FBS), and determine the shape of the FBS that produces the best flow match at the engine inlet. The 2D version of this problem is illustrated in Figure 1 (see [1,2,5,6]).

The underlying mathematical model is based on conservation laws for mass, momentum and energy. For inviscid flow, we have that

$$\frac{\partial}{\partial t}\mathbf{Q} + \frac{\partial}{\partial x}\mathbf{F}_1 + \frac{\partial}{\partial y}\mathbf{F}_2 = 0 \tag{1}$$

where

$$\mathbf{Q} = \begin{pmatrix} \rho \\ m \\ n \\ E \end{pmatrix}, \quad \mathbf{F}_1 = \begin{pmatrix} m \\ mu + P \\ mv \\ (E+P)u \end{pmatrix}, \quad \text{and} \quad \mathbf{F}_2 = \begin{pmatrix} n \\ nu \\ nv + P \\ (E+P)v \end{pmatrix}. \tag{2}$$

The velocity components u and v, the pressure P, the temperature T, and the Mach number M are related to the conservation variables, i.e., the components of the vector $\mathbf{Q}$, by

$$u = \frac{m}{\rho}, \quad v = \frac{n}{\rho}, \quad P = (\gamma - 1)\left(E - \tfrac{1}{2}\rho(u^2 + v^2)\right),$$
$$T = \gamma(\gamma - 1)\left(\frac{E}{\rho} - \tfrac{1}{2}(u^2 + v^2)\right), \quad \text{and} \quad M^2 = \frac{u^2 + v^2}{T}. \tag{3}$$

At the inflow boundary, we want to simulate a free-jet, so that we specify the total pressure P_0, the total temperature T_0, and the Mach number M_0. We also set $v = 0$ at the inflow boundary. If u_I, P_I, and T_I denote the inflow values of the x-component of the velocity, the pressure, and the temperature, these may be recovered from T_0, P_0 and M_0 by

$$T_I = \frac{T_0}{(1 + \tfrac{\gamma-1}{2}M_0^2)}, \quad P_I = \frac{P_0}{(1 + \tfrac{\gamma-1}{2}M_0^2)^{\frac{\gamma}{\gamma-1}}}, \quad \text{and} \quad u_I^2 = M_0^2 T_I = \frac{M_0^2 T_0}{(1 + \tfrac{\gamma-1}{2}M_0^2)}. \tag{4}$$

The components of **Q** at the inflow may then be determined from (4) through the relations

$$\rho_I = \frac{\gamma P_I}{T_I}, \qquad m_I = \rho_I u_I, \qquad n_I = 0, \qquad \text{and} \qquad E_I = \frac{P_I}{\gamma - 1} + \rho_I \frac{u_I^2}{2}. \tag{5}$$

The forebody is a solid surface, so that the normal component of the velocity vanishes, i.e.,

$$u n_1 + v n_2 = 0 \quad \text{on the forebody}, \tag{6}$$

where n_1 and n_2 are the components of the unit normal vector to the boundary. Note that we impose (6) on the velocity components u and v, and not on the momentum components m and n. Insofar as the state is concerned, it is clear that it does not make any difference whether (6) is imposed on m and n or on u and v, since $m = \rho u$ and $n = \rho v$ and $\rho \neq 0$. It can be shown that it does not make any difference to the sensitivities as well.

Assume that at $x = \beta$ the desired steady state flow $\hat{\mathbf{Q}} = \hat{\mathbf{Q}}(y)$ is given as data on the line (called the Inlet Reference Plane)

$$IRP = \{(x,y) | x = \beta, \sigma \leq y \leq \delta\}.$$

Also, we assume here that the inflow (total) Mach number M_0 can be used as a design (control) variable along with the shape of the forebody. Let the forebody be determined by the curve $\Gamma = \Gamma(x), \alpha \leq x \leq \beta$ and let $p = (M_0, \Gamma(\cdot))$. The problem can be stated as the following optimization problem:

Problem FBS. Given data $\hat{\mathbf{Q}} = \hat{\mathbf{Q}}(y)$ on the IRP, find the parameters $p^* = (M_0^*, \Gamma^*(\cdot))$ such that the functional

$$J(p) = \frac{1}{2} \int_\sigma^\delta \|\mathbf{Q}_\infty(\beta, y) - \hat{\mathbf{Q}}(y)\|^2 dy$$

is minimized, where $\mathbf{Q}_\infty(x,y) = \mathbf{Q}_\infty(x,y,p)$ is solution to the steady state Euler equation

$$G(\mathbf{Q}, p) = \frac{\partial}{\partial x} \mathbf{F}_1 + \frac{\partial}{\partial y} \mathbf{F}_2 = 0.$$

Clearly the statement of the problem is not complete. For example, one should carefully specify the set of admissible curves $\Gamma(\cdot)$ and questions remain about existence, uniqueness and integrability of "the" solution $\mathbf{Q}_\infty$. We will not address these issues in this short note.

Most optimization based design methods require the computation of the derivatives $\frac{\partial}{\partial p}\mathbf{Q}_\infty(x,y,p)$. These derivatives are called sensitivities and various schemes have been developed to approximate the sensitivities numerically (see [3,5,10,11]). A common approach is to use finite differences. In particular, the steady state equation (8) is solved for $\hat{p}$ and again for $\hat{p} + \Delta p$ and then $\frac{\partial}{\partial p}\mathbf{Q}_\infty(x,y,\hat{p})$ is approximated by $\frac{[\mathbf{Q}_\infty(x,y,\hat{p}+\Delta p) - \mathbf{Q}_\infty(x,y,\hat{p})]}{\Delta p}$. This method is often costly and can introduce large errors. Another approach is to first derive an equation (the sensitivity equation) for $\mathbf{Q}' = \frac{\partial}{\partial p}\mathbf{Q}_\infty(x,y,p)$ and then numerically solve this equation. We shall illustrate this approach for the forebody design problem and present a comparison of the two methods.

3 Sensitivities with Respect to the Inflow Mach Number

First, we consider the design parameter M_0^2. Thus, we will derive equations for the sensitivity

$$\mathbf{Q}' \equiv \frac{\partial \mathbf{Q}}{\partial M_0^2} \equiv \begin{pmatrix} \rho' \\ m' \\ n' \\ E' \end{pmatrix}, \tag{7}$$

where

$$\rho' \equiv \frac{\partial \rho}{\partial M_0^2}, \qquad m' \equiv \frac{\partial m}{\partial M_0^2}, \qquad n' \equiv \frac{\partial n}{\partial M_0^2}, \qquad \text{and} \qquad E' \equiv \frac{\partial E}{\partial M_0^2}. \tag{8}$$

The differential equation system (1) has no explicit dependence on the design parameter M_0^2, so that equations for the components of $\mathbf{Q}'$ are easily determined by formally differentiating (1) with respect to M_0^2. The result is the system

$$\frac{\partial \mathbf{Q}'}{\partial t} + \frac{\partial \mathbf{F}'_1}{\partial x} + \frac{\partial \mathbf{F}'_2}{\partial y} = 0, \tag{9}$$

where

$$\mathbf{F}'_1 = \begin{pmatrix} m' \\ mu' + m'u + P' \\ mv' + m'v \\ (E+P)u' + (E'+P')u \end{pmatrix} \quad \text{and} \quad \mathbf{F}'_2 = \begin{pmatrix} n' \\ nu' + n'u \\ nv' + n'v + P' \\ (E+P)v' + (E'+P')v \end{pmatrix}, \tag{10}$$

and where,

$$u' = \frac{\partial u}{\partial M_0^2}, \qquad v' = \frac{\partial v}{\partial M_0^2}, \qquad P' = \frac{\partial P}{\partial M_0^2}, \qquad \text{and} \qquad T' = \frac{\partial T}{\partial M_0^2}, \tag{11}$$

and where, through (3), the sensitivities (8) and (11) are related by

$$\begin{aligned} u' &= \frac{1}{\rho} m' - \frac{m}{\rho^2} \rho', & P' &= (\gamma - 1)\left(E' - \tfrac{1}{2}\rho'(u^2 + v^2) - \rho(uu' + vv')\right), \\ v' &= \frac{1}{\rho} n' - \frac{n}{\rho^2} \rho', & \text{and} \quad T' &= \gamma(\gamma-1)\left(\frac{1}{\rho}E' - \frac{E}{\rho^2}\rho' - (uu' + vv')\right) \end{aligned} \tag{12}$$

Note that (9) is of the same form as (1), with a different flux vector. In particular, (9) is in conservation form. As a result of the fact that (9) is *linear* in the primed variables, and that by (12) u', v', and P' are linear in the components of $\mathbf{Q}'$, (9) is a linear system in the sensitivity (7), *i.e.*, in the components of $\mathbf{Q}'$.

Now, we need to discuss the boundary conditions for $\mathbf{Q}'$. Except for the inflow conditions, all boundary conditions are independent of the design parameter M_0^2. Thus, the latter may be differentiated with respect to M_0^2 to obtain boundary conditions for the sensitivities. For example, at the forebody where (6) holds, we simply would have that

$$u' n_1 + v' n_2 = 0 \quad \text{on the forebody}. \tag{13}$$

Similar operations yield boundary conditions for the sensitivities along symmetry lines, other solid surfaces, and at the outflow boundary. Note that if instead of (6), one interprets the no penetration condition as one on the momentum, *i.e.*, $mn_1 + nn_2 = 0$ on the forebody, then instead of (13) we would have that

$$m' n_1 + n' n_2 = 0 \quad \text{on the forebody} \tag{14}$$

which is seemingly different from (13). However, (6) and (12) can be used to show that

$$m' n_1 + n' n_2 = \rho(u' n_1 + v' n_2) + \rho'(u n_1 + v n_2) = \rho(u' n_1 + v' n_2) \tag{15}$$

so that, since $\rho \neq 0$, (13) and (14) are identical.

The inflow boundary conditions for the sensitivities may be determined by differentiating (4) and (5) with respect to the design parameter M_0^2. Note that this parameter appears explicitly in the right-hand-sides of the equations in (4) and (5). Without difficulty, one finds from (5) that

$$\rho_I' = \frac{\gamma}{T_I} P_I' - \frac{\gamma P_I}{T_I^2} T_I', \qquad m_I' = \rho_I u_I' + u_I \rho_I' \tag{16}$$
$$n_I' = 0, \quad \text{and} \quad E_I' = \frac{1}{\gamma - 1} P_I' + \frac{1}{2} u_I^2 \rho_I' + \rho_I u_I u_I',$$

where, from (4),

$$T_I' = -\left(\frac{\gamma-1}{2}\right) \frac{T_0}{(1+\frac{\gamma-1}{2}M_0^2)^2}, \qquad P_I' = -\left(\frac{\gamma}{2}\right) \frac{P_0}{(1+\frac{\gamma-1}{2}M_0^2)^{\frac{2\gamma-1}{\gamma-1}}}, \quad \text{and} \tag{17}$$
$$u_I' = \frac{\sqrt{T_I}}{2M_0} + \frac{M_0}{2\sqrt{T_I}} T_I' = \frac{\sqrt{T_0}}{2M_0(1+\frac{\gamma-1}{2}M_0^2)^{\frac{3}{2}}} \left(1 + (\gamma-1)M_0^2\right).$$

4 Sensitivities with Respect to the Forebody Design Parameters

We assume that the forebody is described in terms of a finite number of design parameters which we denote by P_k, $k = 1, \ldots, K$, and that the forebody may be described by the relation

$$y = \Phi(x; P_1, P_2, \ldots, P_K), \quad \alpha \leq x \leq \beta \tag{18}$$

We express the dependence of the state variable $\mathbf{Q}$ on the coordinates and the design parameters by $\mathbf{Q} = \mathbf{Q}(t, x, y; M_0^2, P_1, P_2, \ldots, P_K)$. We have already seen what equations can be used to determine the sensitivity of the state with respect to M_0^2, i.e., for $\mathbf{Q}'$. We now discuss what equations can be used to determine the sensitivities with respect to the forebody design parameters P_k, $k = 1, \ldots, K$, i.e., for

$$\mathbf{Q}_k \equiv \frac{\partial \mathbf{Q}}{\partial P_k} \equiv \begin{pmatrix} \rho_k \\ m_k \\ n_k \\ E_k \end{pmatrix}, \tag{19}$$

where

$$\rho_k \equiv \frac{\partial \rho}{\partial P_k}, \qquad m_k \equiv \frac{\partial m}{\partial P_k}, \qquad n_k \equiv \frac{\partial n}{\partial P_k}, \quad \text{and} \quad E_k \equiv \frac{\partial E}{\partial P_k}, \qquad k = 1, \ldots, K. \tag{20}$$

System (1) has no explicit dependence on the design parameters P_k, so that equations for the components of $\mathbf{Q}_k$ are easily determined by differentiating (1) with respect to P_k, $k = 1, \ldots, K$. This produces the systems, $k = 1, \ldots, K$, given by

$$\frac{\partial \mathbf{Q}_k}{\partial t} + \frac{\partial \mathbf{F}_{k1}}{\partial x} + \frac{\partial \mathbf{F}_{k2}}{\partial y} = 0, \tag{21}$$

where

$$\mathbf{F}_{k1} = \begin{pmatrix} m_k \\ mu_k + m_k u + P_k \\ mv_k + m_k v \\ (E+P)u_k + (E_k + P_k)u \end{pmatrix} \quad \text{and} \quad \mathbf{F}_{k2} = \begin{pmatrix} n_k \\ nu_k + n_k u \\ nv_k + n_k v + P_k \\ (E+P)v_k + (E_k + P_k)v \end{pmatrix}, \tag{22}$$

and where,
$$u_k = \frac{\partial u}{\partial P_k}, \qquad v_k = \frac{\partial v}{\partial P_k}, \qquad P_k = \frac{\partial P}{\partial P_k}, \qquad \text{and} \qquad T_k = \frac{\partial T}{\partial P_k}. \tag{23}$$

Moreover, by (3), the sensitivities (20) and (23) are related by
$$\begin{aligned} u_k &= \frac{1}{\rho}m_k - \frac{m}{\rho^2}\rho_k, & P_k &= (\gamma - 1)\left(E_k - \tfrac{1}{2}\rho_k(u^2 + v^2) - \rho(uu_k + vv_k)\right) \\ v_k &= \frac{1}{\rho}n_k - \frac{n}{\rho^2}\rho_k, & \text{and} \qquad T_k &= \gamma(\gamma - 1)\left(\frac{1}{\rho}E_k - \frac{E}{\rho^2}\rho_k - (uu_k + vv_k)\right), \end{aligned} \tag{24}$$

for $k = 1, \ldots, K$.

All boundary conditions except the one on the forebody also do not depend on the forebody design parameters P_k, $k = 1, \ldots, K$. For example, consider the inflow boundary conditions (4)-(5). Differentiating these with respect to P_k, $k = 1, \ldots, K$ yields that
$$\rho_{kI} = m_{kI} = n_{kI} = E_{kI} = T_{kI} = P_{kI} = u_{kI} = v_{kI} = 0, \tag{25}$$

at the inflow boundary. Now, consider the boundary condition (6) on the forebody. We have that on the forebody
$$\frac{n_1}{n_2} = -\frac{\partial \Phi}{\partial x}. \tag{26}$$

Combining (6) and (26) we have that
$$u\frac{\partial \Phi}{\partial x} - v = 0 \tag{27}$$

along the forebody or, displaying the full functional dependence on the coordinates and design parameters, we have at a point (x, y) on the forebody, and at any time t,
$$u\left(t, x, y = \Phi(x; P_1, P_2, \ldots, P_K); M_0^2, P_1, P_2, \ldots, P_K\right)\frac{\partial \Phi}{\partial x}(x; P_1, P_2, \ldots, P_K) \\ - v\left(t, x, y = \Phi(x; P_1, P_2, \ldots, P_K); M_0^2, P_1, P_2, \ldots, P_K\right) = 0. \tag{28}$$

We can proceed to differentiate (28) with respect any of the forebody design parameters P_k, $k = 1, \ldots, K$. The result is that, along the forebody for $k = 1, \ldots, K$,
$$u_k \frac{\partial \Phi}{\partial x} - v_k = -\left(\frac{\partial u}{\partial y}\right)\left(\frac{\partial \Phi}{\partial P_k}\right)\left(\frac{\partial \Phi}{\partial x}\right) - u\frac{\partial}{\partial x}\left(\frac{\partial \Phi}{\partial P_k}\right) + \left(\frac{\partial v}{\partial y}\right)\left(\frac{\partial \Phi}{\partial P_k}\right), \tag{29}$$

where u, v, and their derivatives are evaluated at the forebody $(x, y = \Phi(x))$.

If an iterative scheme is used to find a steady state solution of this system ((21), (25), (29)), then we assume that present guesses for the state variables u and v and their derivatives $\partial u/\partial y$ and $\partial v/\partial y$ and for the design parameters M_0^2 and P_k, $k = 1, \ldots, K$, are known. It follows that the right-hand-side of (29) is known as well and equation (29), the boundary conditions along the forebody for the sensitivities with respect to the forebody design parameters, is merely an inhomogeneous version of (27), the boundary condition along the forebody for the state.

Let us now specialize to the type of forebodies considered by Huddleston, [5,6], *i.e.*
$$\Phi(x; P_1, P_2, \ldots, P_K) = \sum_{k=1}^{K} P_k \phi_k(x), \tag{30}$$

where $\phi_k(x)$, $k=1,\ldots,K$, are prescribed functions, e.g., Bezier curves. In this case,

$$\frac{\partial \Phi}{\partial P_k} = \phi_k(x) \quad \text{and} \quad \frac{\partial}{\partial x}\left(\frac{\partial \Phi}{\partial P_k}\right) = \frac{d\phi_k}{dx}(x), \tag{31}$$

and

$$\frac{\partial \Phi}{\partial x} = \sum_{k=1}^{K} P_k \frac{d\phi_k}{dx}(x). \tag{32}$$

Combining (29)-(32), one obtains that, at any point $(x, \Phi(x))$ on the forebody and for each $k=1,\ldots,K$,

$$\left(\sum_{j=1}^{K} P_j \frac{d\phi_j}{dx}\right) u_k - v_k = -\left(\frac{\partial u}{\partial y}\right)\left(\sum_{j=1}^{K} P_j \frac{d\phi_j}{dx}\right)\phi_k - u\frac{d\phi_k}{dx} + \left(\frac{\partial v}{\partial y}\right)\phi_k. \tag{33}$$

For forebodies of the type (30), (33) gives the the boundary conditions along the forebody for the sensitivities with respect to the forebody design parameters P_k, $k=1,\ldots,K$. It is now clear that, given guesses for the state variables u and v and their derivatives $\partial u/\partial y$ and $\partial v/\partial y$ and for the design parameters M_0^2 and P_k, $k=1,\ldots,K$, then the right-hand-side of (33) is known.

5 Computing Sensitivities using an Existing Code for the State

Suppose one has available a code to compute the state variables, i.e., to find approximate solutions of (1) along with boundary and initial conditions. In principle, it is an easy matter to amend such a code so that it can also compute sensitivities.

First, let us compare (1) with (9). If one wishes to amend the existing state code that can handle (1) so that it can treat (9) as well, one has to change the definitions of the flux functions from those given in (2) to those given in (10). Note that the solution for the state is needed in order to evaluate the flux functions of (10).

Next, note that (9) and (22) are identical differential equations. Thus, the changes made to the code in order to treat (9) can also be used to treat (22). In fact, as long as the differential equation and any other part of the problem specification do not explicitly depend on the design parameters, the analogous relations will be the same for all the sensitivities.

The only changes that vary from one sensitivity calculation to another are those that arise from conditions in which the design parameters appear explicitly. In our example, for the sensitivity with respect to M_0^2, one must change the portion of the code that treats the inflow conditions (4)-(5) so that it can instead treat (16)-(17). The only changes needed to accomplish this are to the data of the inflow conditions. In the problem considered here, the nature (i.e. what variables are specified) of the boundary conditions at the inflow, and everywhere else, is not affected. Note that for the sensitivity with respect to M_0^2 the boundary condition (13) on the forebody is the same as that for the state, given by (6).

For the sensitivities with respect to the forebody design parameters, the inflow boundary conditions simplify to (25), i.e., they become homogeneous. The boundary condition at the forebody is now given by (29) or (33). Once again, the nature of the boundary conditions is unchanged from that for the state, and only the data that is specified is different. For the inflow boundary conditions, we may still specify the same conditions for the sensitivities, but now they would be homogeneous. The boundary conditions along the forebody change only in that they become inhomogeneous, (compare (27) and (33)).

In summary, to change a code for the state so that it also handles the sensitivities, one must redefine the flux functions in the differential equations, and the data in the boundary conditions. The changes necessary in the code to account for any particular relation that does not explicitly involve the design parameters are independent of which sensitivity one is presently considering.

The previous remarks are concerned only with the changes one must effect in a state code in order to handle the fact that one is discretizing a different problem when one considers the sensitivities. We have seen that these changes are not major in nature. However, there are additional changes that may be needed when one attempts to solve the discrete equations. In the numerical results presented below we use the finite difference code "PARC" (see [2,5]) to solve the state and sensitivity equations. However, the following comments apply equally well to other CFD codes of this type.

Since we are interested in the steady design problems, the time derivative in (1) is considered only to provide a means for marching to a steady state. Now, suppose that at any stage of a Gauss-Newton, or other iteration, we have used PARC to find an approximate steady state solution of (1) plus boundary conditions. In order to do this, one has to solve a sequence of linear algebraic systems of the type

$$\left(I + \Delta t\, \mathbf{A}(\mathbf{Q}_h^{(n)})\right)\mathbf{Q}_h^{(n+1)} = \left(\mathbf{Q}_h^{(n)} + \Delta t\, \mathbf{B}(\mathbf{Q}_h^{(n)})\right), \qquad n = 0,1,2,\ldots, \tag{34}$$

where the sequence is terminated when one is satisfied that a steady state has been reached and where $\mathbf{Q}_h^{(n)}$ denotes the discrete approximation to the state $\mathbf{Q}$ at the time $t = n\Delta t$. We denote this steady state solution for the approximation to the state by $\mathbf{Q}_h$. One problem of the type (34) is solved for every time step. In (34), the matrix $\mathbf{A}$ and vector $\mathbf{B}$ arise from the spatial discretization of the fluxes and the boundary conditions. Both of these depend on the state at the previous time level.

Having computed a steady state solution by (34), the task at hand is to now compute the sensitivities. We will focus on $\mathbf{Q}'$, the sensitivity with respect to the inflow Mach number. Analogous results hold for the sensitivities with respect to the forebody design parameters. Recall that given a state, the sensitivity equations are linear in the sensitivities. Therefore, if one is interested in the steady state sensitivities, instead of (9) one may directly treat its stationary version

$$\frac{\partial \mathbf{F}'_1}{\partial x} + \frac{\partial \mathbf{F}'_2}{\partial y} = 0. \tag{35}$$

Since (35) is linear in the components of $\mathbf{Q}'$, one does not need to consider marching algorithms in order to compute a steady sensitivity. One merely discretizes (35) and solves the resultant linear system, which has the form

$$A'(\mathbf{Q}_h)\mathbf{Q}'_h = \mathbf{B}'(\mathbf{Q}_h), \tag{36}$$

where $\mathbf{Q}'_h$ denotes the discrete approximation to the steady sensitivity. The matrix A' and vector $\mathbf{B}'$ differ from the A and $\mathbf{B}$ of (34) because we have discretized different differential equations and boundary conditions. Note that A' and $\mathbf{B}'$ in (36) depend only on the steady state $\mathbf{Q}_h$ and thus (36) is a *linear system of algebraic equations* for the discrete sensitivity $\mathbf{Q}'_h$.

The cost of finding a solution of (36) is similar to that for finding the solution of (34) for a single value of n, *i.e.*, for a single time step. The differences in the assembly of the coefficient matrices and right-hand-sides of (34) and (36) are minor. Thus, in theory at least, *one can obtain a steady sensitivity in the same computer time it takes to perform one time step in a state calculation.* If one wants to obtain all the sensitivities, *e.g.*, $K+1$ in our example, one can do so at a cost similar to, *e.g.*, $K+1$ time steps of the state calculation. This is very cheap compared to the multiple state calculations necessary in order to compute sensitivities through the use of difference quotients.

In practice, these "optimal" estimates of speed up are rarely achieved. Moreover, it is important to note that finite difference (FD) and sensitivity equation (SE) methods do not necessarily produce the same results. Since the ultimate goal is to find useful and cheap gradients for optimization, the most important issue is whether or not the SE method combined with an optimization algorithm produces a convergent optimal design as fast as possible. We have tested this scheme on the forebody design problem with excellent results.

6 A Numerical Example

In order to illustrate the use of the SE method in computing sensitivities, we used the PARC code as described above to find approximate solutions of the sensitivity equations and compared the results to the finite difference method. In Figure 2 we show the approximations of $\frac{\partial}{\partial P_1}m(x,y,\hat{M}_0^2,\hat{P}_1,\hat{P}_2)$ for $\hat{M}_0 = 2, \hat{P}_1 = .1$ and $\hat{P}_2 = .15$ where the forebody is described by two Bezier parameters (P_1, P_2). Both pictures are "converged" estimates. Note that there are considerable differences between the FD method and the SE method. Moreover, in Figure 3 we see that not only do the FD and SE methods produce different sensitivities, the value of the step size ΔP_1 can greatly influence the FD approximations. Finally, we note that the SE method ran 4 to 5 times faster than the FD method. Also, although space prohibits a discussion of the optimization problem here, we have used the SE method in a trust region optimization scheme to produce an optimal forebody design for the 2D problem in [5,6]. These results will appear in a forthcoming paper.

7 Conclusions

The problem of computing accurate sensitivities in problems involving solutions to parameterized partial differential equations is an important part of optimal design. The goal is to find derivatives of solutions of partial differential equations with respect to various parameters (including domain shapes) and to use these derivatives in some type of optimization scheme. In almost all practical problems, solutions must be obtained by numerical approximations. This fact leads to "black box" methods for optimal design. In its most basic form, a black box method produces approximate solutions that are then differentiated (by finite differences, automatic differentiation, etc.). The sensitivity equation (SE) method presented here is based on first deriving partial differential equations for the derivatives and then approximating these equations numerically. Both approaches produce numerical approximations of the sensitivities. However, the (SE) method can often reduce computational effort, speed up the calculations and, provided that accurate computational schemes can be devised for the sensitivity equations, the derivatives can be computed with the same degree of accuracy as the state. The 2D optimal forebody simulator problem is an excellent problem for illustrating these points. The numerical results presented here show that the (SE) method is potentially applicable to real problems and, at the same time, raises many interesting theoretical and practical questions.

References

[1] D. Beale and M. Collier, *Validation of a Free-jet Technique for Evaluating Inlet-Engine Com-*

patibility, AIAA Paper 89-2325, AIAA/ASME/SAE/ASEE 25th Joint Propulsion Conference, Monterey, CA, July 1989.

[2] G. Cooper and W. Phares, *CFD Applications in an Aerospace Engine Test Facility*, AIAA Paper 90-2003, AIAA/ASME/SAE/ASEE 26th Joint Propulsion Conference, Orlando, FL, July 1990.

[3] P. Frank and G. Shubin, *A Comparison of Optimization-Based Approaches for a Model of Computational Aerodynamics Design Problem*, Journal of Computational Physics 98, (1992), pp. 74-89.

[4] J. Haslinger and P. Neittaanmäki, *Finite Element Approximation for OSD: Theory and Application*, John Wiley & Sons (1988).

[5] D. Huddleston, *Development of a Free-Jet Forebody Simulator Design Optimization Method*, AEDC-TR-90-22, Arnold Engineering Development Center, Arnold AFB, TN, December, 1990.

[6] D. Huddleston, *Aerodynamic Design Optimization Using Computational Fluid Dynamics*, Ph.D. Dissertation, University of Tennessee, Knoxville, Tennessee, December, 1989.

[7] O. Pironneau, *Optimal Shape Design for Elliptic Systems*, Springer Series in Computational Physics (1983).

[8] G.R. Shubin and P.D. Frank, *A Comparison of the Implicit Gradient Approach and the Variational Approach to Aerodynamic Design Optimization*, Applied Mathematics and Statistics 7, July 1986, pp. 1074-1080, also AIAA Paper 85-0020.

[9] S. Ta'asan, *One Shot Methods for Optimal Control of Distributed Parameter Systems I: Finite Dimensional Control*, ICASE Report No. 91-2, NASA Langley Research Center, Hampton, VA, 1991.

[10] A.C. Taylor III, G.W. Hou and V.M. Korivi, *A Methodology for Determining Aerodynamic Sensitivity Derivatives With Respect to Variation of Geometric Shape*, Proceedings of the AIAA/ASME/ASCE/AHS/ASC 32nd Structures, Structural Dynamics, and Materials Conference, April 8-10, Baltimore, MD, AIAA Paper 91-1101.

[11] A.C. Taylor III, G.W. Hou and V.M. Korivi, *Sensitivity Analysis, Approximate Analysis and Design Optimization for Internal and External Viscous Flows*, AIAA Paper 91-3083, AIAA Aircraft Design Systems and Operations Meeting, Baltimore, MD, September 1991.

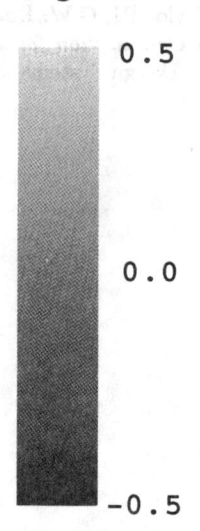

Sensitivity of U-Momentum
with respect to First Bezier Parameter

Sensitivity Equation Method

Values taken at:
Inl. Mach # = 1.7
Bez. P. #1 = 0.10
Bez. P. #2 = 0.15

Finite Difference Method

Legend

0.5

0.0

-0.5

Figure 2

Absolute Difference of U-Mom. Sensitivities with respect to First Bezier Parameter obtained using F.D. and S.E. Methods

```
Values at: Inlet Mach # = 1.7
           Bezier Parameter #1 = 0.10
           Bezier Parameter #2 = 0.15
```

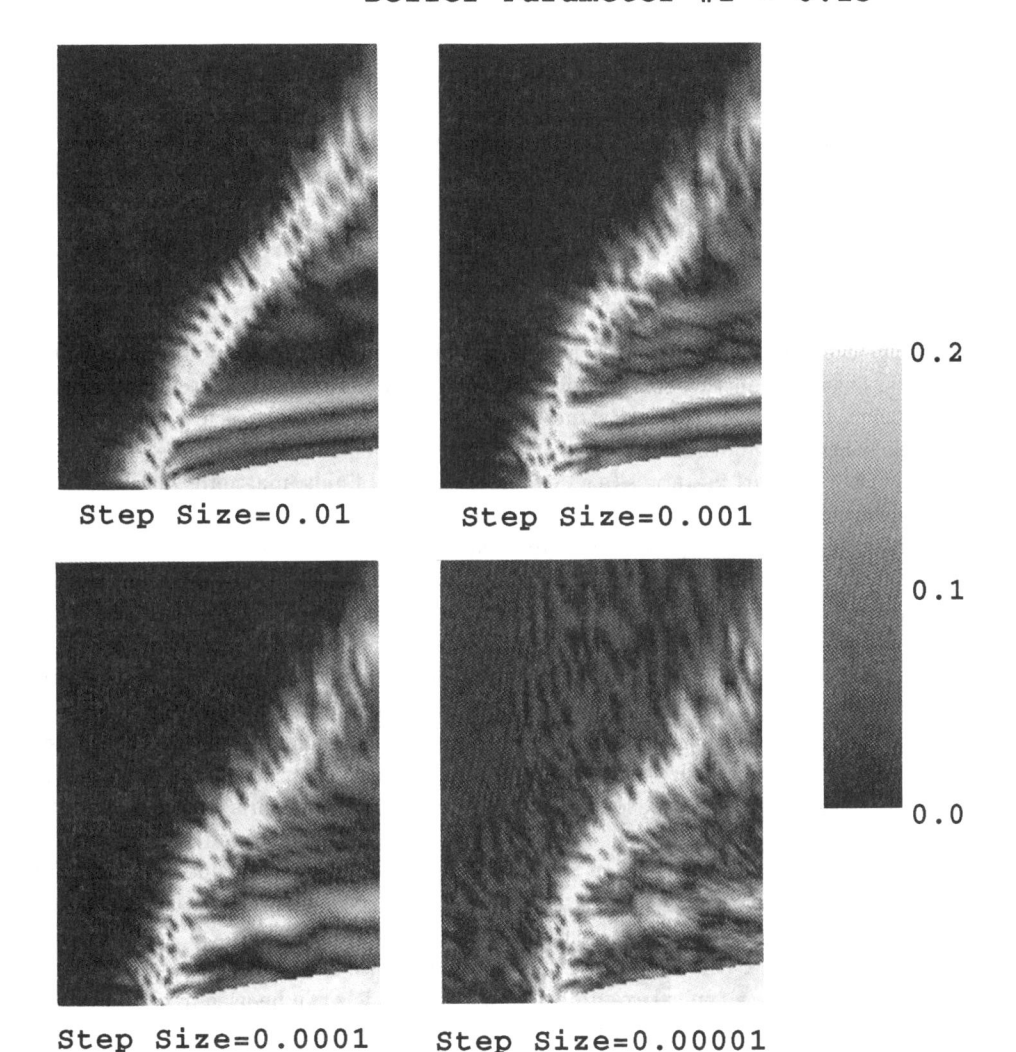

Step Size=0.01   Step Size=0.001

Step Size=0.0001   Step Size=0.00001

Figure 3

Chapter 3
Models for Control in Smart Material Structures *

H.T. Banks †   R.C. Smith ‡

Abstract

Models for piezoceramic actuation of bending moments and extensional/ compressional in-plane forces in structures are outlined in the context of distributed parameter beam models. The resulting actuators are then modeled as the control in a structure/fluid interaction example from acoustics. Computational results for the nonlinear closed loop control system are given.

1 Introduction

An emerging technology involving smart materials offers new challenges and opportunities to scientists, especially in the area of development of sophisticated actuators and sensors in the context of adaptive material structures. Of special importance are issues related to feedback control, in both conceptual design and implementation. There are a large number of classes of smart materials (e.g., electrorheological fluids, magnetostrictives, shape memory alloys), but our focus here is on piezoceramic devices, how to model these, and how they may be used in a typical structure/fluid interaction system.

We concentrate on models in terms of mathematical expressions relating moments and forces to applied voltages which can be used for control purposes. As we shall see below, these lead to unbounded control input operators (and unbounded observation operators in a similar development for sensors). Control theorists and engineers have considered unbounded operator inputs in the context of distributed parameter systems for some years now. However, the related mathematical and engineering theories and developments were most often motivated by idealizations of actual physical actuators. In many cases, careful modeling of actuator (proof-mass, solenoids, etc.) dynamics results in a hybrid distributed parameter system (e.g., see [2]) with bounded operators describing input terms. In this presentation we explain that this is *not* the situation for bonded or imbedded piezoceramic patches employed as actuators. Careful modeling arguments lead to actuator coefficients involving the Dirac delta "function" δ and its "derivative" δ'.

Our goal here is to outline modeling concepts to be used in control via piezoceramic patch actuators and give a typical example of how these models can be used in computations for feedback systems. In this case, we use an example from structural acoustics to illustrate

---

*The research of H.T.B. was supported in part by the Air Force Office of Scientific Research under grant AFOSR-90-0091. This research was also supported by the National Aeronautics and Space Administration under NASA Contract Numbers NAS1-18605 and NAS1-19480 while H.T.B. was a visiting scientist and R.C.S. was in residence at the Institute for Computer Applications in Science and Engineering (ICASE), NASA Langley Research Center, Hampton, VA 23681.

†Center for Research in Scientific Computation, North Carolina State University, Raleigh, NC 27695

‡ICASE, NASA Langley Research Center, Hampton, VA 23681

these new concepts which offer great promise in the development of truly adaptive or smart material structures.

In our presentation we shall concentrate on actuator aspects of piezoceramic devices. For a complete smart structure model, one must also consider piezoceramic sensors or so-called self-sensing actuators (e.g., see [9]). For example, in the noise suppression example detailed below, we assume that acoustic pressure in the cavity and beam displacements are sensed for feedback. For a smart material system, one would use piezoceramic sensors and cavity pressure sensors to construct a state estimator for feedback.

2 Piezoceramic Patch Models

To begin our considerations, we outline quantitative models for excitation and control of structures via piezoceramic patch pairs. To illustrate specific models, we consider a beam of length ℓ, thickness h and width 1 as depicted in Figure 1.

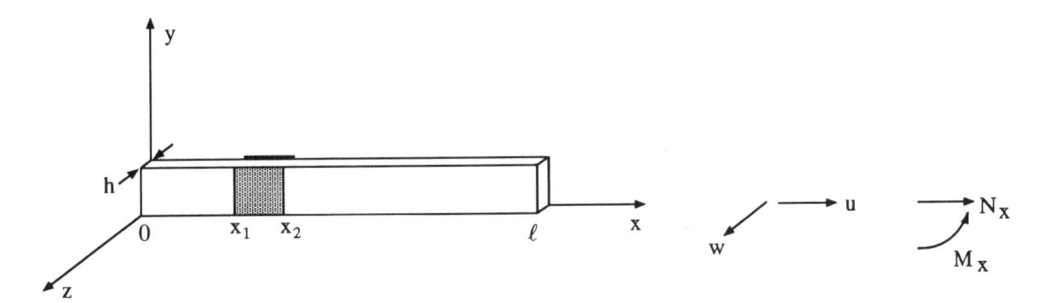

FIG. 1. *Cantilever beam with piezoceramic patches.*

We assume that patch pairs are bonded to the beam at $x_1 \leq x \leq x_2$ along the beam which is cantilevered with fixed end at $x = 0$ and free end at $x = \ell$. Each piezoceramic patch is inherently an elctro-mechanical transducer which, when excited by an electric field, induces a strain in the material in the axial direction of the beam if the patch is poled across the thickness as depicted in Figure 2. We further assume that we have two patches with identical polarization, each of which can be excited independently with an applied voltage to produce elongation or contraction (see, e.g. [8, 10, 11] along with the extensive reference list of [6] for more specific details). If the patches are excited "out-of phase" (with equal magnitude voltages which are in opposite sense across the polarizations as depicted Figure 3a), one produces contraction in one patch and elongation in the other. This induces surface strains in the beam as shown in Figure 3b. The net result on the beam is a pure bending moment about the neutral axis as illustrated in Figure 3c.

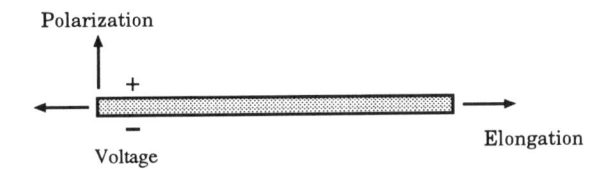

FIG. 2. *Induced piezoceramic patch strain.*

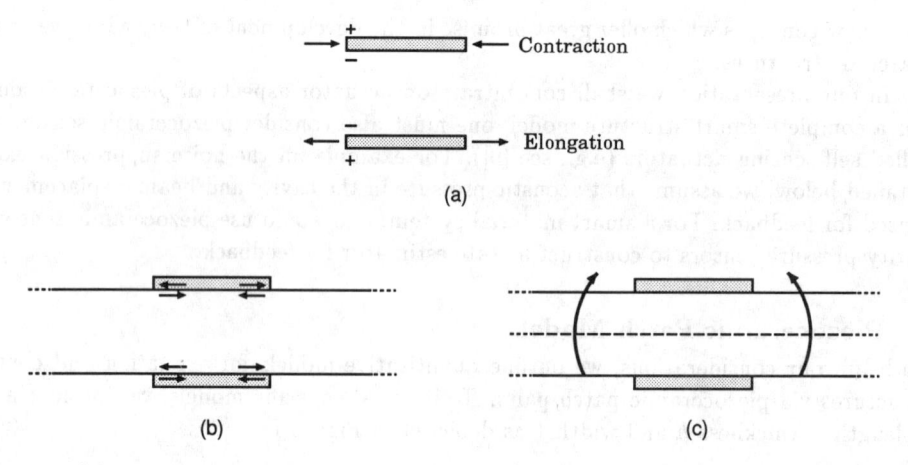

FIG. 3. *(a) Out-of-phase patch excitation; (b) Surface strains and stresses; (c) Pure bending moment about neutral axis of beam.*

The patches may also be excited "in-phase", as depicted schematically in Figure 4, resulting in in-plane compressional or extensional forces along the neutral axis of the beam. Of course, one may excite the patches with independent voltages, resulting , in general, in coupled bending and in-plane deformation of the beam.

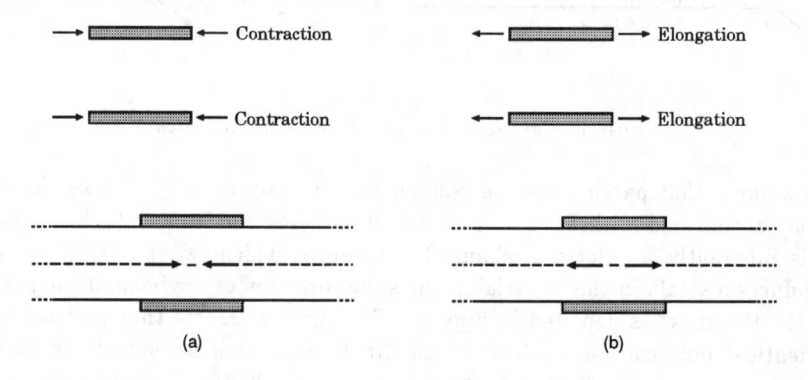

FIG. 4. *In-phase patch excitation; (a) compressional forces, and (b) extensional forces along the neutral axis.*

Quantitative expressions for excitation in the varied ways mentioned above can be derived as special cases of the general patch/thin shell interaction models developed in [6]. We consider transverse and axial displacements, denoted by $w(t,x)$ and $u(t,x)$, respectively, of the beam in Figure 1. Referring to [6], we find that for an undamped (assumed here for simplicity; internal damping can be readily added) beam we have the dynamic equations of motion given by

$$\rho h(x)\frac{\partial^2 u}{\partial t^2} - \frac{\partial}{\partial x}\left(Eh(x)\frac{\partial u}{\partial x}\right) = -S_{1,2}(x)\frac{\partial}{\partial x}[N_x]_{pe}$$

(1)

$$\rho h(x)\frac{\partial^2 w}{\partial t^2} + \frac{\partial^2}{\partial x^2}\left(EI(x)\frac{\partial^2 w}{\partial x^2}\right) = \hat{q}_n + \frac{\partial}{\partial x}\left(-\frac{\partial}{\partial x}[M_x]_{pe}\right)$$

for $0 \leq x \leq \ell$, $t > 0$. For patch pairs with edges at x_1 and x_2 (see Figure 1), the density is

$$\rho h(x) = \rho_b h + 2\rho_{pe}T\chi_{pe}(x) \quad , \quad \chi_{pe}(x) = \begin{cases} 1 & , \quad x_1 \leq x \leq x_2 \\ 0 & , \quad \text{otherwise} \end{cases}$$

where ρ_b is the mass density (in mass per unit volume) of the beam, and ρ_{pe} and T are the mass density and thickness of the patch, respectively (throughout this section, we are assuming that the beam and patches have width 1). Similarly, the terms Eh and EI are given by

$$Eh(x) = E_b h + 2E_{pe}T\chi_{pe}(x)$$
$$EI(x) = E_b \frac{h^3}{12} + E_{pe}\left[hT^2 + \frac{1}{2}h^2T + \frac{2}{3}T^3\right]\chi_{pe}(x)$$

where E_b and E_{pe} are the Young's moduli for the beam and patch, respectively.

The force components are given in terms of $\hat{q}_n$, the total external normal (transverse) surface load, and $[N_x]_{pe}$, the piezoceramic induced line forces in the x-direction (see [6] for a discussion concerning the relationship between the induced line forces $[N_x]_{pe}$, having units of force, and the longitudinal force $S_{1,2}\frac{\partial}{\partial x}[N_x]_{pe}$ having units of force per unit length along the neutral axis). The indicator function appearing in the longitudinal force has the values 1 for $x < (x_1 + x_2)/2$, 0 for $x = (x_1 + x_2)/2$, and -1 for $x > (x_1 + x_2)/2$, and determines the direction of the applied force. The quantity $[M_x]_{pe}$ represents the piezoceramic induced line moments (about the neutral axis) about the x-axis and has units of moment.

These dynamic equations must be coupled with appropriate boundary conditions which, in the configuration depicted in Figure 1, are given by

$$u(t,0) = 0 \quad , \quad \frac{\partial u}{\partial x}(t,\ell) = 0 \, ,$$

$$w(t,0) = \frac{\partial w}{\partial x}(t,0) = 0 \quad , \quad \frac{\partial^2 w}{\partial x^2}(t,\ell) = \frac{\partial}{\partial x}\left(EI(x)\frac{\partial^2 w}{\partial x^2}\right)(t,\ell) = 0 \, ,$$

for $t \geq 0$, as well as initial conditions $w(0,x) = \phi(x)$, $\frac{\partial w}{\partial t}(0,x) = \psi(x)$, $u(0,x) = \eta(x)$ and $\frac{\partial u}{\partial t}(0,x) = \gamma(x)$.

As is shown in [6], the external moments and forces resulting from the excitation of the patches can be written as

(2)
$$[N_x]_{pe} = -E_{pe}d_{31}\left[H(x - x_1) - H(x - x_2)\right]S_{1,2}(x)(V_1 + V_2)$$
$$[M_x]_{pe} = -E_{pe}\frac{d_{31}}{2}[h + T]\left[H(x - x_1) - H(x - x_2)\right](V_1 - V_2)$$

where V_1 and V_2 are the voltages into the patches (note that this allows for different voltages into the individual patches in the pair). The piezoelectric charge coefficient d_{31} relates the applied electric field with the resulting mechanic strain.

A variety of responses can be elicited through the manner in which the voltages V_1 and V_2 are chosen. If we excite the patches *out-of-phase*, for example, with $V_1 = -V_2 = V$, we have

(3)
$$[N_x]_{pe} = 0$$
$$[M_x]_{pe} = -E_{pe}d_{31}[h + T][H(x - x_1) - H(x - x_2)]V$$

which, of course, results in pure bending of the beam. If we excite the patches *in-phase*, e.g., with $V_1 = V_2 = V$, we have

(4)
$$[N_x]_{pe} = -2d_{31}E_{pe}[H(x-x_1) - H(x-x_2)]S_{1,2}(x)V$$
$$[M_x]_{pe} = 0,$$

resulting in deformation along the neutral axis. Other activation scenarios can be produced by varying the voltages into the individual patches with the ensuing external resultants following directly from (2). For example, *single patch activation in patch pairs* can be obtained with the choice $V_1 = V, V_2 = 0$. Hence the formulation (1) with external resultants given by (2) provides a great deal of flexibility in describing the beam dynamics resulting from a variety of patch geometries and excitation strategies.

3 Structural Acoustics Applications

To demonstrate the applicability of piezoceramic patches in a structural acoustics setting, we consider the problem of reducing structure-borne noise levels inside an acoustic cavity. This example is motivated by the physically important problem of reducing noise inside a fuselage which is due to structural vibrations caused by low frequency high amplitude exterior acoustic fields produced by the engines. Thus the general formulation of the problem consists of an exterior noise source which is separated from an interior cavity by an elastic boundary. This boundary transmits noise or vibrations from the exterior field to the interior cavity via fluid/structure interactions. Control is implemented via piezoceramic patches which are bonded to the elastic boundary of the cavity, and through applied voltages, can be excited so as to produce pure bending moments and/or extensional forces. In this way, we can take advantage of the natural "feedback" loop which is due to the coupling between the structural vibrations and the acoustic fields.

3.1 Nonlinear Model

Although the underlying physical applications are three dimensional in nature, many of the issues concerning modeling techniques, the identification of physical parameters, and the development of feasible control strategies can be studied in 2-D geometries, and it is in this reduced setting that we initially investigated these problems. For this work, we consider a simplified but typical 2-D domain $\Omega(t)$ which is separated from a periodic exterior perturbing force f by an elastic beam at one end (see Figure 5). This force, which can be considered to be due to an exterior noise source, causes vibrations in the beam which in turn generates unwanted noise in the cavity.

As motivated by experiments being designed for corresponding 3-D problems, the boundaries on three sides of the variable cavity $\Omega(t)$ are taken to be "hard" walls thus leading to zero normal velocity boundary conditions. It is also assumed that the perturbable boundary $\Gamma_0(t)$ consists of an impenetrable fixed-end Euler-Bernoulli beam with Kelvin-Voigt damping. As discussed in [5], the overall fluid/structure interaction model is nonlinear due to the variable domain as well as the nonlinear velocity and pressure couplings between the acoustic and structural responses. Under an assumption of small displacements which is inherent in the Euler-Bernoulli theory, the variable domain $\Omega(t)$ can be approximated by the fixed domain $\Omega \equiv [0,a] \times [0,\ell]$ (see Figure 6). The fully nonlinear coupling terms, however, are retained throughout the following discussion.

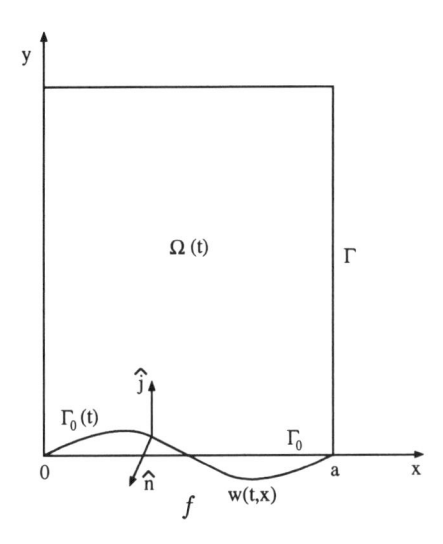

FIG. 5. *The 2-D domain.*

For control of structural vibrations and the acoustic pressure field in this model, s piezoceramic patches are attached to the beam as shown in Figure 6. As illustrated in this figure, the patches are excited in a manner so as to produce pure bending moments.

In terms of the velocity potential ϕ (so that $p = \rho_f \phi_t$ is the acoustic pressure) and the transverse beam displacements w, the approximate nonlinear controlled model for the coupled system is then given by

$$\phi_{tt} = c^2 \Delta \phi \qquad (x,y) \in \Omega\ , t > 0\ ,$$

$$\nabla \phi \cdot \hat{n} = 0 \qquad (x,y) \in \Gamma\ , t > 0\ ,$$

$$\nabla \phi(t, x, w(t,x)) \cdot \hat{n} = w_t(t,x) \qquad 0 < x < a\ , t > 0\ ,$$

(5)
$$\rho w_{tt} + \frac{\partial^2}{\partial x^2}\left(EI\frac{\partial^2 w}{\partial x^2} + c_D I \frac{\partial^3 w}{\partial x^2 \partial t}\right) = -\rho_f \phi_t(t, x, w(t,x)) + f(t,x) \qquad \begin{array}{l} 0 < x < a\ , \\ t > 0\ , \end{array}$$
$$+ \frac{\partial^2}{\partial x^2}\left(\sum_{i=1}^{s} \mathcal{K}_i^B u_i(t)\left[H(x - x_{i1}) - H(x - x_{i2})\right]\right)$$

$$w(t,0) = \frac{\partial w}{\partial x}(t,0) = w(t,a) = \frac{\partial w}{\partial x}(t,a) = 0 \qquad t > 0\ ,$$

$$\phi(0,x,y) = \phi_0(x,y)\ , \quad w(0,x) = w_0(x)$$
$$\phi_t(0,x,y) = \phi_1(x,y)\ , \quad w_t(0,x) = w_1(x)\ .$$

Here ρ_f and c are the equilibrium density of the atmosphere and speed of sound in the cavity. The beam parameters ρ, EI and $c_D I$ represent the linear mass density, stiffness and damping, respectively. Also, H denotes the Heaviside function, $u_i(t)$ is the voltage applied to the ith patch, and $\mathcal{K}_i^B$ is a parameter which depends on the geometry and piezoceramic material properties (see (3)). The combined piezoceramic material parameters $\mathcal{K}_i^B$, $i = 1, \cdots, s$, as well as the beam parameters ρ, EI and $c_D I$ are considered to be unknown and are estimated via the techniques discussed in later sections. It should be noted that the system (5) contains an unbounded control input term since it involves the second derivative of the Heaviside function.

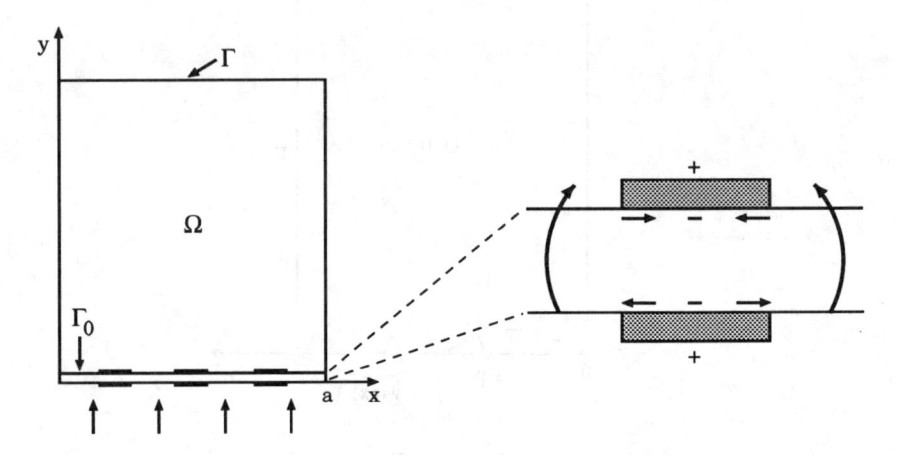

FIG. 6. *Acoustic cavity with piezoceramic patches.*

Variational Formulation

In the system model just developed, the beam and acoustic equations are in strong form which leads to difficulties in both the parameter estimation and control problems (for example, it necessitates the differentiation of discontinuous material parameters, leads to difficulties when implementing approximation schemes, and causes the problems associated with the unbounded control input term). To avoid these difficulties, it is advantageous to formulate the problem in weak or variational form.

In order to pose the problem in a variational form which is conducive to approximation, parameter identification and control, the state is taken to be $z = (\phi, w)$ in the state space $H = \bar{L}^2(\Omega) \times L^2(\Gamma_0)$ with the energy inner product

$$\left\langle \begin{pmatrix} \phi \\ w \end{pmatrix}, \begin{pmatrix} \xi \\ \eta \end{pmatrix} \right\rangle_H = \int_\Omega \frac{\rho_f}{c^2} \phi \xi \, d\omega + \int_{\Gamma_0} \rho w \eta \, d\gamma \ .$$

Here $\bar{L}^2(\Omega)$ is the quotient space of L^2 over the constant functions. The use of the quotient space results from the fact that the potentials are determined only up to a constant.

To provide a class of functions which are considered when defining a variational form of the problem, we also define the Hilbert space $V = \bar{H}^1(\Omega) \times H_0^2(\Gamma_0)$ where $\bar{H}^1(\Omega)$ is the quotient space of H^1 over the constant functions and $H_0^2(\Gamma_0)$ is given by $H_0^2(\Gamma_0) = \{\psi \in H^2(\Gamma_0) : \psi(x) = \psi'(x) = 0 \text{ at } x = 0, a\}$. The V inner product is taken as (here and below we use the notation $D = \frac{\partial}{\partial x}$)

$$\left\langle \begin{pmatrix} \phi \\ w \end{pmatrix}, \begin{pmatrix} \xi \\ \eta \end{pmatrix} \right\rangle_V = \int_\Omega \nabla \phi \cdot \nabla \xi \, d\omega + \int_{\Gamma_0} D^2 w D^2 \eta \, d\gamma \ .$$

The system can be written in first-order form by defining the product spaces $\mathcal{V} = V \times V$ and $\mathcal{H} = V \times H$ and taking the state to be $\mathcal{Z} = (\phi, w, \dot{\phi}, \dot{w})$. Note that the state now contains a multiple of the pressure since $p = \rho_f \dot{\phi}$. As discussed in [5], integration in combination with the use of Green's theorem then yields the first-order variational form

(6)
$$\begin{aligned}
\int_\Omega \frac{\rho_f}{c^2}(\dot\phi)_t \xi d\omega &+ \int_{\Gamma_0} \rho(\ddot w)_t \eta d\gamma \\
&+ \int_\Omega \rho_f \nabla\phi \cdot \nabla\xi d\omega + \int_{\Gamma_0} EID^2 w D^2 \eta d\gamma \\
&+ \int_{\Gamma_0} \left\{ c_D I D^2 \dot w D^2 \eta + \rho_f \left[\dot\phi(w)\eta - \dot w \xi \right] \right\} d\gamma \\
&= \int_{\Gamma_0} \sum_{i=1}^s \mathcal{K}_i^B u_i(t)(H_{i1} - H_{i2}) D^2 \eta d\gamma + \int_{\Gamma_0} f \eta d\gamma
\end{aligned}$$

for all (ξ, η) in V. Again, a more complete discussion and motivation concerning the formulation of the first-order system in weak form is given in [5].

3.2 System Approximation

To define an approximating subspace for the first-order system, suitable bases must be chosen for the beam and cavity discretizations. Cubic splines are used as a basis for the beam since they satisfy the smoothness requirement as well as being easily implemented when adapting to the fixed-end boundary conditions and patch discretizations. Letting $\{B_i^n\}_{i=1}^{n-1}$ denote the cubic splines which have been modified to satisfy the boundary conditions (see [1, 5] for details), the corresponding $n-1$ dimensional beam approximating subspace is given by $H_b^n = \text{span}\{B_i^n\}_{i=1}^{n-1}$ and the approximate beam solution is taken to be

$$w^N(t, x) = \sum_{i=1}^{n-1} w_i^N(t) B_i^n(x).$$

A tensored Legendre basis is used for the cavity discretization. Let $P_i^a(x)$ and $P_i^\ell(y)$ denote the standard Legendre polynomials that have been scaled by transformation to the intervals $[0, a]$ and $[0, \ell]$, respectively. The basis functions $\{B_{ij}^m\}$ for the cavity are then defined as

$$B_{ij}^m(x, y) = P_i^a(x) P_j^\ell(y) \quad \text{for} \quad i = 0, 1, \cdots, m_x, \quad j = 0, 1, \cdots, m_y, \quad i + j \neq 0,$$

where $m = (m_x + 1) \cdot (m_y + 1) - 1$. The condition $i + j \neq 0$ eliminates the constant function thus guaranteeing that the set of functions is suitable as a basis for the quotient space. The m dimensional cavity approximating subspace is taken to be $H_c^m = \text{span}\{B_i^m\}_{i=1}^m$ and the approximate cavity solution is given by

$$\begin{aligned}
\phi^N(t, x, y) &= \sum_{i=1}^m \phi_i^N(t) B_i^m(x, y) \\
&= \sum_{j=0}^{m_y} \sum_{\substack{i=0 \\ i+j \neq 0}}^{m_x} \tilde\phi_{ij}^N(t) P_i^a(x) P_j^\ell(y).
\end{aligned}$$

The approximating state space is then taken to be $H^N = H_c^m \times H_b^n$ where $N = m+n-1$, and the product space for the first order system is $\mathcal{H}^N = H^N \times H^N$. By restricting the infinite dimensional system (6) to $\mathcal{H}^N \times \mathcal{H}^N$, one obtains the nonlinear finite dimensional system

$$M^N \dot{y}^N(t) = \tilde{\mathcal{A}}^N \left(y^N(t) \right) + \tilde{B}^N u(t) + \tilde{F}^N(t)$$

$$M^N y^N(0) = \tilde{y}_0^N$$

or equivalently

(7)
$$\dot{y}^N(t) = \mathcal{A}^N \left(y^N(t) \right) + B^N u(t) + F^N(t)$$

$$y^N(0) = \tilde{y}_0^N \ .$$

Explicit descriptions of the mass and stiffness matrices M^N and $\tilde{\mathcal{A}}^N \left(y^N(t) \right)$ as well as detailed definitions of the control matrix $\tilde{B}^N$ and the force vector $\tilde{F}^N(t)$ can be found in [1, 4]. The vector $y^N(t) = (\phi_1^N(t), \cdots, \phi_m^N(t), w_1^N(t), \cdots, w_{n-1}^N(t), \dot{\phi}_1^N(t), \cdots, \dot{\phi}_m^N(t), \dot{w}_1^N(t), \cdots, \dot{w}_{n-1}^N(t))^T$ contains the $2N \times 1$ approximate state coefficients while $u(t) = (u_1(t), \cdots u_s(t))^T$ contains the s control variables. As detailed in [5], the nonlinearity in the matrix $\tilde{\mathcal{A}}^N \left(y^N(t) \right)$ manifests itself in the dependence of the matrix on the unknown coefficients $\{w_j(t)\}$.

3.3 Parameter Estimation Problems

The goal of the finite dimensional parameter estimation problem is to determine estimates of the "true" material parameters $\rho, EI, c_D I$ and $\mathcal{K}_i^B$, $i = 1, \cdots, s$, given data measurements z from some observable subspace Z of the approximate state space. To pose this mathematically, we let $q = (\rho, EI, c_D I, \mathcal{K}_1^B, \cdots, \mathcal{K}_s^B)$ and assume that $q \in Q$ where Q denotes an admissible parameter space. The finite dimensional parameter estimation problem is to then seek $\bar{q} \in Q$ which minimizes

(8)
$$J(w^N, z; q) = \left| \mathcal{C}_2 \left[\mathcal{C}_1 \left\{ w^N(t_i, q) \right\} - \{z(t_i)\} \right] \right|^2$$

given pointwise temporal measurements $\{z(t_i)\}$ at given points on the beam. Note that this minimization is performed subject to w^N satisfying the approximating coupled system equations (hence the coefficients $\left\{w_i^N(t)\right\}$ of w^N must satisfy (7)). Depending on the experimental apparatus, the data observations may consist of position, velocity, or acceleration measurements at points on the beam. In the first case, the operator $\mathcal{C}_1$ is simply the identity whereas in the latter two cases, it is $\mathcal{C}_1 = \frac{d}{dt}$ and $\mathcal{C}_1 = \frac{d^2}{dt^2}$, respectively. The form of the operator $\mathcal{C}_2$ depends upon whether one is performing the identification procedures in the time domain or in the frequency domain. In the time domain, $\mathcal{C}_2$ is the identity whereas it is the Fourier transform for identification in the frequency domain.

In order to develop a feasible scheme for estimating the material parameters ρ, EI and $c_D I$, we must make some assumptions regarding their spatial behavior. Because the beam and patches are considered to be homogeneous as well as uniform in width and thickness, it is reasonable to assume that the density, stiffness and damping parameters of the combined beam/piezoceramic patches are piecewise constant in nature (see for example, [7]). A suitable partition is then taken to be $\{x_k\} = \{0, a\} \bigcup \{\alpha_{ij}\}_{j=1,2}^{i=1,\cdots,s}$ where the $2s$ points $\{\alpha_{ij}\}$ are the endpoints of the s piezoceramic patches (see (5)). Finally, we assume that the parameters have the form

$$\rho(x) = \sum_{k=1}^{2s+1} c_k \tilde{B}_k(x) \quad , \quad c_1 = c_3 = \cdots = c_{2s+1}$$

(9)
$$EI(x) = \sum_{k=1}^{2s+1} \tilde{c}_k \tilde{B}_k(x) \quad , \quad \tilde{c}_1 = \tilde{c}_3 = \cdots = \tilde{c}_{2s+1}$$

$$c_D I(x) = \sum_{k=1}^{2s+1} \hat{c}_k \tilde{B}_k(x) \quad , \quad \hat{c}_1 = \hat{c}_3 = \cdots = \hat{c}_{2s+1}$$

$$\mathcal{K}_i^B(x) = \bar{c}_i \tilde{B}_i(x) \quad , \quad i = 1, \cdots, s$$

where the piecewise constant basis functions are defined by $\tilde{B}_k(x) \equiv H(x-x_{k-1})-H(x-x_k)$. The coefficient constraints $c_1 = c_3 = \cdots = c_{2s+1}$, and so on, result from the uniformity of the beam in areas not covered by patches. The admissible parameter space Q is obtained by restricting the set of linear basis combinations to those which satisfy the coefficient constraints as well as the physical constraints $\rho > 0$, $EI > 0$ and $c_D I \geq 0$ on $(0, a)$.

Example 1: Numerical Parameter Estimation

To test the parameter estimation methodology, we considered the system (5) in which the bounding end beam has bonded to it a centered piezoceramic patch covering 1/3 of its length (see Figure 7). The beam was assumed to have length, width and thickness $.6\,m, .1\,m$ and $.005\,m$, respectively, and the Young's modulus and beam density were taken to be $E = 7.1 \times 10^{10}\,N/m^2$ and $\rho_b = 2700\,kg/m^3$. This yielded stiffness $EI = 73.96\,Nm^2$ and linear mass density $\rho = 1.35\,kg/m$ for the beam where no patch is attached. The damping parameter for the beam was chosen to be $c_D I = .001\,kg\,m^3/sec$. The cavity was assumed to have dimensions $.6\,m$ by $1\,m$ and the speed of sound and atmospheric density were taken to be $c = 343\,m/sec$ and $\rho_f = 1.21\,kg/m^3$, respectively.

FIG. 7. *Acoustic chamber with one centered 1/3 length patch.*

The patch was assumed to have thickness $\mathcal{T} = .0005\,m$, and the Young's modulus and density were taken to be $E_{pe} = 6.3 \times 10^{10}\,N/m^2$ and $\rho_{pe} = 7650\,kg/m^3$ which are reasonable for a patch made from G-1195 piezoceramic material. Hence both the stiffness coefficient and the density in the region of the combined beam and patch (Region 2) were greater than that of the beam (Regions 1 and 3). We also assumed that the damping coefficient were slightly larger in Region 2 than on the outer regions.

For testing purposes, the true values were chosen as specified in Table 1. Due to the uniformity of the beam, only values in Regions 1 and 2 are reported in the tables since the values in Region 3 are constrained to be equal to those in Region 1.

TABLE 1. *True values of the material parameters.*

	$\rho\,(kg/m)$	$EI\,(Nm^2)$	$c_D I\,(kg\,m^3/sec)$
Region 1	1.35	73.96	.001
Region 2	2.115	125.4	.00125

As detailed in [5], the natural frequencies for the system were numerically determined by simulating an impact to the center of the beam, and it was found that the first four system responses occur at $62.9, 179.7, 342.1$ and 397.9 hertz (strictly speaking, these are the frequencies for the linearized system). These values can be compared to the frequencies $65.9, 181.3, 343.9$ and 387.8 hertz which were obtained for the system in which the beam was devoid of patches (see [3]).

To demonstrate the recovery of the physical parameters ρ, EI and $c_D I$, the patch parameter $\mathcal{K}^B$ was set to zero and the forcing function was taken to be

$$f(t,x) = \begin{cases} \sin(120\pi t) + \sin(360\pi t) + \sin(800\pi t) &, \quad 0 \le t \le 1/60 \\ 0 &, \quad 1/60 < t \le 8/60 \end{cases}$$

which initially excites the first, third and fourth system modes and then allows the oscillations to begin dying away due to the damping in the beam. A force of this type can be generated by an acoustic source with the above frequencies.

The parameter estimation was performed with data which was generated by calculating the acceleration of the central point of the beam at 498 uniformly distributed points throughout the time interval $[0, 8/60]$ (hence $C_1 = \frac{d^2}{dt^2}$ in (8)). The acceleration was determined by using a second-order central difference on the displacements which were observed at the points $(.3, t_k)$, $t_k = k \cdot \frac{8}{60(500)}$, $k = 2, \cdots, 499$. These displacements were obtained by solving the nonlinear finite dimensional system (7) using $m_x = m_y = 12$ and $n = 18$ basis functions. Due to the relatively small number of frequencies being matched, all identification procedures were performed in the time domain which implies that C_2 in (8) was taken to be the identity.

The initial guesses for the parameters are given in Table 2 while the estimates obtained with noisefree data are reported in Table 3. To obtain these results, $m_x = m_y = 10$ and $n = 16$ basis functions were used and the optimization was performed via a Levenberg-Marquardt routine. These results demonstrate that very accurate estimates for the material parameters can be obtained if the data is indeed noisefree (the accuracy of these results may be slightly enhanced by the fact that we used the same numerical scheme to generate the data as was used to solve the inverse problem, although a larger number of basis functions was used to generate the data so as to obtain a more accurate representation). Numerical results demonstrating the method for data containing various levels of noise as well as illustrating the estimation of all four parameters (that is, including $\mathcal{K}^B$) can be found in

[5]. Even with noise in the simulated data, one is able to obtain reasonable estimates of the parameters with our methods.

TABLE 2. *Initial guesses for the material parameters.*

	$\rho\,(kg/m)$	$EI\,(Nm^2)$	$c_D I\,(kg\,m^3/sec)$
Region 1	1.4	75.0	.0001
Region 2	2.0	127.0	.0001

TABLE 3. *Estimated values of the material parameters obtained with no noise added to the data.*

	$\rho\,(kg/m)$	$EI\,(Nm^2)$	$c_D I\,(kg\,m^3/sec)$
Region 1	1.34999996	73.95994605	.00100002
Region 2	2.11500038	125.40017944	.00124993

3.4 Control Problems

As noted previously, the infinite dimensional coupled system (6) and corresponding finite dimensional problem (7) are nonlinear due to the nature of the coupling terms. In order to determine a feedback law for this system, the following strategy was adopted. A linearized system corresponding to the infinite dimensional variational problem (6) was determined by assuming small beam displacements and approximating the nonlinear pressure and momentum coupling terms by corresponding linear terms (this linearization approximation is justified by the observation that for physically reasonable input forces, the beam displacements are of the order 10^{-5} m for the geometries of interest). The finite dimensional approximate linearized system was determined in a manner similar to that described in Section 2.3 for the nonlinear system, and the Riccati feedback gains for this approximate system were then computed by way of a periodic LQR theory (see [1]). Finally, these linear gains were fed back into the nonlinear problem to create a stable nonlinear closed loop control system.

To illustrate this control strategy, the linearized system and control problem are outlined, and examples demonstrating the behavior of the approximate feedback functional gains and controlled states are presented (a comprehensive discussion of the linear control problem can be found in [1, 3]). The linear gains are then applied to the nonlinear problem of interest and the ensuing control results are briefly discussed (a more detailed discussion of the nonlinear results is given in [4]).

Linear Control Problem

As discussed in [1], the approximation of the nonlinear coupling terms by the corresponding linear expressions leads to the linear finite dimensional system

$$M^N \dot{y}^N(t) = \tilde{A}^N y^N(t) + \tilde{B}^N u(t) + \tilde{F}^N(t)$$
$$M^N y^N(0) = \tilde{y}_0^N$$

or equivalently the finite dimensional Cauchy equation

(10)
$$\dot{y}^N(t) = A^N y^N(t) + B^N u(t) + F^N(t)$$
$$y^N(0) = y_0^N$$

(this latter system can be compared to the corresponding nonlinear system (7)). The components of the linear stiffness matrix can be found in [1].

The periodic finite dimensional control problem is then to find $u \in L^2(0,\tau)$ which minimizes
$$J^N(u) = \frac{1}{2}\int_0^\tau \left\{\langle Q^N y^N(t), y^N(t)\rangle_{\mathbf{R}^N} + \langle Ru(t), u(t)\rangle_{\mathbf{R}^s}\right\} dt, \quad N = m + n - 1$$
where y^N solves (10), τ is the period, R is an $s \times s$ diagonal matrix and $r_{ii} > 0, i = 1, \cdots, s$ is the weight on the controlling voltage into the ith patch.

The nonnegative definite matrix Q^N can be chosen so as to emphasize the minimization of particular state variables as well as to create windows that can be used to decrease state variations of certain frequencies. From energy considerations as discussed in [1], an appropriate choice for Q^N in this case is
$$Q^N = M^N \mathcal{D}$$
where M^N is the mass matrix, and the diagonal matrix $\mathcal{D}$ is given by
$$\mathcal{D} = \text{diag}\left[d_1 I^m, d_2 I^{n-1}, d_3 I^m, d_4 I^{n-1}\right].$$
Here I^k, $k = m, n-1$, denotes a $k \times k$ identity and the parameters d_i are chosen to enhance stability and performance of the feedback.

The optimal control is then given by
$$u^N(t) = R^{-1}(B^N)^T \left[r^N(t) - \Pi^N y^N(t)\right]$$
where Π^N is the solution to the algebraic Riccati equation
$$(A^N)^T \Pi^N + \Pi^N A^N - \Pi^N B^N R^{-1}(B^N)^T \Pi^N + Q^N = 0. \tag{11}$$
For the regulator problem with periodic forcing function $F^N(t)$, $r^N(t)$ must satisfy the linear differential equation
$$\begin{aligned}\dot{r}^N(t) &= -\left[A^N - B^N R^{-1}(B^N)^T \Pi^N\right]^T r^N(t) + \Pi^N F^N(t) \\ r^N(0) &= r^N(\tau)\end{aligned} \tag{12}$$
while the optimal trajectory is the solution to the linear differential equation
$$\begin{aligned}\dot{y}^N(t) &= \left[A^N - B^N R^{-1}(B^N)^T \Pi^N\right] y^N(t) + B^N R^{-1}(B^N)^T r^N(t) + F^N(t) \\ y^N(0) &= y^N(\tau).\end{aligned} \tag{13}$$

To highlight the contribution of the feedback functional gains to the control law, it is noted that with suitable assumptions and the definition $z^N = (\phi^N, w^N)$, the optimal control for the finite dimensional linear periodic problem is given by
$$\begin{aligned}u^N(t) &= -\langle f^N, z^N(t)\rangle_V - \langle g^N, \dot{z}^N(t)\rangle_H + R^{-1}(B^N)^T r^N(t) \\ &= -\left[\int_\Omega k_1^N \cdot \nabla \phi^N d\omega + \int_{\Gamma_0} k_2^N D^2 w^N d\gamma\right] - \left[\int_\Omega \frac{\rho_f}{c^2} k_3^N \dot{\phi}^N d\omega + \int_{\Gamma_0} \rho k_4^N \dot{w}^N d\gamma\right] \\ &\quad + R^{-1}(B^N)^T r^N(t).\end{aligned}$$

Here f^N and g^N are the finite dimensional gains, and k_1^N through k_4^N are the approximating functional gains. We observe that k_1^N, k_2^N, k_3^N and k_4^N can be thought of as the approximate acoustic velocity, beam bending moment, pressure, and beam velocity gains, respectively. These functions can be expressed as

$$k_1^N(x,y) = \sum_{i=1}^{m} k_{1,i}^N \nabla B_i^m(x,y) \quad , \quad k_2^N(x) = \sum_{i=1}^{n-1} k_{2,i}^N D^2 B_i^N(x)$$

$$k_3^N(x,y) = \sum_{i=1}^{m} k_{3,i}^N B_i^m(x,y) \quad , \quad k_4^N(x) = \sum_{i=1}^{n-1} k_{4,i}^N B_i^N(x)$$

where the vector

$$K^N = \left[k_{1,1}^N, \cdots, k_{1,m}^N, k_{2,1}^N, \cdots, k_{2,n-1}^N, k_{3,1}^N, \cdots, k_{3,m}^N, k_{4,1}^N, \cdots, k_{4,n-1}^N \right]$$

is given by $K^N = R^{-1} \left(B^N \right)^T \Pi^N \left(M^N \right)^{-1}$. The matrix Π^N again satisfies the algebraic Riccati equation (11).

Example 2: Linear Control

To illustrate the effects of feedback control on the linearized system, we considered the case in which the beam is subjected to the uniform (in space) forcing function

$$f(t,x) = 2.04 \left[\sin(150\pi t) + \sin(790\pi t) \right].$$

As noted in [3], this input excites the first and fourth system modes which have natural frequencies of 65.9 and 387.8 hertz, respectively. Physically, this models a multiple frequency exterior noise source with a root mean square (rms) sound pressure level of 120 dB.

The beam and cavity dimensions and parameters were taken to be the same as those used in the parameter estimation example except that here the combined structure consisting of the beam and piezoceramic patch was assumed to have uniform density, stiffness and damping coefficients along its length. This was done to simplify the ensuing computations and can be physically justified by assuming that the patches are imbedded in the beam and that the patch material differs only slightly in mass density, stiffness, and damping properties from the beam material.

Control was implemented via a single centered piezoceramic patch covering 1/6 of the beam length, and the results reported below were obtained with the choices $d_1 = d_2 = d_4 = 1$, $d_3 = 10^4$ and $R = 10^{-6}$ for the quadratic cost functional parameters. The parameter d_3 was chosen to have larger magnitude so as to more heavily penalize large pressure variations.

The functional gains k_2^N, k_3^N and k_4^N are plotted in Figure 8. Recall that k_2^N and k_4^N are the bending and beam velocity gains, respectively, and hence are defined on the interval $[0, .6]$ for this example. The pressure gain k_3^N is defined on $[0, .6] \times [0, 1]$ and Figure 8 contains plots of the functional values along the edge of the cavity at $(x, 0)$. In each case, the functional gains obtained with a variety of choices for the number of basis functions are plotted to show the behavior as the number of basis functions is increased. Although not completely definitive, the bending gains appear to be converging strongly as the basis number is increased. It is more difficult to determine the manner of convergence of the pressure and beam velocity gains (as might be expected when one carefully examines the functional spaces in which the problem is posed). Numerical tests indicate at least a weak convergence of these gains, and work is continuing to determine whether or not the convergence is actually strong in the function spaces of interest.

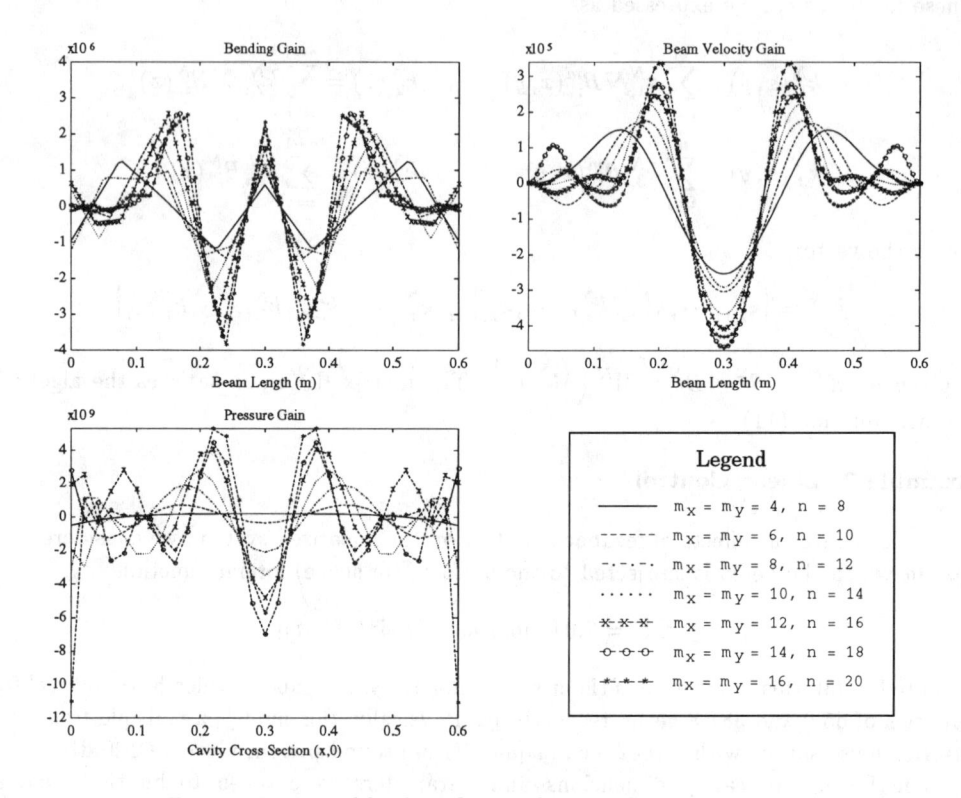

FIG. 8. *Functional bending, beam velocity and pressure gains.*

To demonstrate the effect of the feedback gains on the linear system, the uncontrolled and controlled beam displacements and acoustic pressures at the spatial points $X = .3$ and $(X, Y) = (.3, .1)$ are plotted in Figure 9. From numerical tests, it was found that for the above forcing function, the uncontrolled and controlled pressure oscillations could be resolved with the choices $m_x = m_y = 10$ and $n = 16$ which results in a total of 120 cavity and 15 beam basis functions. The time interval of interest was taken to be $[0, 9/75]$ and the controlling voltage obtained with control starting at time $T = 0$ is plotted in Figure 10.

From Figure 9, it can be seen that when control is started at $T = 0$, the solutions are periodic and are maintained at a level which is about 12% of that found in the uncontrolled case. By calculating the rms pressures, it was determined that at the point $(X, Y) = (.3, .1)$, the uncontrolled sound pressure level is 97.75 dB whereas the controlled sound pressure in this case is maintained at a level of 78.93 dB.

As illustrated in [3], the results in Figure 9 are representative of those obtained throughout the cavity (in fact, the response in the controlled case decreases even more as one moves deeper into the cavity), and they demonstrate that the pressure and beam displacements are significantly reduced and maintained at a very low level of magnitude in spite of the periodic forcing function.

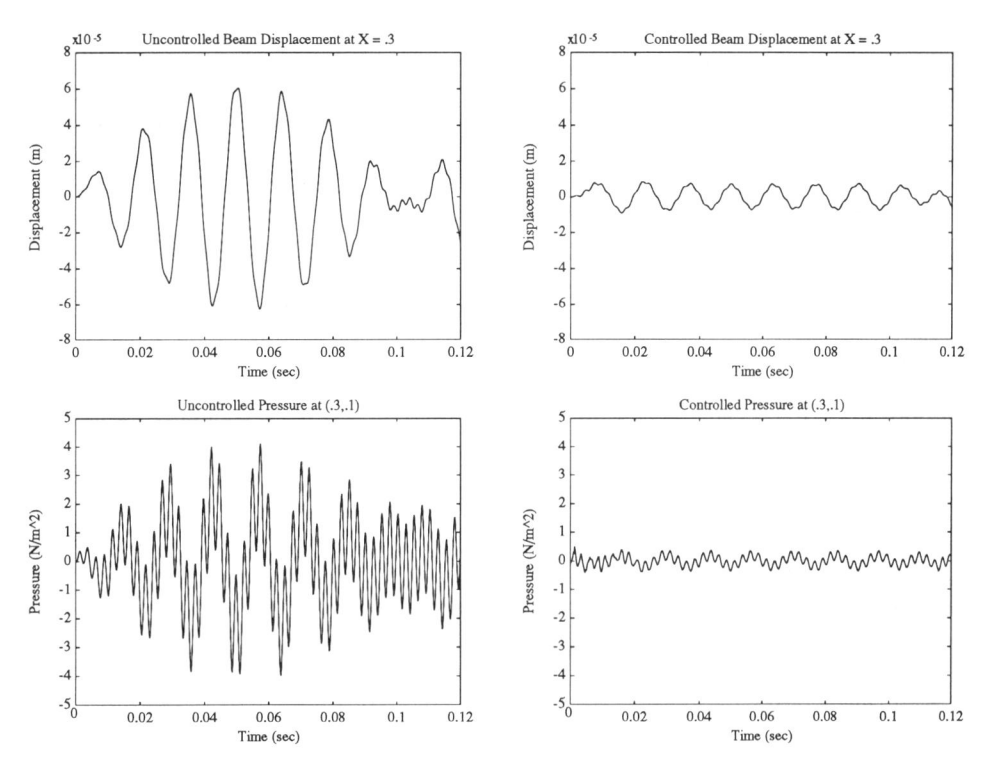

FIG. 9. *Uncontrolled and controlled beam displacements and pressures for the linearized system at the points $X = .3$ and $(X, Y) = (.3, .1)$. Control is started at $T = 0$.*

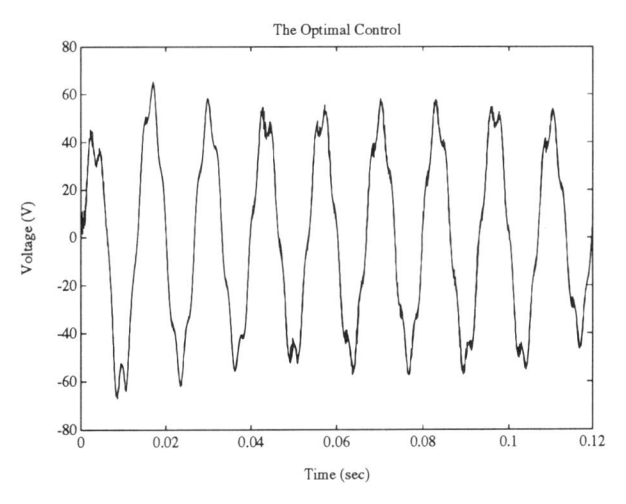

FIG. 10. *The optimal control $u(t)$ for the linearized system when control is started at $T = 0$.*

Nonlinear Control Problem

The previous discussion and example demonstrated the application of the LQR methodology to the linearized periodic system. To extend these results to the nonlinear system of interest, the linear gains were calculated and fed back into the nonlinear system (7), thus yielding the suboptimal control

$$u^N(t) = R^{-1}(B^N)^T \left[r^N(t) - \Pi^N y^N(t) \right] \tag{14}$$

and the closed loop system

$$\dot{y}^N(t) = \mathcal{A}^N\left(y^N(t)\right) - B^N R^{-1}(B^N)^T \Pi^N y^N(t) + B^N R^{-1}(B^N)^T r^N(t) + F^N(t) \tag{15}$$
$$y^N(0) = y^N(\tau).$$

The Riccati matrix Π^N and tracking vector $r^N(t)$ are solutions to (11) and (12) which arise when formulating the corresponding LQR problem.

Example 3: Nonlinear Control

The effects of feeding the linear gains back into the nonlinear system are demonstrated by considering the problem from Example 2, but in this case retaining the fully nonlinear coupling terms. For the parameter choices listed in the previous example, the uncontrolled trajectories were determined using (7) (with $u(t) = 0$) and are plotted in Figure 11. The same figure contains the plots of the controlled beam displacements and pressure oscillations as determined by (15).

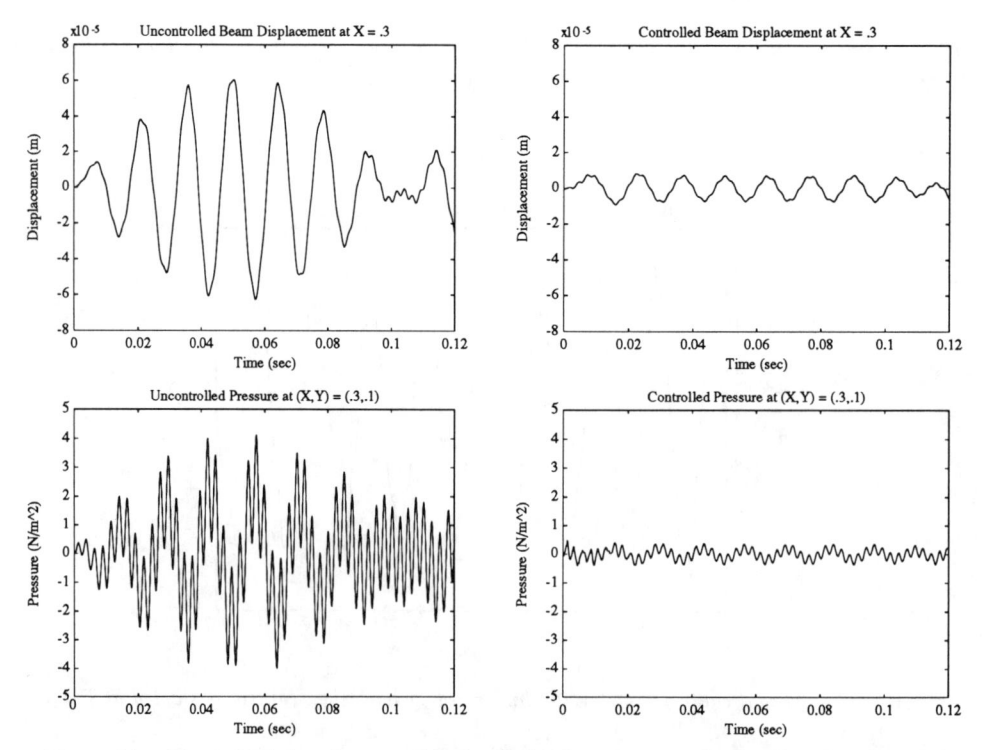

FIG. 11. *Uncontrolled and controlled beam displacements and pressures for the nonlinear problem at the points $X = .3$ and $(X, Y) = (.3, .1)$. Control is started at $T = 0$.*

As demonstrated by results from this example as well as the more extensive set of examples in [4], the strategy of using the gains from the linearized system to control the nonlinear system of interest is very effective for this problem. This is partly due to the weakness of the nonlinearity. In fact, by comparing the nonlinear results in Figure 11 with the corresponding linear ones in Figure 9, it can be seen that the two are virtually identical. Qualitatively, the linear and nonlinear results are the same to within graphical accuracy, and while very small quantitative differences can be detected, these differences do not affect the decibel levels obtained in either the uncontrolled or controlled cases. The controlling voltage $u^N(t)$ as given by (14) was also virtually identical to that obtained in the linear simulation (see Figure 10), and its plot is not repeated here. The high degree of similarity between the linear and nonlinear results can be explained by the fact that the beam displacements are very small and hence the linearized coupling terms quite accurately approximate the true nonlinear expressions.

4 Conclusion

In this paper we have outlined quantitative modeling concepts for use of piezoceramic devices as actuators and illustrated these ideas by example from structural acoustics. In this example, it is seen how one might use such control devices to develop a noise suppression system based on feedback. Our presentation represents the first steps in the development of a theory of control via smart materials wherein one senses and actuates via the material itself in a dynamic feedback setting.

Acknowledgements

The authors would like to thank H.C Lester and R.J. Silcox of the Acoustics Division, NASA Langley Research Center, for numerous discussions concerning the structural acoustics example in this work.

References

[1] H.T. Banks, W. Fang, R.J. Silcox and R.C. Smith, *Approximation Methods for Control of Acoustic/Structure Models with Piezoceramic Actuators*, Journal of Intelligent Material Systems and Structures, 4(1) (1993), pp. 98-116.

[2] H.T. Banks, S. Gates, G. Rosen and Y. Wang, *The Identification of a Distributed Parameter Model for a Flexible Structure*, SIAM Journal of Control and Optimization, 26 (1988), pp. 743–762.

[3] H.T. Banks, R.J. Silcox and R.C. Smith, *The Modeling and Control of Acoustic/Structure Interaction Problems via Piezoceramic Actuators: 2-D Numerical Examples*, ICASE Report 92-17, submitted to ASME Journal of Vibration and Acoustics.

[4] H.T. Banks and R.C. Smith, *Feedback Control of Noise in a 2-D Nonlinear Structural Acoustics Model*, preprint.

[5] ———, *Parameter Estimation in a 2-D Acoustic/Structure Interaction Model with Fully Nonlinear Coupling Conditions*, preprint.

[6] H.T. Banks, R.C. Smith and Y. Wang, *The Modeling of Piezoceramic Patch Interactions with Shells, Plates and Beams*, ICASE Report 92-66, to appear in Quarterly of Applied Mathematics.

[7] H.T. Banks, Y. Wang, D.J. Inman and J.C. Slater, *Variable Coefficient Distributed Parameter System Models for Structures with Piezoceramic Actuators and Sensors*, Proc. of the 31st Conf. on Decision and Control, Tucson, AZ, Dec. 16-18, 1992, pp. 1803–1808.

[8] E.F. Crawley, J. de Luis, N.W. Hagood and E.H. Anderson, *Development of Piezoelectric Technology for Applications in Control of Intelligent Structures*, Applications in Control of Intelligent Structures, American Controls Conference, Atlanta, June 1988, pp. 1890–1896.

[9] J.J. Dosch, D.J. Inman and E. Garcia, *A Self-Sensing Piezoelectric Actuator for Collocated Control*, Journal of Intelligent Material Systems and Structures, 3 (1992), pp. 166–185.

[10] J.L. Fansan and T.K. Caughey, *Positive Position Feedback Control for Large Space Structures*, AIAA Paper No. 87-0902, Proceedings of the 28th AIAA/ ASME/ASCE/AHS Structures, Structural Dynamics and Materials Conference, AIAA Dynamics Specialists Conference, Part IIb, Monterey, CA, April 9-10, 1987, pp. 588–598.

[11] H. Hagood, W. Chung and A. von Flotow, *Modelling of Piezoelectric Actuator Dynamics for Active Structural Control*, Journal of Intelligent Material Systems and Structures, 1 (1990), pp. 327–354.

Chapter 4
Bootstrap Methods for Inference in Least Squares Identification Problems

B.G. Fitzpatrick [*]   G. Yin [†]

Abstract

We consider the estimation of error distributions in least squares identification of distributed parameter systems. In particular, we examine consistency and asymptotic normality of empirical estimates of the error distribution, and we develop a bootstrap method for inference, including confidence intervals and tests for normality. We apply these techniques to identification of material parameters in vibration problems for flexible beams.

1 Introduction

When applying mathematical models in engineering and scientific problems, one is confronted with the task of choosing model parameters based on experimental observations. Typically, such parameter identification problems can be handled effectively with nonlinear least squares estimation. In many cases the observations contain various sources of error, a situation which inevitably leads to a discussion of statistical techniques for quantifying uncertainty. A major difficulty in understanding the impact of observation error on system identification and control is the determination of the probability distribution of these errors. In this paper we discuss the estimation of this error distribution within the context of least squares identification.

There has been a great deal of research activity focused on the statistical analysis of identification problems. One approach (see [3, 4, 11]) involves consistency and asymptotic normality of parameter estimators, as well as ANOVA-type test statistics. These techniques have been very useful in several applications (see [3, 4] for details), but they are somewhat limited in applicability to situations in which the errors are independent, identically distributed, and additive. To examine data for applicability of these methods, we need statistical techniques for testing these assumptions.

An interesting question along these lines involves testing observation errors for spatial and/or temporal uniformity. For example, noise levels in accelerometers used in vibration experiments are often reported as percentages, leading us to a multiplicative (rather than additive) error model. In another example (see [2]), a size structured model for larval striped bass populations was considered. Here a distributed model was fit to observations using least squares techniques, and the above-mentioned ANOVA statistics were used to

---

[*] Department of Mathematics and Center for Research in Scientific Computation, North Carolina State University, Raleigh, NC 27695-8205. Research of this author was supported in part by the Air Force Office of Scientific Research under grant AFOSR-91-0021.

[†] Department of Mathematics, Wayne State University, Detroit, MI 48202. Research of this author was supported in part by the National Science Foundation under grant DMS-9022139.

study changes in mortality rates. These tests seemed to work well only when data from fish of smaller sizes was omitted from the least squares fit, suggesting that the errors were not identically distributed over the size structure. Formally applied Kolmogorov-Smirnov tests rejected at all levels the hypothesis of i.i.d. errors; however, such tests are not applicable in cases in which parameters are estimated, as the results below will show.

For inference problems arising in identification and control applications, Bayesian techniques (see [1, 12]) show great promise; however, precise knowledge of the probability distribution of the observation error is required.

In this paper, we examine empirical distributions which can be used to estimate the measurement error distribution. We discuss methods for inference concerning parameter estimates in least squares as well as testing hypotheses about the error distribution. Since the errors themselves are not observable, the empirical distribution we construct requires some careful analysis. Thus, we discuss some of the asymptotic results contained in [13]: these results provide the theoretical foundation for inference techniques. Once an estimator for the error distribution has been computed, one can use this estimator in a Monte Carlo fashion to simulate data and test hypotheses about the parameters as well as the error distribution. We describe the implementation of a bootstrap method for building confidence sets around parameter estimates and developing tests concerning the validity of proposed statistical error models. We illustrate the bootstrap techniques using as an example some least squares estimation problems for an Euler-Bernoulli beam.

In Section 2 below, we discuss the least squares estimation problem, illustrating the ideas with a flexible beam example. We construct the error distribution estimator in Section 3, and we discuss the asymptotic behavior as the number of observations increases. The bootstrap technique is described in Section 4, and in Section 5 we give results from bootstrap computations for an Euler-Bernoulli beam example.

2 Empirical Distributions in Least Squares Identification

An important aspect of many applied problems is assessing the fit of a model to observed data. One particular question that arises is the impact of measurement uncertainty on parameter estimates obtained from such a fit. For an example problem, we consider an experiment in which a flexible beam, clamped at one end and free at the other, is deformed to an initial deflection and then released to vibrate freely. In our example, we assume that an accelerometer is placed at the tip of the beam, so that we obtain a sequence of acceleration measurements over a period of time. To model the beam we use the Euler-Bernoulli beam equation

$$\rho \frac{\partial^2 u}{\partial t^2}(x,t) + \gamma \frac{\partial u}{\partial t}(x,t) + \frac{\partial^2}{\partial x^2}\left[EI\frac{\partial^2 u}{\partial x^2}(x,t) + c_D I \frac{\partial^3 u}{\partial x^2 \partial t}(x,t)\right] = 0, \qquad t > 0, \qquad 0 < x < \ell,$$

where $u(x,t)$ denotes the displacement of the beam at position x along its axis and at time t, ρ is the linear mass density, γ is the viscous damping coefficient, EI is the stiffness, and $c_D I$ is the Kelvin-Voigt damping coefficient. The initial displacement $u(x,0)$ is known, and the beam is initially at rest, so that $\partial u/\partial t$ is initially 0. The beam is assumed to be clamped at $x = 0$ and free at $x = \ell$, conditions which determine the boundary conditions. We set $q = (\gamma, c_D I, EI)$, to denote the vector of parameters for the problem, and we include explicitly the parameter dependence of the solution with the notation $u(x,t;q)$.

We denote by $\{\hat{u}_{tt}(\ell, t_k)\}_{k=1}^n$ the collection of acceleration measurements. If the model equation provides a good description of the dynamics, one might expect that $\hat{u}_{tt}(\ell, t_k) \approx$

$u_{tt}(\ell, t_k; q^*)$, for some value of q^*. The identification task is to estimate the parameter vector q^* from these observations. The least squares approach involves minimizing the cost functional

$$J_n(q) = \frac{1}{n} \sum_{k=1}^{n} |\widehat{u}_{tt}(\ell, t_k) - \frac{\partial^2 u}{\partial t^2}(\ell, t; q)|^2.$$

Statistical tasks center on the relationship of the model to the data and the impact of this relationship on the parameter estimates.

The general least squares statistical problem that we consider here can be described as follows. We have a collection of observations $\{Y_k\}$ with

(1) $$Y_k = f(x_k, q^*) + \varepsilon_k, \ 1 \leq k \leq n,$$

where $\{x_k\}$ (with $x_k \in X \subset \mathbb{R}^r, 1 \leq k \leq n$) is a collection of settings at which the observations are made, $f(x, q)$ is a parameterized function, called the model function, and $\{\varepsilon_k\}$ denotes the measurement error. The function f typically arises from a differential equation model describing the system ($f = u_{tt}$ in the above example). As above, the idea here is that the "true parameter" q^* is unknown: we wish to estimate it by fitting the observations $\{Y_k\}$ to f. In what follows, we assume q^* lies in a known set Q_{ad}, referred to as the admissible parameter set. This set incorporates various constraints on the parameters, which in general depend on the specific application. For example, we would require $EI > 0$, and $\gamma, c_D I \geq 0$ in our flexible beam. To estimate the parameter q^*, we minimize the mean square objective function

(2) $$J_n(q) = \frac{1}{n} \sum_{k=1}^{n} (Y_k - f(x_k, q))^2,$$

and we denote by q_n a minimizer over Q_{ad}. The factor of $1/n$ is introduced to scale the functional for purposes of analyzing large sample ($n \to \infty$) behavior.

At this point, we review some results on least squares estimators. These results will provide the foundation for our study of the empirical distribution. To proceed, we first make the following assumptions:

(A1) The sequence $\{\varepsilon_k\}$ is a sequence of independent and identically distributed (i.i.d.) random variables with $E\varepsilon_k = 0$, $E\varepsilon_k^2 = \sigma^2 < \infty$, and distribution function F.

(A2) The set Q_{ad} is a compact subset of a complete, separable metric space Q. The set X is a compact subset of $\mathbb{R}^r$. The function $f: Q \to C(X)$ is continuous, where $C(X)$ denotes the space of continuous functions defined on X.

(A3) The settings $\{x_k\} \subset X$ are chosen in such a way that there exists a finite measure μ on X, such that for each bounded and continuous function h, $\frac{1}{n}\sum_{k=1}^{n} h(x_k) \xrightarrow{n} \int_X h \, d\mu$.

(A4) The functional $J^*(q) = \sigma^2 + \int_X (f(x, q^*) - f(x, q))^2 d\mu$ has a unique minimizer $q^* \in$ Int $Q_{ad} \subset$ Int Q, where Int G denotes the interior of the set G.

(A5) The set Q is a subset of $\mathbb{R}^r$, and that for each x, $f(x, \cdot)$, is twice continuously differentiable. In addition, the matrix

$$\mathcal{J} = 2 \int_X \frac{\partial f(x, q^*)}{\partial q} \frac{\partial f'(x, q^*)}{\partial q} d\mu(x) = \frac{\partial^2 J^*(q^*)}{\partial q^2}$$

is positive definite.

For motivations and applicability of these conditions, we refer the readers to the papers [3, 4, 11]. Note that continuity of f and compactness of Q_{ad} guarantee the existence of a minimizer q_n of J_n. The following results, which can be found in [3, 4, 11], will be used heavily in the sequel.

PROPOSITION 2.1 . *Under the above conditions (A1)—(A4)*
(1) *for each $q \in Q$, $P(\lim_n J_n(q) = J^*(q)) = 1$ and the convergence is uniform on each compact subset of Q.*
(2) *If, additionally, we assume that (A5) holds, then we have $P(\lim_n q_n = q^*) = 1$.*
(3) *$\sqrt{n}(q_n - q^*) \xrightarrow{n} N(0, 2\sigma^2 J^{-1})$ in distribution, where J is given in (A5) and where $N(0, S)$ denotes a normal random vector with mean 0 and covariance S.*

In the proof of this result, the following representation is crucial for asymptotic normality (see [3, 4, 11] for details):

$$\begin{aligned}\sqrt{n}(q_n - q^*) &= -\sqrt{n}\left(\frac{\partial^2 J_n(\bar{q}_n)}{\partial q^2}\right)^{-1}\frac{\partial J_n(q^*)}{\partial q} \\ &= \frac{2}{\sqrt{n}}J^{-1}\sum_{k=1}^{n}\frac{\partial f(x_k, q^*)}{\partial q}\varepsilon_k + o_P(1),\end{aligned} \quad (3)$$

where $\bar{q}_n$ is a vector with each of its components between the corresponding components of q^* and q_n, and $o_P(1)$ denotes a random variable that tends to 0 in probability. We shall make use of this representation in the next section.

In fact, more far reaching convergence and rate of convergence results have been obtained: interested readers are referred to [15] for the corresponding functional limit results which exploit the 'dynamic' behavior of the estimation sequence. The results given above, however, will provide enough information on the parameter estimators for the purposes of empirical distribution studies.

3 Residuals and Empirical Distributions

It would seem that to develop a more accurate distribution theory for the parameter estimators one would need to know more about the error distribution F. In many cases this distribution is not known. There is a well-developed theory for estimating the distribution of a random sample; however, it requires that one observe the random variables whose distribution is to be estimated (see, for example, [5, 6, 7, 8, 9]). Unfortunately, the errors $\{\varepsilon_k\}$ are not observable. To overcome this difficulty, we estimate the errors by the residuals

$$\varepsilon_k^n = Y_k - f(x_k, q_n),$$

where q_n minimizes J_n over Q_{ad}. Since Y_k is the actual observation at time k and q_n is the parameter estimator, the residual sequence $\{\varepsilon_k^n\}$ is effectively an approximating sequence of the observation errors $\{\varepsilon_k\}$. In order to study the distribution of ε_k, we define an empirical distribution function

$$\begin{aligned}\hat{F}_n(t,) &= \frac{1}{n}\sum_{k=1}^{n} I_{\{\varepsilon_k^n \leq t\}} \\ &= \frac{1}{n}(\# \text{ of } \varepsilon_k^n \leq t),\end{aligned} \quad (4)$$

which is based on the estimated errors, instead of the unobservable "true" errors. The empirical distribution of the actual errors is given by

$$F_n(t) = \frac{1}{n}\sum_{k=1}^{n} I_{\{\varepsilon_k \leq t\}} \quad (5)$$

As we have mentioned, the empirical distribution function F_n has large sample convergence properties that are well understood, but the results rely on the fact that the sequence $\{\varepsilon_k\}$ is an independent, identically distributed sequence. The major difficulty in analyzing $\hat{F}_n$ is that the random variables in the sequence $\{\varepsilon_k^n\}$ are neither independent nor identically distributed. We study in this section some adaptations of well-known results on empirical distributions, combining the results of the previous section with the classical theory of empirical distributions.

For the distribution function defined in (4), we first wish to derive a result analogous to the well-known Glivenko-Cantelli's Theorem, which is an almost sure convergence result. From Taylor's theorem and the assumptions and results of the previous section, we have that
$$\varepsilon_k^n = \varepsilon_k - \tau_k^n,$$
where τ_k^n is given by

(6) $$\tau_k^n = \frac{\partial f'(x_k, \bar{q}_n)}{\partial q}(q_n - q^*)$$

where $\bar{q}_n$ is a vector with each of its components between the corresponding components of q^* and q_n. The results outlined in the previous section imply that $\tau_k^n = o_s(1)$, uniformly in $k < n$, where $o_s(1) \xrightarrow{n} 0$ w.p.1. Thus, $\{\varepsilon_k^n\}$ behaves asymptotically like $\{\varepsilon_k\}$. We see in the following theorem a quantification of the asymptotic behavior. The proof of this result can be found in [13].

THEOREM 3.1. *Assume that (A1) — (A4) hold. Then,*
$$P\left(\limsup_n \sup_t |\hat{F}_n(t, \tilde{q}_n) - F(t, q^*)| = 0\right) = 1.$$

The above theorem shows us that the empirical distributions of the actual and estimated errors behave the same in an almost sure sense. We see below, in a weak convergence result, that the limiting probability laws of the scaled estimated and actual error sequences are quite different. We begin with the scaled sequence

(7) $$\hat{W}_n(t) = \sqrt{n}(\hat{F}_n(t) - F(t)),$$

whose limiting distribution we seek. This distributional result can be viewed as a rate of convergence result. In addition, it provides an approach to estimating distributions of statistics which can be used to test for biases in data and goodness of fit of hypothesized error distributions.

The desired result requires one more assumption:

(A6) The sequence $\{\varepsilon_k\}$ has a common density function $\pi_\varepsilon(\cdot)$ which is continuous on $(-\infty, \infty)$, and $E\varepsilon_1^4 < \infty$.

We thus restrict ourselves to continuous random variables with finite fourth moment. A typical example satisfying (A6) is a Gaussian distribution. In view of this assumption,

(8) $$F(t) = P(\varepsilon_k \leq t) = \int_{-\infty}^t \pi_\varepsilon(s)ds \text{ and } \frac{\partial F(t)}{\partial t} = \pi_\varepsilon(t).$$

To analyze the asymptotic behavior of $\hat{W}_n(\cdot)$, we first from (6) note that

(9) $$\hat{W}_n(t) = \frac{1}{\sqrt{n}}\sum_{k=1}^n \left([I_{\{\varepsilon_k \leq t\}}] - F(t)\right) + \frac{1}{\sqrt{n}}\sum_{k=1}^n \left(F(t + \tau_k^n)) - F(t)\right)$$
$$+ \frac{1}{\sqrt{n}}\sum_{k=1}^n \left((I_{\{\varepsilon_k \leq t + \tau_k^n\}} - F(t + \tau_k^n)) - (I_{\{\varepsilon_k \leq t\}} - F(t))\right).$$

A similar technique has been used by Durbin (see [7, 8]) in the case of i.i.d. observations when F is evaluated using an estimated parameter. Again, we remark that the nature of our observations will not allow the direct application of the results in [7, 8], but the details of the analysis bear several similarities.

Now, in [13] it was shown that the third term on the right hand side of (9) tends to 0 in probability; hence, the asymptotic distribution of the limit is determined by the first two terms. The first term tends to a stretched Brownian bridge, a fact which is part of the classical theory of empirical distributions (see for example [5]). The second term requires some analysis, involving Taylor's theorem (using (A6) for F) and the results of Section 2. Some straightforward (but tedious) calculations based on (3) above show that $E(\hat{W}_n) \to 0$, and that the asymptotic covariance is given by

$$
\begin{aligned}
C(t_1, t_2) = & \min(F(t_1), F(t_2)) - F(t_1)F(t_2) \\
& + 2\pi_\varepsilon(t_2)\left(\int_X \frac{\partial f'(x, q^*)}{\partial q} d\mu \mathcal{J}^{-1} \int_X \frac{\partial f(x, q^*)}{\partial q} d\mu \left(E\varepsilon_1\left(I_{\{\varepsilon_1 \leq t_1\}} - F(t_1, q^*)\right)\right)\right) \\
& + 2\pi_\varepsilon(t_1)\left(\int_X \frac{\partial f'(x, q^*)}{\partial q} d\mu \mathcal{J}^{-1} \int_X \frac{\partial f(x, q^*)}{\partial q} d\mu \left(E\varepsilon_1\left(I_{\{\varepsilon_1 \leq t_2\}} - F(t_2, q^*)\right)\right)\right) \\
& + 8\sigma^2 \pi_\varepsilon(t_1)\pi_\varepsilon(t_2)\left(\int_X \frac{\partial f'(x, q^*)}{\partial q} d\mu \mathcal{J}^{-1} \int_X \frac{\partial f(x, q^*)}{\partial q} d\mu\right).
\end{aligned}
\tag{10}
$$

In proving weak convergence of the empirical distribution process, one first establishes the tightness (relative compactness) of the sequence and then identifies the limit process through the finite dimensional distributions, using standard central limit theorem arguments together with the computed mean and covariance. The details of the arguments leading to the following theorem can be found in [13].

THEOREM 3.2 . $\hat{W}_n(\cdot)$ *converges weakly to a zero mean Gaussian process* $W(\cdot)$, *whose covariance function is given by (10).*

Also discussed in [13] are results for two sample, as well as general multiple sample, problems. We now turn our attention to Monte Carlo and bootstrapping methods for using the empirical distribution in inference.

4 Bootstrap Inference Methods

In Section 3 above, we defined the empirical distribution and examined some of its large sample properties. The asymptotic distribution of the scaled empirical process allows one to test hypotheses concerning the actual error distribution, F. Once one has found the distribution F, one can proceed with a Monte-Carlo based approach to compute confidence intervals for the parameters. There are, however, some drawbacks to this approach. In particular, one may test several distribution functions and find them all rejected by the hypothesis test. In such an event, one may still use the empirical distribution to perform inference, via bootstrapping.

The bootstrap procedure begins with the empirical distribution $\hat{F}_n$. One obtains a collection of simulated data by generating a random sample from $\hat{F}_n$. That is, we use a random number generator to obtain simulated errors, $\{\varepsilon_1^1, \ldots, \varepsilon_n^1\}, \{\varepsilon_1^2, \ldots, \varepsilon_n^2\}, \ldots, \{\varepsilon_1^B, \ldots, \varepsilon_n^B\}$. Here B denotes the number of simulated experiments we are performing. We then construct the bootstrap observations

$$Y_k^b = f(x_k, q_n) + \varepsilon_k^b, \qquad 1 \leq k \leq n, \tag{11}$$

for each b with $1 \leq b \leq B$. At this point we proceed as in the original least squares problem: the observations Y_k^b are known, but the values of the sampled errors are not. We consider a collection of least squares problems

$$J_n^b(q) = \frac{1}{n} \sum_{k=1}^{n} (Y_k^b - f(x_k, q))^2, \quad (12)$$

and we denote by q_n^b a minimizer of J_n^b. One may then compute an empirical distribution from these parameter estimates for the purposes of finding confidence intervals and performing hypothesis tests.

Another use of the bootstrap is to perform distribution tests. For example, a common goodness of fit test is the Kolmogorov-Smirnov test, which uses the statistic $T_n = \sqrt{n} \sup_t |\hat{F}_n(t) - F^0(t)|$, where F^0 is the hypothesized error distribution. For the case in which the empirical distribution is constructed from an i.i.d. sample, the limiting distribution of T_n is well understood, and is independent of any unknown parameters. In many situations, the limiting distribution of T_n is very difficult to obtain, due to the complicated nature of the covariance given in (10). Hence, a Monte Carlo approach may be necessary.

One possible bootstrap test involves sampling from the hypothesized distribution; that is, our simulated errors $\{\varepsilon_1^1, \ldots, \varepsilon_n^1\}, \{\varepsilon_1^2, \ldots, \varepsilon_n^2\}, \ldots, \{\varepsilon_1^B, \ldots, \varepsilon_n^B\}$ are generated from F^0 rather than the empirical distribution. From each observation set $\{Y_1^b, \ldots, Y_n^b\}$ we construct a set of residuals

$$\tilde{\varepsilon}_k^b = Y_k^b - f(x_k, q_n^b), \quad (13),$$

using the least squares estimator q_n^b obtained from that sample through minimizing (12). Next, we construct bootstrap empirical distributions

$$F_n^b(t) = \frac{1}{n} \sum_{k=1}^{n} I_{[\varepsilon_k^b \leq t]},$$

and from these, test statistics T_n^b as above. The distribution function for T_n can then be approximated by the empirical distribution of the collection $\{T_n^b\}_{b=1}^B$. To test the hypothesis that the original error distribution is F^0, we determine a threshold based on a given significance level with the empirical distribution of the collection $\{T_n^b\}_{b=1}^B$, and we test the hypothesis that F^0 is the error distribution by comparing the statistic T^n with that threshold.

5 An Example

To illustrate the use of the empirical distribution, we consider the vibration experiment discussed in Section 2. We begin with a collection of data for a 2 meter long beam. This data was generated by solving the model equation via Galerkin approximation using 12 Legendre polynomials adjusted to match the clamped boundary conditions. The data was obtained by solving the equation with $\rho = .03$, and with $\gamma^* = .002, c_D I^* = .005$, and $EI^* = 109.237$, and adding random noise sampled from different distributions. The time period of the experiment is $[0, 1]$. Figure 1 shows the acceleration at the free tip of the beam before the noise was added. Figures 2,3, and 4 show Gaussian, gamma, and proportional Gaussian noise added to the signal. The sample size n is 1000. The Gaussian errors were generated with mean 0 and standard deviation $\sigma = 10.0$ (roughly 10% error). The gamma errors were shifted and scaled to have mean 0 and standard deviation 10, as well. The proportional errors were generated with mean 0 and standard deviation .1; these errors are used to generate data of the form

$$\widehat{u}_{tt}(t_k, 1) = u_{tt}(t_k, 1; q^*)(1 + \varepsilon_k).$$

The data in these three plots is the data that we use for the numerical experiments below.

Our first step is to solve the least squares problem and to generate empirical distributions. Then for the purposes of solving the least squares problem we approximate the solution of the Euler-Bernoulli equation with 10 Legendre polynomials. We use the Minpack routine LMDIF1, an implementation of the Levenberg-Marquardt algorithm, to perform the minimization. These calculations were carried out on a Sun Microsystems SPARCstation 1. In Figure 5, we see that the empirical distribution compares well with the Gaussian distribution, and in Figure 6, we see that the empirical distribution fits the Γ quite well. A Gaussian fit is also provided for comparison. As Figure 7 shows, the empirical distribution obtained from least squares fit to the "proportional error" data did not compare well at all to a Gaussian, even when we tuned the variance for a best fit. Of course, the theory outlined above is not valid in this case, and if we knew before computing the estimates that the errors were proportional, we should use that information in defining the residuals. The point of this part of the example is to see if our test can detect a problem with proportional data.

To compute the bootstrap quantities, we implemented the least squares algorithm and empirical distributions on the Oak Ridge National Laboratory 128 processor Intel iPSC/860 hypercube using either 32 or 64 processors, depending on the work load of the machine. The bootstrapping task is inherently parallelizable, but random numbers must be generated with care to insure independence across processors. We developed random number generators based on those of [10, 14], specially designed for MIMD computers. We used a bootstrap sample size of $B = 512$. Figure 8 gives the bootstrap distribution of γ_n, with a comparison to the Gaussian distribution predicted by the theory of [3, 4, 11]. Bootstrap distributions for EI and $c_D I$ gave similar results. Finally, Figure 9 shows the empirical distribution for the collection $\{T_n^b\}_{b=1}^B$, as well as the Kolmogorov Smirnov distribution. Note that the distributions agree very well. Further computations indicated that for the parameter values that we used, the terms in the covariance additional to the Brownian bridge in (10) were quite small (on the order of 10^{-7}). The second derivative is large, and the integral $\int_X \partial f(x; q^*)/\partial q$ in (10) is small. We ran another test with $(\rho^*, \gamma^*, c_D I^*, EI^*) = (1, 1, 1, 10)$, and we obtained Figure 10 for the empirical distribution for the collection $\{T_n^b\}_{b=1}^B$. This comparison shows a definite difference from the Kolmogorov-Smirnov distribution.

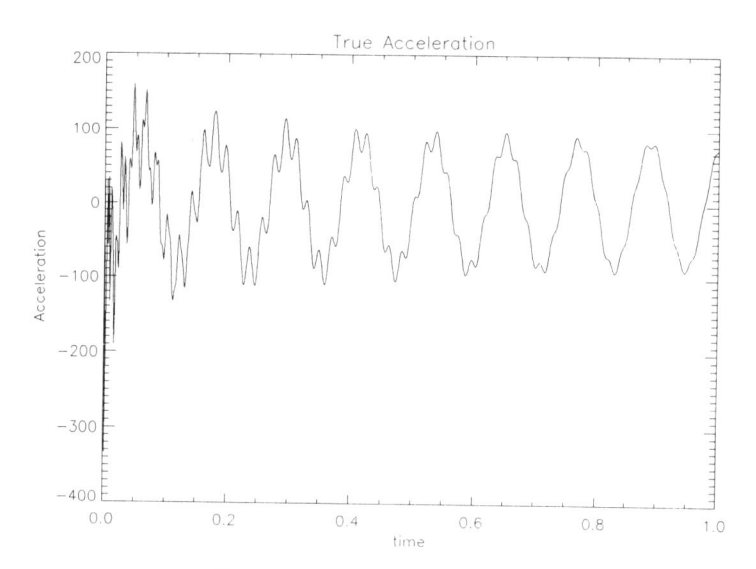

FIG. 1. *True tip acceleration.*

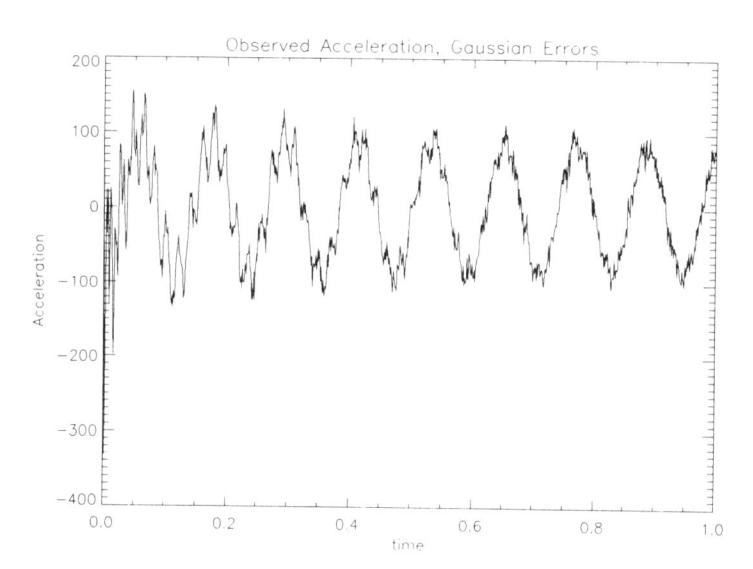

FIG. 2. *Observed tip acceleration with Gaussian noise.*

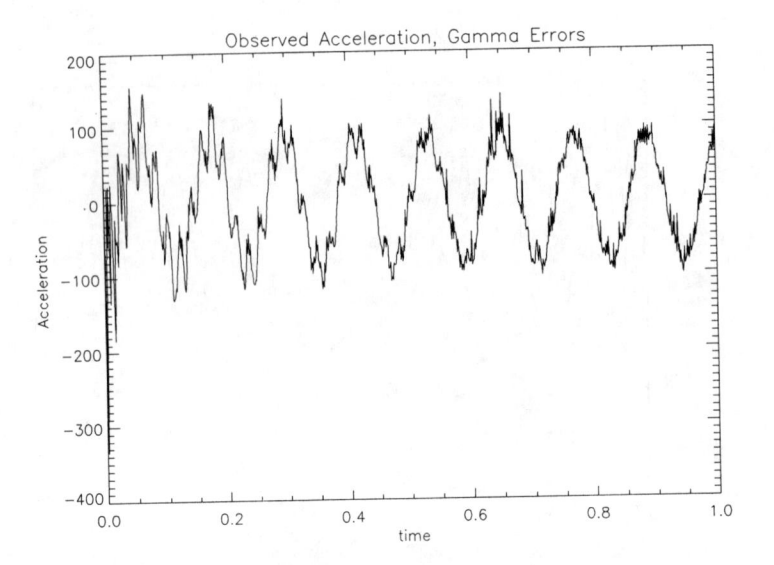

FIG. 3. *Observed tip acceleration with shifted gamma noise.*

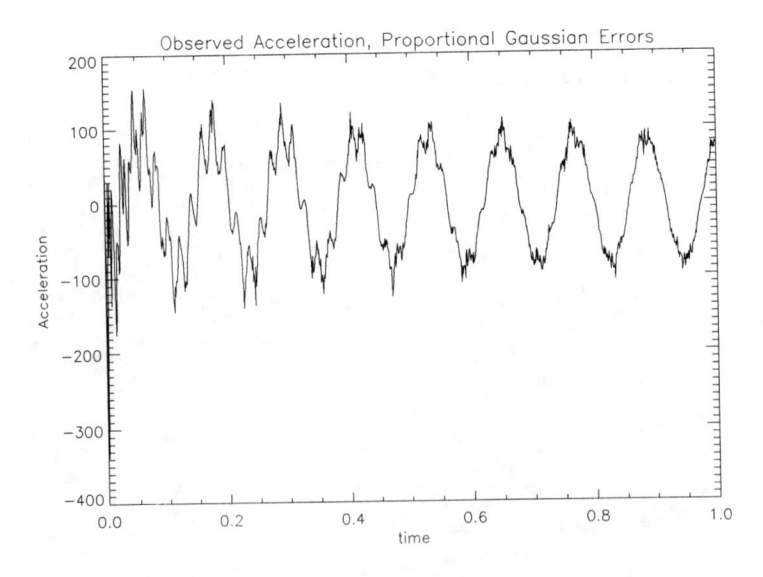

FIG. 4. *Observed tip acceleration with proportional Gaussian noise.*

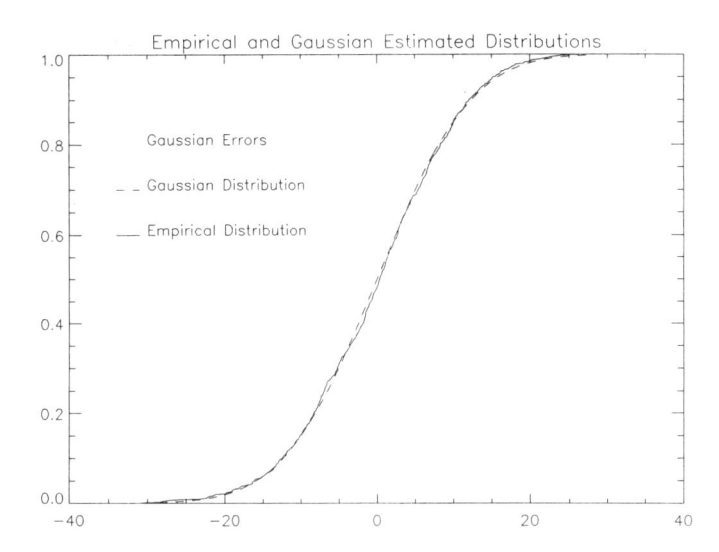

FIG. 5. *Empirical and Gaussian distributions from Gaussian noise example.*

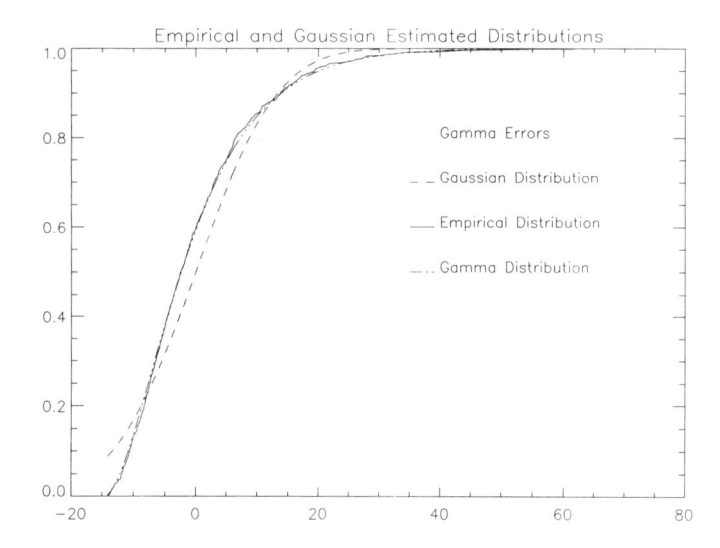

FIG. 6. *Empirical, Gaussian, and gamma distributions from shifted gamma noise example.*

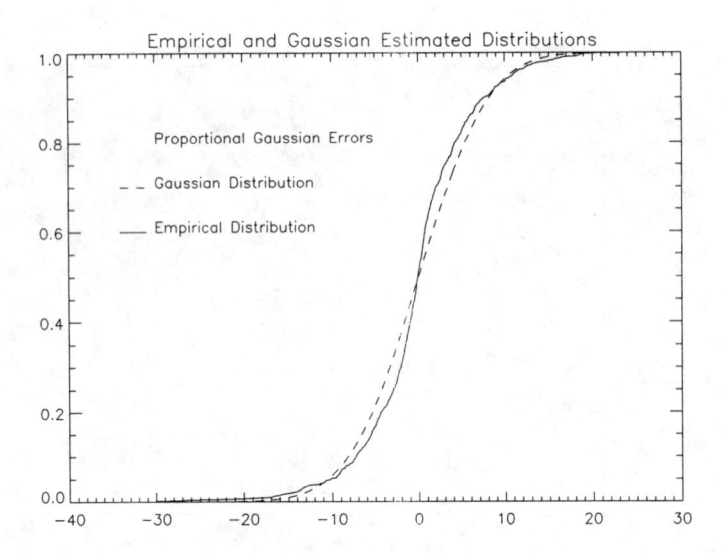

FIG. 7. *Empirical and Gaussian distributions from proportional Gaussian noise example.*

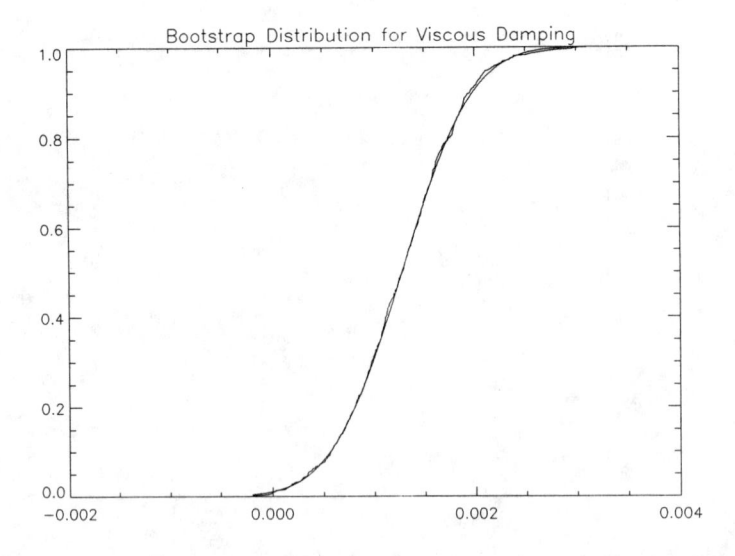

FIG. 8. *Comparison of bootstrap empirical distribution and Gaussian distribution for $\hat{\gamma}_n$.*

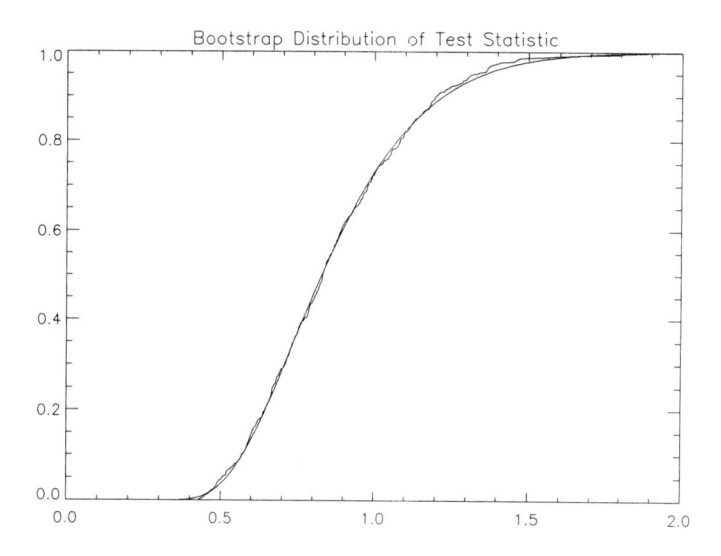

FIG. 9. *Comparison of bootstrap distribution of T_n and the Kolmogorov distribution.*

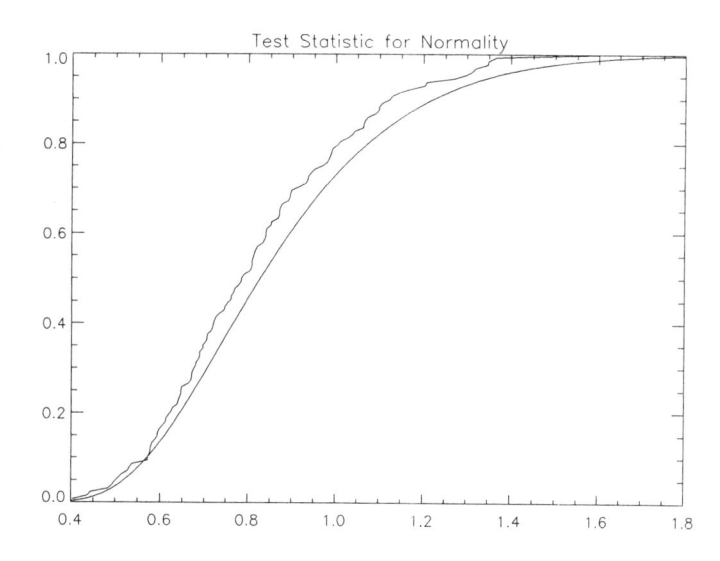

FIG. 10. *Comparison of bootstrap distribution of T_n and the Kolmogorov distribution.*

6 Concluding remarks

We have concentrated on the convergence and the rate of convergence of empirical processes of residuals from nonlinear least squares estimation problems. These processes can be used to answer many statistical questions of interest in application problems. We are particularly interested in testing for normality of the measurement errors and in determining whether or not collections of errors are identically distributed.

One of the difficulties with the above results is the nature of the limiting Gaussian process. The covariance depends on the unknown parameter, and it also depends on derivatives of the model function with respect to the parameter. In the applications mentioned in the introduction, the model function is (a function of) the solution of a partial differential equation, so that approximating the covariance of $\hat{W}_n$ is quite involved. To surmount this difficulty, we have examined the use of bootstrap methods for these problems. Our computations so far seem quite promising, especially in their efficient use of parallelism in computing.

References

[1] M. D. Aczon, H. T. Banks, and B. G. Fitzpatrick, *The Linear Regulator Problem for Systems with a Distribution of Parameters*, Proc. 31st IEEE CDC, Tucson, AZ, Dec. 16–18, 1992, pp. 1168-1171.

[2] H. T. Banks, L. Botsford, F. Kappel, and C. Wang, *Estimation of Growth and Survival in Size Structured Cohort Data: an Application to Larval Striped Bass (*Morone Saxatilis*)*, J. Math. Bio. 30 (1991), pp. 125-150.

[3] H. T. Banks and B. G. Fitzpatrick, *Inverse Problems for Distributed Systems: Statistical Tests and ANOVA*, in Proceedings of the International Symposium on Mathematical Approaches to Environmental and Ecological Problems, Lecture Notes in Biomath. 81, Springer-Verlag, New York, 1989.

[4] H.T. Banks and B.G. Fitzpatrick, *Statistical tests for model comparison in parameter estimation problems for distributed systems*, J. Math. Bio. 28 (1990), pp. 501-527.

[5] P. Billingsley, *Convergence of Probability Measures*, wiley, New York, 1968.

[6] R.M. Dudley, *A course on empirical processes*, Lecture Notes Math. 1097, Springer-Verlag, New York, 1984.

[7] J. Durbin, *Weak convergence of the sample distribution function when parameters are estimated*, Ann. Statist. 1 (1973), pp. 279-290.

[8] J. Durbin, *Kolmogorov-Smirnov tests when parameters are estimated*, in Lecture Notes Math. 566, pp. 33-44, Springer-Verlag, New York, 1976.

[9] S.N. Ethier and T.G. Kurtz, *Markov Processes, Characterization and Convergence*, Wiley, New York, 1986.

[10] G. S. Fishman, *Multiplicative Congruent Random Number Generators with Modulus 2^β: an Exhaustive Analysis for $\beta = 32$ and a Partial Analysis for $\beta = 48$*, Math. Comp. 54 (1990), pp. 341-344.

[11] B. G. Fitzpatrick, *Statistical methods in parameter identification and model selection*, Ph. D. Dissertation, Division of Applied Mathematics, Brown University, Providence, RI., 1988.

[12] B. G. Fitzpatrick, *Bayesian Analysis in Inverse Problems*, Inverse Problems, 7 (1991), pp. 675-702.

[13] B. G. Fitzpatrick and G. Yin, *Empirical Distributions in Least Squares Estimation for Distributed Parameter Systems*, to appear.

[14] O. E. Percus and M. H. Kalos, *Random Number Generators for MIMD Parallel Processors*, J. Parallel Distrib. Comp. 6 (1989), pp. 477-497.

[15] G. Yin and B.G. Fitzpatrick, *On invariance principles for distributed parameter identification algorithms*, Informatica 3 (1992), pp. 98-118.

Chapter 5
MATLAB-Software for Parameter Estimation in Two-Point Boundary Value Problems

M. Kroller[*]   K. Kunisch[*†]

Abstract

MATLAB-based software solving certain parameter estimation problems is described. It can be used to study how different discretizations and various regularization terms influence the behavior of the algorithms solving the inverse problem.

In this note we summarize our efforts to develop a MATLAB code for the estimation of parameters in two point boundary value problems. It is intended to be used by anybody who wishes to experience some of the features of an illposed inverse problem without writing the codes himself and who has MATLAB software available.

Numerical techniques to solve infinite dimensional inverse problems depend - among other criteria - strongly on the type of discretization that is chosen, on the noise level in the data and on the type and size of the regularization term that is used. Such choices can be made using the menu that is provided with the code.

The equation under investigation is

$$\begin{cases} -(au_x)_x + cu = f \text{ on } (0,1), \\ u(0) = u(1) = 0, \end{cases} \qquad (1)$$

and the task consists in the determination of one of the coefficients a or c, or of the input f, from knowledge of an observation z corresponding to the state variable u. More precisely, the user can compare different variants of an algorithm to solve one of the following three problems:

---

[*]Partially supported by a grant of the Bundesministerium fr Wissenschaft und Forschung, Austria.

[†]Partially support through the Christian Doppler Laboratory on "Parameter Identification and Inverse Problems".

1. Given an observation z for u, identify f by the output least squares (OLS) method:

$$\min \frac{1}{2}|u(f) - z|^2_{L^2} + \frac{\beta}{2}|f|^2 \text{ over } f,$$

where $u(f)$ is the solution to $-\Delta u = f$ on $(0,1)$ and $u(0) = u(1) = 0$.

The solution of the minimization problem in 1. is based on solving the normal equations (necessary optimality condition).

2. Given an observation z for u identify a using the augmented Lagrangian method (ALM) to solve the constrained minimization problem

$$\min \frac{1}{2}|u - z|^2_{L^2} + \frac{\beta}{2}|a|^2$$

subject to (a, u) satisfying $(au_x)_x + f = 0$, $u \in H^1_0(0,1)$.

For a description of the ALM we refer to [IK, IKK] and the brief explanation further below.

3. Given an observation z for u identify c using the augmented Lagrangian method to solve the constrained minimization problem

$$\min \frac{1}{2}|u - z|^2_{L^2} + \frac{\beta}{2}|c|^2$$

subject to (c, u) satisfying $-\Delta u + cu = f$, $u \in H^1_0(0,1)$.

In 1.-3. above L^2 stands for $L^2(0,1)$ and Δ denotes the Laplacian from $H^1_0(0,1)$ to $L^2(0,1)$. The norm in the regularization terms denotes either the L^2-norm or the H^1-seminorm. In the latter case $|f|$ stands for $|Df|_{L^2}$ with D denoting differentiation, and analogously for a and c.

With regards to discretization of the state variable u the user can choose between approximation by linear splines or by cubic Hermite splines, and discretization of the unknown parameters a respectively c or f can be carried out by either piecewise constant functions or by linear splines. With respect to the grid one can choose between either using the same grid for the state variable as well as for the parameter variable or alternatively one can take the grid for the state discretization twice as fine as that for the unknown parameter. Thus, if the grid for u is given by $\{\frac{i}{N+1}\}_{i=0}^{N+1}$, then that for the coefficient is $\{\frac{j}{M+1}\}_{j=0}^{M+1}$ with $N = M$ or $N = 2M + 1$.

The observation z was produced in the following manner. The solution u corresponding to the "true" coefficient was evaluated on an equidistant grid $\{x_i\}_{i=1}^{N_D} \subset (0,1)$ (the default value in $N_D = 43$) and uniformly distributed random numbers $\delta_i \in [-\delta, \delta]$ were added to u at the gridpoints to obtain

$$z(x_i) = u(x_i) + \delta_i, \text{ for } i = 1, \cdots, N_D, \text{ and } z(0) = z(1) = 0.$$

Then a linear interpolating spline is passed through $z(x_i)$ to obtain the "observation" z. The noise level δ as well as the regularization parameter β are chosen by the user.

The augmented Lagrangian algorithm to solve 2. proceeds as follows:

(i)   set $k > 0 (k = 1$ is default$), n = 0, \lambda_0 = 0, u_0 = z$(or $u_0 = 0$), choose $\beta \geq 0$.
(ii) minimize $L_k(a, u_n, \lambda_n)$ over a to obtain a_{n+1},
(iii) minimize $L_k(a_{n+1}, u, \lambda_n)$ over u to obtain u_{n+1},
      if convergence is achieved then stop, otherwise
(iv) $\lambda_{n+1} = \lambda_n + ke(a_{n+1}, u_{n+1})$, set $n = n + 1$ and goto (ii).

Here we put

$$L_k(a, u, \lambda) = \frac{1}{2}|u - z|_{L^2}^2 + <D\lambda, De(a,u)>_{L^2} + \frac{k}{2}|De(a,u)|_{L^2}^2 + \frac{\beta}{2}|a|^2,$$

and

$$e(a, u) = (-\Delta)^{-1}((au_x)_x + f).$$

The minimization of the quadratic subproblems in (ii) and (iii) is carried out by solving the necessary optimality conditions.

The algorithms also provide for a presmoothing of the data. In this case z in 1. - 3. is replaced by the solution to

$$\min_{z^\epsilon} \epsilon |z^\epsilon|_{H_0^1}^2 + \sum_{i=1}^{N_D} |z^\epsilon(x_i) - z(x_i)|^2,$$

with $\epsilon > 0$.

Additional internal variables controlling the flow of the algorithm are described in the README.DOC file. The code was tested under 386-MATLAB, version 3.5k, on a 486-PC and under PROMATLAB on a DEC-station.

One of the main goals of the software that we prepared is to provide for the possibility to make easy comparisons. Therefore the menu asks the user whether he would like to prepare only one, or rather two or four graphs simultaneously on the screen.

Typical comparisons that can be carried out with the prepared software are described next. The specifications for the problems to be solved are:

for 1.: $f(x) = 1 + x^2$,

the resulting unperturbed state variable is given by $u(x) = -\frac{x^4}{12} - \frac{x^2}{2} + \frac{7x}{12}$.

for 2.: $f(x) = -((1+x^2)(e^x \sin \pi x)_x)_x$, $a(x) = 1 + x^2$,

the resulting unperturbed state variable is given by $u(x) = e^x \sin \pi x$,

for 3.: $f(x) = -(e^x \sin \pi x)_{xx} + (1+x^2)e^x \sin \pi x$, $c(x) = 1 + x^2$,

the resulting unperturbed state variable is given by $u(x) = e^x \sin \pi x$.

The graphs in the plots below are labeled in the follow way:

x-axis: noise level δ, size of regularization parameter β, (iteration number in case of ALM).

y-axis: searched for parameter (a or c or f), L^2-error between true and numerical coefficient.

headline: type of parameter regularization (L^2-norm or H^1-seminorm), method for solving inverse problem (OLS or ALM), discretization of u, discretization of parameter, relative mesh size (state vs. parameter discretization), value for M.

Solid lines represent the "true" infinite dimensional parameter, the dashed lines show the numerical solution. No effort has been made yet to optimize the codes with respect to speed of convergence.

Comparison 1 and 2. The first comparison indicates that it is harder to estimate c than f, and that it is harder to estimate f than a, see Plot 1. A heuristic reasoning can be given as follows. The mappings $f \to u(f) = (-\Delta)^{-1}f$, $c \to u(c) = (-\Delta + c)^{-1}f$, and $a \to u(a) = -\int_0^x \frac{1}{a}(F - \bar{F})ds$, where $F(s) = \int_0^s f(\tau)d\tau$, $\bar{F} = \int_0^1 \frac{1}{a}Fds / \int_0^1 \frac{1}{a}ds$, are all smoothing, with the first two of them involving two integrations and the last one only one. If the domain and the range of the above three mappings are endowed with the same norms, then these mappings allow no continuous inverse, and in this respect the inversion of the first two of them is more illposed than that of the last mapping. We note, however, that not only the smoothing property of these mappings leads to difficulties in determining the unknown parameter, but also the fact that a and c appear in bilinear form together with u_x, respectively u, in (1). In neighborhoods of the singular set $S_a = \{x : u_x(x) = 0\}$ for the problem of identifying a, and $S_c = \{x : u(x) = 0\}$ for the problem of identifying c, it is difficult to estimate the unknown coefficients, and it is impossible in the interior of the singular sets.

While the data in Plot 1 are noise free, noise is added to the data for the results of Plot 2, where again the result for a is better than that for f and that for f is better than that for c. The absolute noise level δ is chosen such that the signal-to-noise ratio is approximately the same for all three subplots of Plot 2.

Comparison 3. (Plot 3) Here regularization by discretization and regularization using a Tikhonov regularization term are compared. Subplot 1 shows the (usless) result if $N = M$. The numerical difficulties are due to the fact that as a consequence of using the same grid for coefficient and state variable and due to Dirichlet boundary conditions, the number of unknowns in the discretization for f exceeds the number of state variables (number of equations) by one. Subplot 2 (top right) and subplot 3 (bottom left) give the result if a coarser mesh is taken for the discretization of f or alternatively a regularization term is used. In subplot 4, both a coarser mesh and a Tikhonov regularization term are used simultaneously. This gives the smallest L^2-error.

Comparison 4. (Plot 4) The effect of the singular set is demonstrated and it is shown how regularization can be used to suppress it. For the problem of identifying a the singular set is a singleton given by the unique solution of $\sin \pi x + \pi \cos \pi x = 0$ in $(0, 1)$ which is approximately .598. In subplots 1 and 2 regularization by L^2-norm and H^1-seminorm are compared while a is discretized by step functions. L^2-norm regularization does not eliminate the lack of information in the neighborhood of the singular set, while the H^1-seminorm solution does. Of course, in view of the interpretation of the regularized solutions as least (semi-) norm solutions, these results should be expected [EKN]. In subplots 3 and 4 the comparison between L^2-norm and H^1-seminorm is repeated with linear spline discretization of a. Due to the more global nature of the linear spline basis functions than that of step functions, the result in subplot 3 is better than that in subplot 1.

Comparison 5. (Plot 5) Here the results for identifying c with a good choice of the regularization parameter are compared to the results when overregularization occurs. Both L^2-norm and H^1-seminorm regularization results are given. Note that the maximum error occurs at the singular set $\{0,1\}$ in all subplots.

Comparison 6. (Plot 6) Different combinations for the discretization of a and u are compared. For these and the following examples the data are perturbed by random noise.

Comparison 7. (Plot 7) For this comparison for identifying a in the presence of noisy data, we first determined a best β value among the values $\beta = 10^{-k}, k = 1, 2, \cdots$, and then we compared the results for L^2-norm and H^1-seminorm regularization.

Comparison 8. (Plot 8) This series of plots shows that there exists an optimal β-value! For a systematic choice of the best regularization parameter we refer to [SEK] and the references given these.

Remark. The software that we described in this note can be obtained by sending an E-Mail message to KROLLER@EDVZ.UNI-GRAZ.ADA.AT .

References

[EKN]   H.W. Engl, K. Kunisch and A. Neubauer: Tikhonov Regularization for the Solution of Nonlinear Illposed Problems, Inverse Problems, 5 (1989), 523-540.

[IK]   K. Ito and K. Kunisch: A Hybrid Method Combining the Output Least Squares and the Equation Error Approach for the Estimation of Parameters in Elliptic Systems, SIAM Journal on Control and Optimization, 28 (1990) 113-136.

[IKK]   K. Ito, M. Kroller and K. Kunisch: A Numerical Study of the Augmented Lagrangian Method for the Estimation of Parameters in Elliptic Systems, SIAM J. on Sci. and Stat. Comput., 12 (1991), 844-910.

[SEK]   O. Scherzer, H.W. Engl and K. Kunisch: Optimal A-Posteriori Parameter Choice for Tikhonov Regularization for Solving Nonlinear Ill-posed Problem, to appear in SIAM J. Numerical Analysis.

PLOT 1

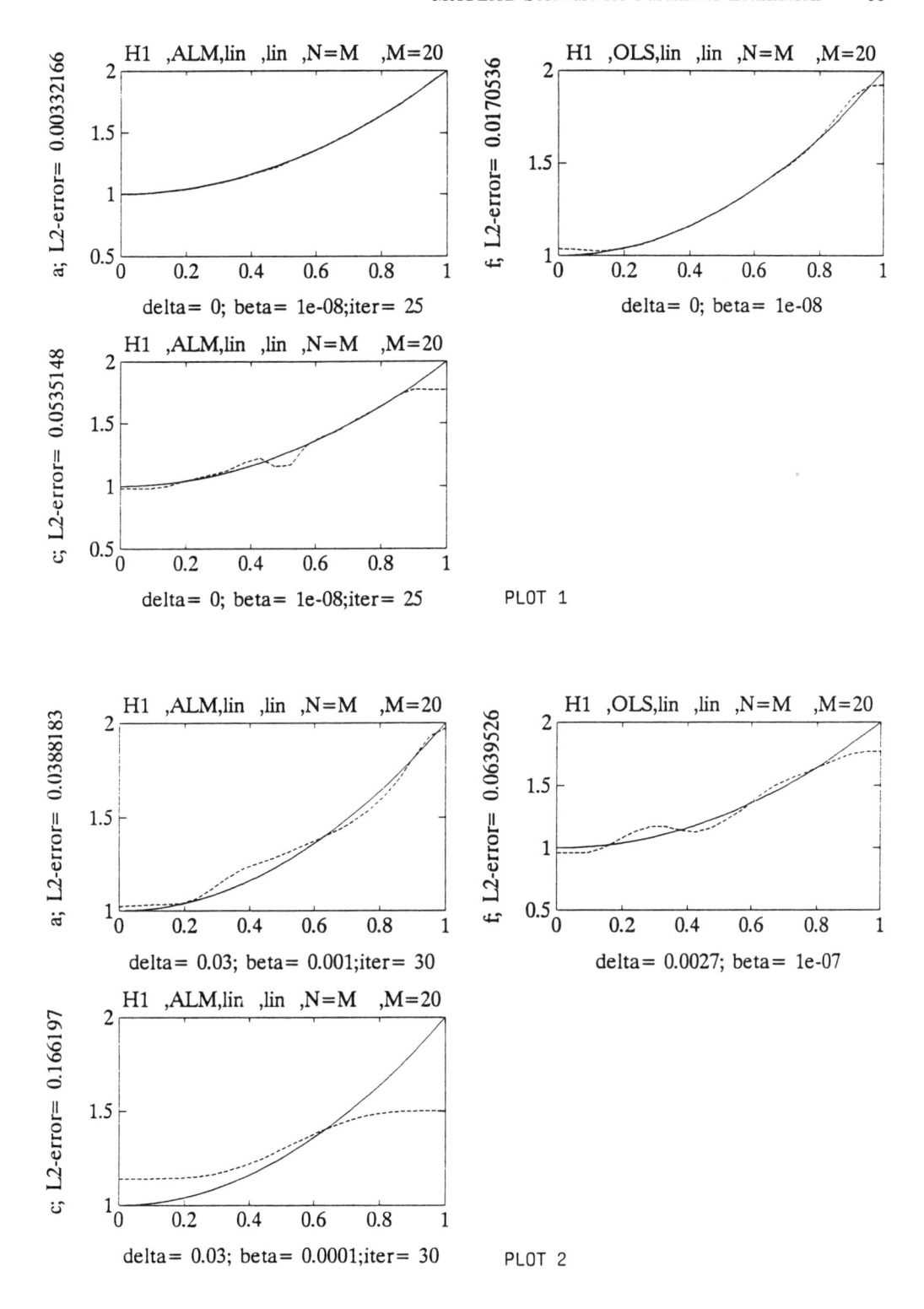

PLOT 2

PLOT 3

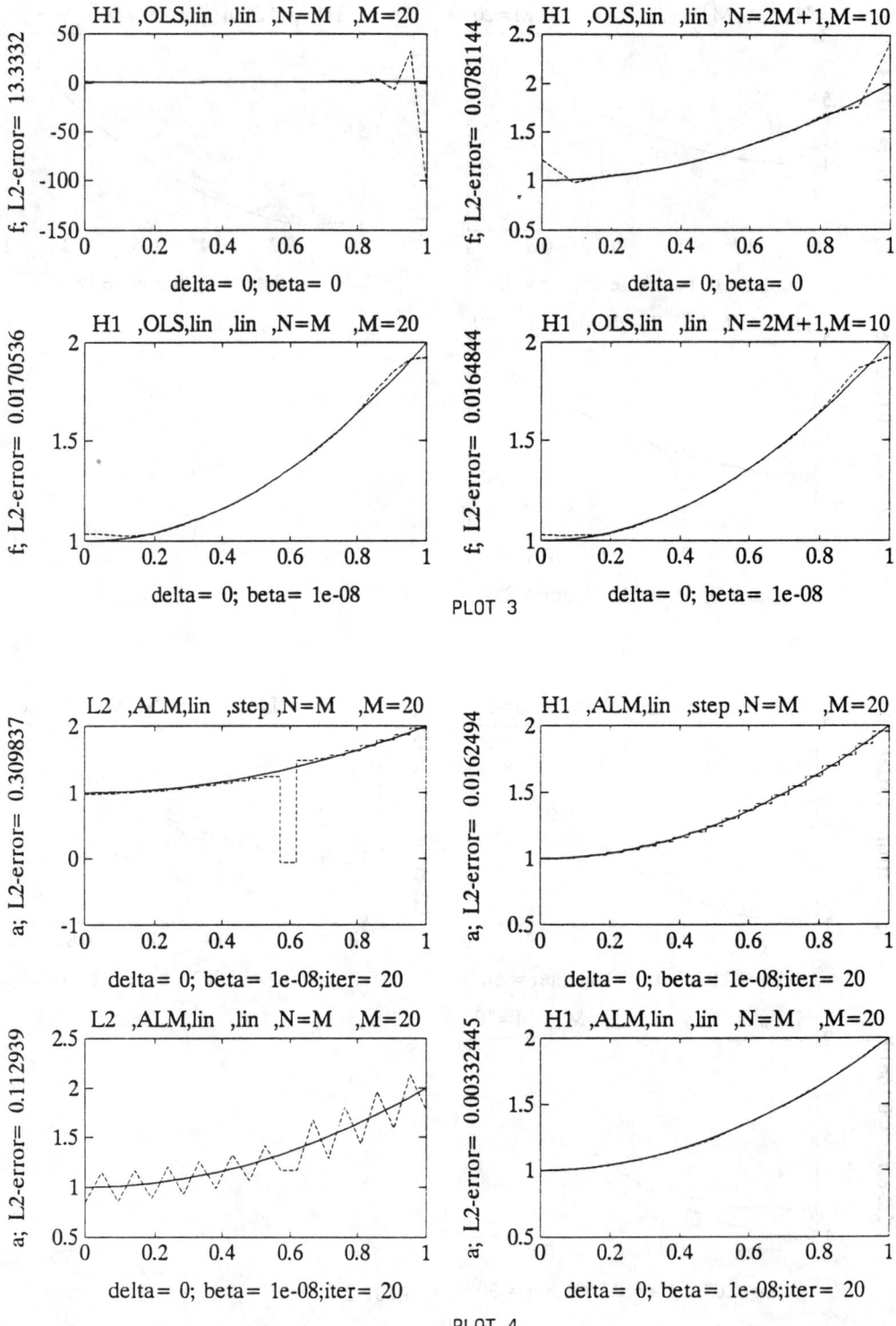

PLOT 4

PLOT 5

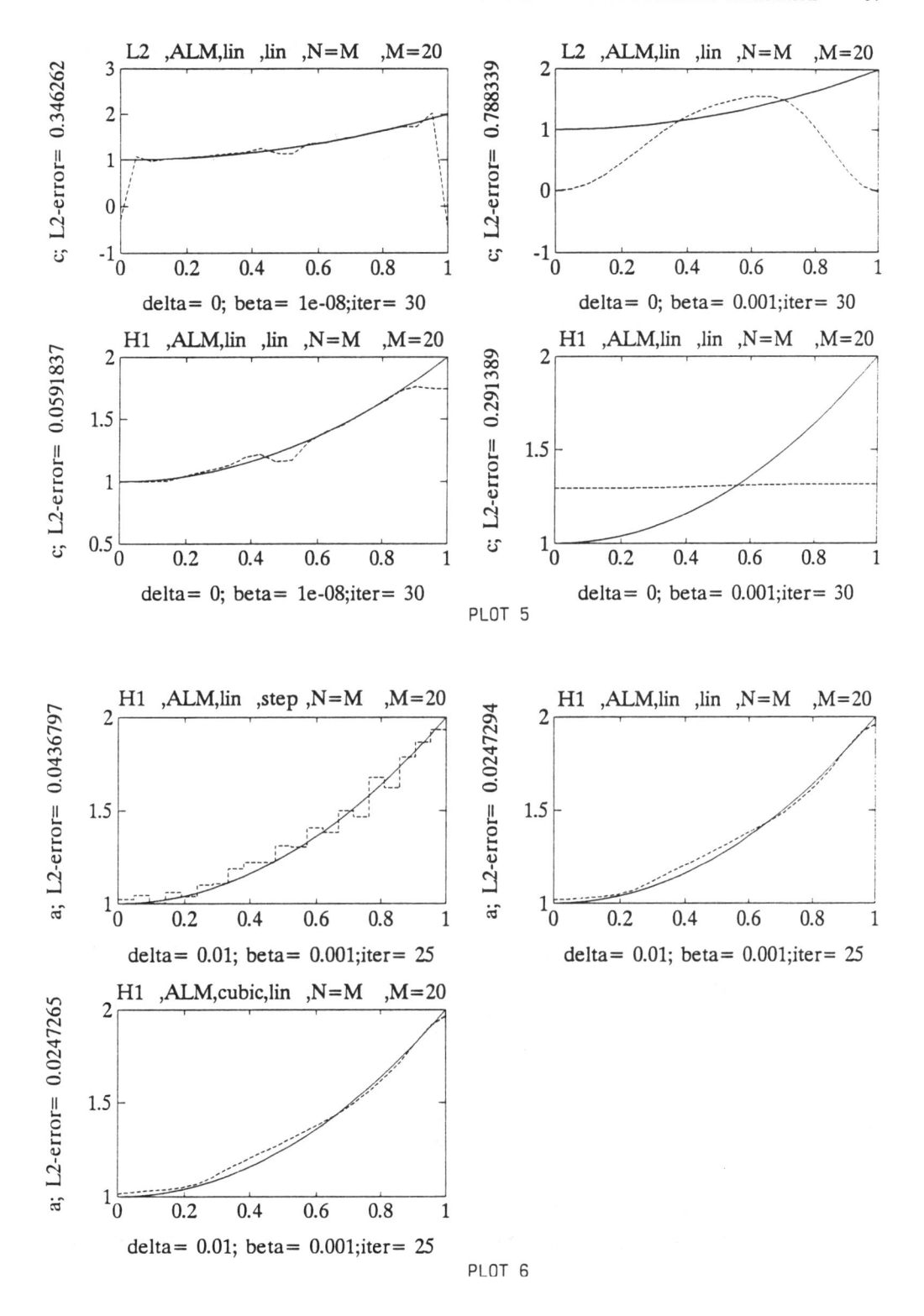

PLOT 6

PLOT 7

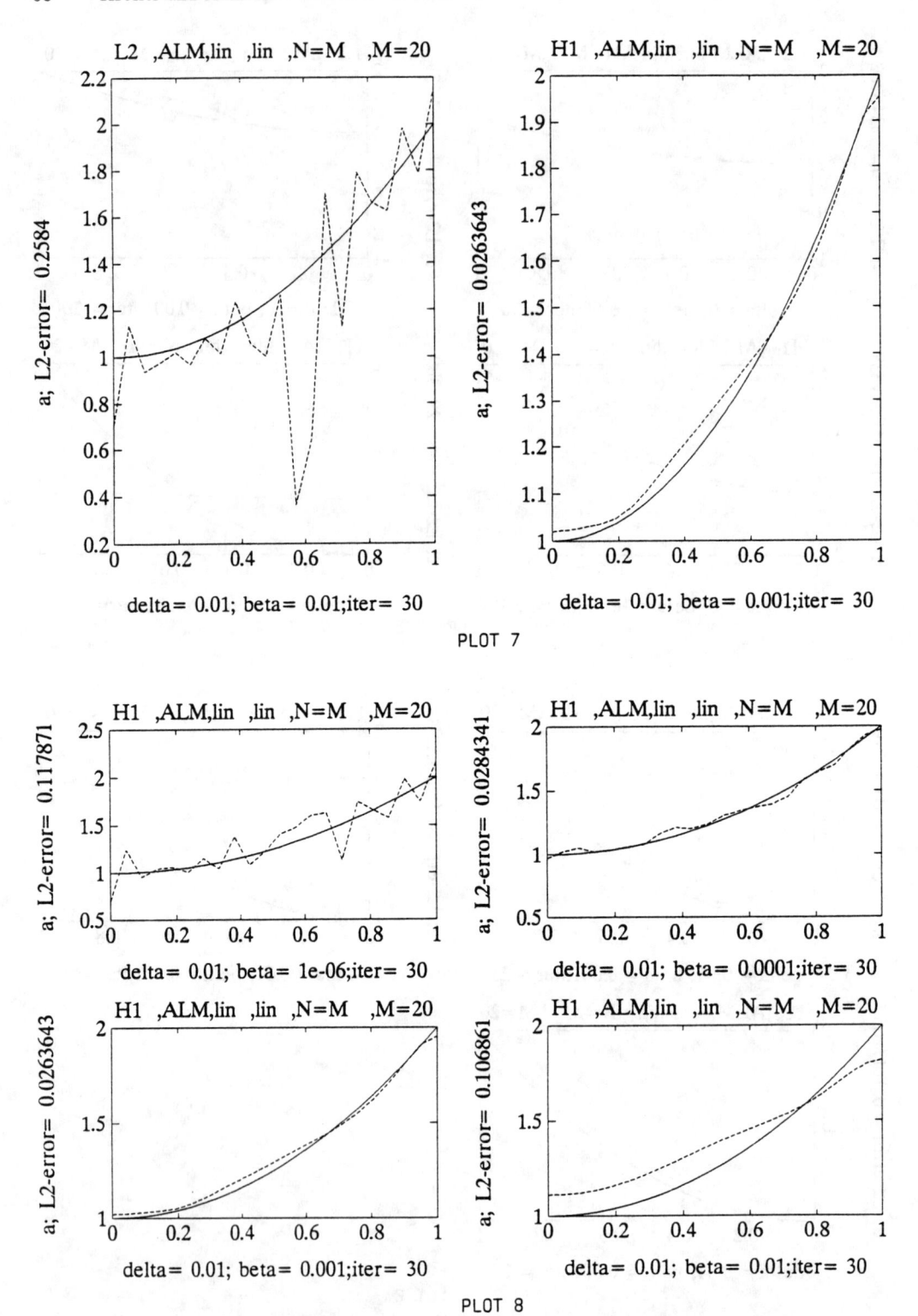

PLOT 8

Chapter 6
Some Stability Estimates for the Identification of Conductivity in the One – Dimensional Heat Equation*

Giovanni Crosta†

Abstract

The stability of position – dependent conductivity in one spatial dimension is considered. This inverse problem is assumed throughout to have at least one measurable, bounded and strictly positive solution. Since conductivity satisfies an ordinary differential equation (ODE), uniqueness conditions may result from information of local or non – local type. Local information corresponds to a Cauchy datum, which can be supplied either at a regular or at a critical point; at the latter temperature is stationary (singular Cauchy problem). Non – local information is supplied as the domain average of thermal flux at a given instant of time. The main purpose of the paper is to provide a unified view over the stability estimates pertaining to the unique solution. Some additional restrictions (regularization) are imposed on the temperature data. If uniqueness is due to a regular Cauchy problem, L^∞– estimates are obtained. Singular problems, on the other hand, yield L^r–estimates, $1 \leq r < \infty$. Non local conditions are treated similarly. The unifying device is the defect equation, an ODE for conductivity differences in a space of distributions. Estimates are arrived at by suitably integrating said ODE. Some examples and counterexamples are provided.

Introduction

One of the best known inverse problems is the identification of position dependent conductivity $a(.)$, the leading coefficient appearing in the one – dimensional heat equation

$$(0.1) \qquad (au_x)_x = u_t + f \quad \text{in } Q := (x_0, x_1) \times (t_0, t_1) ,$$

from knowledge of the (thermal) *potential* u and of the source term f in the whole interval at one or more instants of time.

---

*This work and the related activities have been funded by the Italian Ministry of University and Scientific Research (*MURST 60%*, from 1990 onwards). Related travel support by the *Frequent Flyer Program* of Delta Air Lines, Inc., Atlanta (GA) and by the former *FT WorldPass Program* of Pan American World Airways, Inc., New York (NY) is gratefully acknowledged. The *Dipartimento di Scienze dell' Informazione*, Universita' degli Studi di Milano has made the word processing facilities available. F. Dal Fabbro (Milan) and M. Hazewinkel (Amsterdam) are thanked for constructive criticism about a preliminary version of this paper.

†Dipartimento di Scienze dell' Informazione, Universita' degli Studi, via Comelico 39, I 20135 Milan (Italy). e_mail: crosta@imiucca.csi.unimi.it ; FAX: +39 (2) 55 00 63 73.

This problem has been extensively considered before: the reader may refer e.g., to Banks and Kunisch [2] and to the literature quoted therein. Cannon [4] also provides many references. This paper aims at unifying the types of stability estimates, which follow from different uniqueness conditions.

Eq. 0.1, to be understood in the distribution sense, is, $\forall\, t$, a first order ODE w.r. to $a(.)$. Throughout the paper a measurable, bounded and strictly positive a is assumed to exist i.e., the $\{u; f\}$ data pair shall give rise to at least one solution $a(u; f)$. Uniqueness and stability will be often established by means of another ODE, the *defect equation*

$$(0.2) \qquad (Bv_x)_x - V_t + (aV_x)_x = 0 ,$$

where $b(v; f)$ is the second conductivity, $V := v - u$ is the difference between the given potentials and $B := b(v; f) - a(u; f)$ is the difference conductivity. The solutions a, resp. B, are unique whenever adequate supplementary information is available, either of local or non-local type.

Section 1 deals with uniqueness. Conditions of local type correspond to Cauchy data for either Eq. 0.1 or 0.2. Cauchy problems can be regular or singular. The latter arise e.g., when u_x vanishes at some point (critical point). Uniqueness relies then on the properties of the set of critical points. The uniqueness condition of non-local type considered herewith is the domain average of thermal flux at a given instant of time.

Section 2 provides the main results i.e., stability estimates based on the same amount of data, which was needed to achieve uniqueness, subject to regularization. If e.g., $v_x(., t)$ vanishes nowhere for some t and its reciprocal is bounded above by a given constant, then a uniform (L^∞) stability estimate for B follows. This is no longer the case if uniqueness is due to the existence of a critical point: integral estimates are obtained, which in particular depend on the regularity of V_x and $\frac{1}{v_x}$.

The stability of those solutions, which are unique because of non-local conditions, is assessed in a similar way.

Estimates for a, resp. B, appearing in the *time independent* counterparts of Eqs. 0.1, resp. 0.2 are special cases, which do not need separate treatment.

Motivation for this work has come from applications. Consider e.g., ground water flow or heat conduction in a layered medium, where only one coordinate (depth) is relevant. This justifies the choice of Eq. 0.1 and of a related initial - BV problem. Potential (water pressure or temperature) can be measured at the top and bottom layers as well as in the interior; source terms can also be determined, whereas local heat flux is harder to measure. Experimental data are then to be inverted, in order to reconstruct $a(.)$.

1 The Uniqueness Conditions

1.1 Notation and Definitions

In the space-time domain $Q := (x_0, x_1) \times (t_0, t_1) \equiv D \times T$ consider the following inverse problem.

PROBLEM 1.1. *Given the thermal potential (temperature) - source term pair $\{u; f\}$ s.t., f satisfies*

$$f \in C^0(\overline{T}; H^{-1}(D)) \tag{1.1}$$

and u satisfies e.g.,

$$u \in \mathcal{U} := C^0(\overline{T}; W^{1,\infty}(D)) \cap C^1(\overline{T}, L^2(D)), \tag{1.2}$$

find a (position dependent) thermal conductivity a(.) complying with the following

$$a \in \mathcal{A}_{ad} := \{ a \mid a \in L^\infty(D), \exists \lim_{x \to x_0^+} a ; \; 0 < \text{ess}\inf_D a := a_L, \; \text{ess}\sup_D a = a_H \}, \tag{1.3}$$

and with the one – dimensional heat equation

$$(au_x)_x =_{d.w.} u_t + f := g, \; \forall t \in T. \tag{1.4}$$

Partial derivatives, denoted by italics subscripts, shall be understood in the distribution sense. The subscript d.w. stands for distribution – wise. Derivatives of $a(.)$ and of other quantities, which only depend on x, are denoted by primes. One – sided limits will be denoted by e.g., $a(x_0^+)$ (see (1.3)), the Lebesgue measure of a set by meas[.] and the closure of a set either by clos{ . } or by an overbar. Norms in the $W_p^s(.)$ spaces, $1 \leq p \leq \infty$ (by abuse of notation), where s is the order of differentiation w.r. to x, will be denoted by $\| \cdot \|_{s,p}$. More notation and properties of distributions are provided by Appendix A.

ASSUMPTION. *Throughout the paper at least one solution to Pbm. 1.1 is assumed to exist.*

REMARK 1.1. Although most of the results which follow refer to $\mathcal{U}$ of (1.2), the latter shall not be understood as a restriction. See § 2.4 in particular.

The following is the device, which will provide unified uniqueness and stability results.
PROPOSITION 1.1. *(The **defect equation**)*
Consider another data pair $\{ v; f \}$ where v, the second potential, satisfies (1.2) and f is the same as above. Assume $\{ v; f \}$ gives rise to a second conductivity b complying with

$$b \in L^\infty(D), \; \text{ess}\inf_D b = a_L, \tag{1.5}$$

and

$$[(bv_x)_x - v_t - f](t) =_{d.w.} 0, \; \forall t \in T. \tag{1.6}$$

Denote the difference conductivity by $B := b - a$ and the difference of the two potentials by $V := v - u$. Then B satisfies the defect equation

$$(Bv_x)_x(t) =_{d.w.} [V_t - (aV_x)_x](t), \; \forall t \in T. \tag{1.7}$$

Eq. 1.7 is a consequence of Eq. 1.4, 1.6 and of the definitions of B and V.
The r.h.s. of Eq. 1.7, the *defect*, is denoted by r.
The solutions to Eq. 1.7 shall in some cases satisfy the following property.

(1.8) $\quad b \in \mathcal{B}_{ad} := \{ B \mid , B \in L^\infty(D) ; \exists\, B(x_0^+), B(x_0^+) = 0 \}$

Uniqueness conditions and therefore the corresponding stability estimates covered by § 2, can be classified according to three criteria.

i) The type of available information, which may be either local or non – local. *Local* information corresponds to a Cauchy datum for either Eq. 1.4 or Eq. 1.7, both of which are, $\forall\, t \in T$, ordinary differential equations (ODEs) w.r. to the unknown quantities, a and resp. B. *Non – local* information originates from knowledge of the domain average of thermal flux at a given instant of time $\tau \in T$ e.g.,

(1.9) $$\frac{1}{|x_1 - x_0|} \int_D a\, u_x(x, \tau) = k(\tau).$$

The *domain average* of a function $h \in C^0(\overline{T}, L^1(D))$ at a given instant τ will be denoted by $\langle h \rangle(\tau)$, hence Eq. 1.9 is rewritten as $\langle a u_x \rangle(\tau) = k(\tau)$.

ii) The Cauchy problem on its turn may be either *regular* or *singular*. The former requires that the coefficient of $a'(.)$, resp. $B'(.)$, in the affected ODE vanish almost nowhere in D, at least at the considered instant. A singular Cauchy problem arises when information is supplied at a critical point $\xi(\tau) \in \overline{D}$ s.t., either

(1.10) $$\lim_{x \to \xi(\tau)_\pm} u_x(x, \tau) = 0$$

in Eq. 1.4 or

(1.11) $$\lim_{x \to \xi(\tau)_\pm} v_x(x, \tau) = 0,$$

in Eq. 1.7, whenever this makes sense. See PROP. 1.4 below.

iii) Finally, information may be available at one specific instant of *time* $\tau \in \overline{T}$, or may be needed in the whole of T in order to yield uniqueness.

A regular Cauchy problem is met in the following circumstances:
C.1) when $a(x_0)$ is known, see PROP. 1.2;
C.2) when a' has a bounded variation ($a' \in \mathcal{BV}(\overline{D})$) and satisfies $a'(\xi) = 0$, $\xi \in \overline{D}$, known; this corresponds to "zoning" in geophysics (see PROP. 1.3);
C.3) when two more regular data pairs $\{u; f\}$, $\{v; h\}$ satisfy at some τ

(1.12) $$\exists\, y_1, y_2 \in \overline{D} \ni \frac{1}{u_x(y_2,\tau)\, v_x(y_1,\tau)} - \frac{1}{u_x(y_1,\tau)\, v_x(y_2,\tau)} \neq 0.$$

This relationship, the *independence condition*, will not be considered herewith.

In principle the identification of a is related to an inverse source problem (see e.g., Isakov [7]): consider $(Bv_x)_x$ as the source term F in Eq. 1.7, rewritten as $V_t - (aV_x)_x = F$. However no uniqueness result applies, because $F = 0 \not\Rightarrow B = 0$.

1.2 Uniqueness due to local conditions

1.2.1 Uniqueness due to regular Cauchy problems for the defect equation

The statement corresponding to C.1 is straightforward.

PROPOSITION 1.2. *Let (1.1) hold, $u, v \in \mathcal{U}$ of (1.2) and let there $\exists \, \tau \in \overline{T} \ni \cdot$*

(1.13) $$\frac{1}{v_x}(\cdot, \tau) \in L^\infty(D)$$

(1.14) $$\exists \, u_x(x_0^+, \tau), v_x(x_0^+, \tau).$$

Moreover let there

(1.15) $$\exists \, a \in \mathcal{A}_{ad},$$

$\exists \, b$ s.t., (1.5) holds and

(1.16) $$B \in \mathcal{B}_{ad}.$$

Uniqueness then consists of the following implication

(1.17) $$\{[V(x, \tau) = 0, \forall \, x \in \overline{D}] \wedge [V_t(\cdot, \tau) =_{a.e.} 0 \text{ in } D]\} \Rightarrow$$
(1.18) $$\Rightarrow B =_{a.e.} 0 \text{ in } D.$$

Proof. Let Eq. 1.17 hold, then the r.h.s. of Eq. 1.7 is the null distribution. The antiderivative of $(Bv_x)_x(\cdot, \tau)$, which is in L^∞ as a consequence of the hypotheses, shall be a constant. Condition 1.13 gives sense to

(1.19) $$B(\cdot) =_{a.e.} \frac{c}{v_x(\cdot, \tau)}.$$

Since both sides of Eq. 1.19 are defined at x_0^+, (1.16) implies $c = 0$, hence Eq. 1.18. □

REMARK 1.2. Following Ch. 8 of Fattorini [5], a uniqueness result for Eq. 1.7 can be established under less restrictive hypotheses on the differential operator and on the (distribution – valued) control term, r. No attempt in this direction is made herewith: conditions (1.2), (1.16) and (1.14) have been chosen in view of the stability estimates to be discussed in § 2.

The other regular Cauchy problem, C.2, obviously needs stronger hypotheses.

PROPOSITION 1.3. (uniqueness *via a stationary conductivity*)
Let the reference conductivity possess a derivative of bounded variation

(1.20) $$a \in \mathcal{A}_2 := \{ a \mid a \in \mathcal{A}_{ad}, a' \in \mathcal{BV}(\overline{D}) \}$$

s.t., the (null measure) set S_a of points where a' is discontinuous satisfies

(1.21) $$S_a \cap \{x_0 ; x_1\} = \emptyset.$$

Denote by E_a the set of points, where a is stationary and let $E_a \neq \emptyset$. Require

(1.22) $$f \in C^0(\bar{T}, H^{-1}(D) \cap C^0(\bar{D} \setminus S_a)) .$$

(1.23) $$u, v \in \mathcal{U} := C^0(\bar{T}, C^2(\bar{D})) \cap C^1(\bar{T}, C^0(\bar{D})) ,$$

(1.24) $$\exists \tau \in \bar{T} \ni u_x(x,\tau), v_x(x,\tau) \neq 0, \forall x \in \bar{D} ,$$

(1.25) $$\exists b \in C^0(\bar{D}), 0 < a_L \leq b, \forall x \in \bar{D} .$$

Then

(1.26) $$B' \in \mathcal{BV}(\bar{D})$$

and condition (1.17) implies (1.18), with equality holding everywhere.

Proof. Given (1.20) and (1.23), Eq. 1.7 holds a.e. and term – wise differentiation is allowed. Since $Bv_{xx} \in C^0(\bar{D})$ and the r.h.s. of Eq. 1.7 is at most in $\mathcal{BV}(\bar{D})$, the same must be for $B' v_x$, whence (1.26).

Uniqueness is proved as in PROP. 1.2: only, the Cauchy datum is available at ξ instead of x_0. Explicitly, the unique a is given by

(1.27) $$a(x) = \left(\frac{g_\xi^{[-1]}}{u_x} \right)(x,\tau) + \left(\frac{g}{u_{xx}} \right)(\xi, \tau) ,$$

where $g_\xi^{[-1]}(\xi, \tau)$ is the antiderivative, which vanishes at ξ. All quotients on the r.h.s. make sense at time τ because of (1.24) and of the stronger hypotheses above. □

REMARK 1.3
i)   If $\mathrm{meas}[E_a] > 0$, then $a'' = 0$ on an interval: the above uniqueness condition continues to hold. Namely, the second derivative of conductivity never comes into play, unlike u_{xx}, which is relevant to PROP. 1.4 below.
ii)   The existence HP. 1.20 contains more information about the data $\{u; f\}$ e.g., $g(\xi) = 0, \forall \xi \in E_a$ and $\mathrm{meas}[E_a] > 0$ together imply $u_{xx} = 0$ in E_a.

1.2.2 Uniqueness due to singular Cauchy problems for Eq. 1.4.

Let

(1.28) $$u_x \in C^0(\bar{D} \setminus S_u) , \forall t \in T$$

s.t., the set $S_u(t)$ of the points where u_x is discontinuous satisfies

(1.29) $$\mathrm{meas}[S_u] = 0 \text{ and } S_u \cap \{x_0 ; x_1\} = \emptyset , \forall t \in T .$$

The set $E_u(t)$ of critical points of $u(.,t)$ can be then defined.

DEFINITION 1.1.

(1.30) $$E_u(t) := \text{clos}\{ \xi_u(t) \mid \xi_u(t) \in \bar{D}; \text{ either } u_x(\xi_u(t), t) = 0 \text{ or } \lim_{x \to \xi_u(t)_\pm} u_x(x(t), t) = 0 \}$$

PROPOSITION 1.4. (uniqueness from singular Cauchy problems). *Given $f \in L^2(T; H^{-1}(D))$, assume there $\exists a \in \mathcal{A}_{ad} \ni u(a; f)$ satisfies (1.28) and $\exists b$, complying with (1.5) s.t.,*

(1.31) $$u(a; f) = u(b; f) \text{ in } Q.$$

If either of the following conditions holds

(1.32) $$\exists \tau \in T \ni \{ E_u(\tau) \neq \emptyset \wedge \text{meas}[E_u(\tau)] = 0 \}$$

or

(1.33) $$\{ E_u(t) \neq \emptyset, \forall t \in T \wedge \text{meas}[\bigcap_{t \in T} E_u(t)] = 0 \}.$$

then

(1.34) $$b =_{a.e.} a \text{ in } D.$$

Proof
i) (necessity of each requirement taken by itself)
Let $E_u(t) = \emptyset, \forall t \in \bar{T}$, then non-uniqueness corresponds to b given by

(1.35) $$b =_{a.e.} a + \frac{c}{u_x},$$

where $c(.) \in C^0(\bar{T})$, independent of x and s.t., b satisfies (1.5). Since u_x is piecewise continuous and never changes its sign in a continuous fashion, it makes sense to multiply both sides of Eq. 1.35 by u_x. From Eq. 1.4, rewritten as a relationship between dual pairings it obviously follows

(1.36) $$\langle au_x + c \mid w_x \rangle = \langle au_x \mid w_x \rangle, \forall w \in \mathcal{W}_{ad}$$

where $\mathcal{W}_{ad}$ is defined in Appendix A. Since there is *no* constraint which forces $c \equiv 0$, the *first* requirement posed on $E_u(.)$ by both HP. 1.32 and 1.33 is thus necessary. The necessity of the *second* requirement is proved in a similar way. Begin considering HP. 1.32 : if $\text{meas}[E_u(\tau)] > 0$, then $\exists \tilde{b}$, no longer related to a by Eq. 1.35, which satisfies (1.5) and $\text{supp}(\tilde{b}-a) \subseteq E(\tau) \neq \emptyset$, because $(\tilde{b}-a)' u_x + (\tilde{b}-a)u_{xx} = 0$ in $E_u(\tau)$ and $u_{xx} = 0$ there. If meas$[\bigcap_{t \in T} E_u(t)] > 0$ in HP. 1.33, then at any $t \in T$ the same argument applies and leads to the same conclusion. Namely, replacing $E_u(\tau)$ by the general $E_u(t)$ yields

(1.37) $$\text{meas}[\text{supp } \tilde{b}] = \text{meas}[\bigcap_{t \in T} E_u(t)] > 0 .$$

ii) (sufficiency of HP. 1.32)
Now let HP. 1.32 hold. Assume that in addition to a there exists b, which also complies with Eq. 1.4. By definition, Eq. 1.36 holds. This time a condition on $c(.)$ will be obtained, however. Namely, after integration of Eq. 1.36 by parts, the value $c(\tau)$ is bounded above by

(1.38) $$| c(\tau) | \leq_{\text{a.e.}} \| b - a \|_{0,\infty} | u_x(x,\tau) | .$$

Choose $\epsilon > 0$, arbitrary, and consider a $\delta(\epsilon)$-neighbourhood $B_\delta(\xi)$ of a point $\xi \in E_u(\tau)$, where u_x is continuous i.e., $B_\delta(\xi) \cap S_u(\tau) = \emptyset$, and where $|u_x(x,\tau)| < \epsilon$. From

(1.39) $$\int_{B_\delta(\xi)} |c(\tau)| \, dx \leq \| b - a \|_{0,\infty} \int_{B_\delta(\xi)} |u_x(x,\tau)| \, dx <$$
$$< \epsilon \| b - a \|_{0,\infty} \text{meas}[B_\delta(\xi)]$$

and from the arbitrariness of ϵ it follows $c(\tau) = 0$ i.e.,

(1.40) $$(b - a)u_x =_{\text{a.e.}} 0 , \text{ hence } (b - a) =_{\text{a.e.}} 0 \text{ in } D .$$

A similar procedure applies to the left (right) neighbourhoods referred to by (1.30), where $u_x(.,\tau)$ is discontinuous as of (1.28), (1.29). The conclusion, Eq. 1.40, is the same.

iii) (sufficiency of HP. 1.33)
Finally consider HP. 1.33. The same procedure applies to the equation $(b - a) u_x =_{\text{a.e.}} c$, $\forall t \in T$. The intermediate results are of course different at each step. Ineq. 1.38 and Eqns. 1.40 now hold a.e. in $D \setminus E_u(t)$. The collection of constraint equations obtained from Eqns. 1.40 at each t imply

(1.41) $$(b - a)u_x = 0 \text{ a.e. in } D \setminus \bigcap_{t \in T} E_u(t) ,$$

i.e., with reference to Eq. 1.37,

(1.42) $$\text{meas}[\text{ supp } \tilde{b}] = \text{meas}[\bigcap_{t \in T} E_u(t)] = 0 ,$$

which completes the proof. □

Identification of Conductivity in the 1D Heat Equation   77

REMARK 1.4. Uniqueness conditions based on the properties of $E_u(t)$ were originally proven by Kitamura and Nakagiri [8] for a smooth { potential, conductivity } pair. The arguments used herewith are different, due to the decreased regularity of the data.

REMARK 1.5. Specifying the closure in (1.30) is essential. Assume it were *not* required. Take $u_x \in C^0(\overline{Q})$ and $u_x(\xi, t) = 0$, $\forall \xi \in \mathbb{Q} \cap \overline{D}$ (the rationals in $\overline{D}$), $\forall t$. Then continuity implies $u_x(t) \equiv 0$, hence the impossibility of achieving uniqueness.

EXAMPLE 1.1 (Application of PROP. 1.4). Let HP. 1.32 hold. It can be easily shown that a is given by the quotient $\frac{1}{u_x}[g_0^{[-1]} - g_0^{[-1]}(\xi_u^{\pm})](\tau)$, where the two limits represented by second summand need not coincide. (In fact one of them may not exist).
Let $D := (-1, 1)$, $u_x(\tau) = (1 + x^2)\theta(x) + x^2\theta(-x)$, where $\theta(.)$ is Heaviside's unit step function and $g_0^{[-1]}(\tau) = x^2\theta(-x) + 2(1 + x^2)\theta(x) - 1$. Clearly $S_u(\tau) = E_u(\tau) = \{0\}$. Now drop the τ-dependence. Since $g_0^{[-1]}(0^+) = 1$ and $g_0^{[-1]}(0^-) = -1$, the above quotient is found to read $\frac{1}{u_x}\left[2(1 + x^2)\theta(x) + x^2\theta(-x)\right]$. It returns the unique $a = 1 + \theta(x)$, which is also discontinuous at $x = 0$.

REMARK 1.6. (On the compatibility between singular and regular Cauchy uniqueness conditions). Let (1.13) and (1.28) simultaneously hold. HP. 1.13 and 1.33 applied to v are clearly incompatible: $\forall t$, either meas$[E_v(t)] = 0$ or meas$[E_v(t)] > 0$. The case meas$[E_v(t)] = 0$ takes to HP. 1.32, whereas meas$[E_v(t)] > 0$ implies $(B')^{[-1]} \not\subset L^\infty(D)$.

HP. 1.13 and 1.32 may hold for the same set of data at different time instants: choose $\epsilon > 0$ and take e.g., $v_x = x + t - \epsilon$ in $\overline{Q} := [0, 1] \times [0, 1]$. Then $\left(\frac{1}{v_x}\right)(t) \in L^\infty(D)$, $\epsilon < t \leq 1$, whereas $E_v(t) = \{\xi_v(t) = \epsilon - t\}$.

REMARK 1.7. It is understood that data on the whole of $\overline{T}$ are required when HP. 1.33 holds. Now assume that, as time elapses, the latter is met at a date earlier than t_1. Then data collection may be halted and the interval be redefined accordingly.

1.3 Uniqueness due to a non – local condition (*self – identifiability*)

The non local condition introduced by Eq. 1.9 leads to the following statement.

THEOREM 1.1. (*self – identifiability*). *Let f comply with (1.1), $u \in \mathcal{H}$. Assume $\exists$ $a(u; f) \in \mathcal{A}_{ad}$ and $\exists \tau \in \overline{T} \ni$ the following properties hold*

(1.43) $$\mathrm{meas}[E_u(\tau)] = 0,$$

(1.44) $\quad \exists \ \{ g_0^{[-1]}(x_0^+, \tau) \}$ and $u_x(x_0^+, \tau)$

(1.45) $\quad < au_x >(\tau) = k_\tau$.

Then the unique $a(\ u;\ f\ )$ is given by

(1.46) $\quad a =_{\text{a.e.}} \left(\dfrac{g_0^{[-1]} - <g_0^{[-1]}> + k_\tau}{u_x} \right)(\tau)$.

Proof. Eq. 1.4 means that au_x is an antiderivative of g

(1.47) $\quad au_x \in \{ g^{[-1]}(t) \}$, $\forall t \in \overline{T}$,

where $\{ g^{[-1]} \} \subset L^\infty$. The continuity hypotheses (1.44) yield

(1.48) $\quad au_x(\tau) = g_0^{[-1]}(\tau) + c_\tau$.

The constant c_τ is determined by averaging Eq. 1.48 over D and making use of Eq. 1.45 i.e., $c_\tau = k_\tau - <g_0^{[-1]}>(\tau)$. Dividing both sides of Eq. 1.48 by u_x , which is allowed by Eq. 1.43, yields Eq. 1.46. $\square$

EXAMPLE 1.2 (Application of THM. 1.1). Choose $\overline{Q} = [-1 + \epsilon, 1 - \epsilon] \times [1,2]$, where $0 < \epsilon \ll 1$, ϵ fixed. Let the data be: $u_x(x, \tau) = [2\ \theta(x) - 1 - x]\tau$, $g(x,\tau) = [2\delta(x) - 1]\ \tau$, where $\delta(.)$ is Dirac's measure, and $k_\tau = 0$.
Note that $E_u(\tau) = \emptyset$ and $S_u(\tau) = \{ 0 \}$: in other words $u_x(., \tau)$ vanishes nowhere in $\overline{D}$, hence changes sign in a discontinuous fashion. Then $g_0^{[-1]} = [\epsilon + 2\theta(x) - 1 - x]\ \tau$ and $<g_0^{[-1]}> = \epsilon\tau$. Eq. 1.46 applies and returns the unique $a = 1$.

2 Stability Estimates

For each of the uniqueness conditions in the previous Section, a stability estimate will be stated under supplementary regularizing hypotheses.

2.1 Estimates for solutions to regular Cauchy problems

When uniqueness is due to a Cauchy datum supplied at a regular point, stability is estimated as follows.

THEOREM 2.1.
Let $u, v \in \mathcal{C}$ s.t., (1.14) are met. In addition let the following uniform bounds hold

(2.1) $\quad \exists\ \tau \in \overline{T} \ni \left|\dfrac{1}{u_x}\right|(\tau), \left|\dfrac{1}{v_x}\right|(\tau) \leq c_v$;

Let a, B comply with (1.15) and (1.16), resp. Then a uniform estimate applies to B

$$(2.2) \qquad \| B \|_{0,\infty} \leq (1 + 2a_H) \, c_v \, \| V \|_{\mathcal{G}(\tau)} ,$$

where

$$(2.3) \qquad \| V \|_{\mathcal{G}(\tau)} := \max_{\bar{D}} |V(\tau)| + \| V_x(\tau) \|_{0,\infty} + \sqrt{|x_1 - x_0|} \, \| V_t(\tau) \|_{0,2} .$$

Proof. The starting point is the defect equation at $t = \tau$, from which the t - dependence is dropped. Eq. 1.7 is integrated term - wise to yield

$$(2.4) \qquad Bv_x =_{a.e.} r_0^{[-1]} + c_\tau .$$

The continuity hypotheses imply both $r_0^{[-1]} =_{a.e.} \int_{x_0}^{x} V_t \, dy - aV_x + (aV_x)(x_0^+)$ and $c_\tau = 0$.

The quotient $\dfrac{r_0^{[-1]}}{v_x}$ is defined a.e. and is estimated from above by means of Ineq. 2.1

$$(2.5) \qquad |B| \leq_{a.e.} \left[\sqrt{|x_1 - x_0|} \, \| V_t \|_{0,2} + 2a_H \| V_x \|_{0,\infty} \right] c_v \leq$$

$$\leq (1 + 2a_H) \, c_v \, \| V \|_{\mathcal{G}(\tau)} ,$$

which leads to Ineq. 2.2. □

REMARK 2.1 (On the role of procedure). Instead of integrating Eq. 1.7 and estimating the quotient by Ineq. 2.5, one may consider estimating $|B|$ by means of Gronwall – Bellman's inequality, which arises naturally in connection with Cauchy problems for evolution equations and the related stability theory (e.g., Haraux [6]), including those obtained from inverse problems: see e.g., Baumeister and Kunisch [3]. Very briefly, this procedure can indeed be worked out, even if Eq. 1.7 is in a distribution space. It consists of the following steps:

i)   Gronwall – Bellman's inequality is extended to measurable functions;

ii)   the growth of B is estimated: this requires the derivative of the product Bv_x to make sense term - wise; moreover the condition $\left(\dfrac{v_{xx}}{v_x}\right)(\tau) \in L^1(D)$ is needed at some stage, which implies $\left(\dfrac{1}{v_x}\right)(\tau) \in AC(\bar{D})$;

iii)   the hypotheses of THM. 2.1 are strengthened by $v_{xx} \in L^2$ s.t., $\| v_{xx} \|_{0,2}(\tau) \leq$
$\leq c_s$ and $a \in \mathcal{A}_{ad} \cap \mathcal{BV}(\bar{D})$ s.t., the total variation of a is no more than c_A.

iv)   The resulting estimate differs from Ineq. 2.2, because the r.h.s. is replaced by

$$(1 + a_H + c_A) c_v \, \| V \|_{\mathcal{G}(\tau)} \exp\left[c_s c_v \sqrt{|x_1 - x_0|} \, \right].$$

The next result applies to the uniqueness condition of PROP. 1.3.

THEOREM 2.2 (stability of the *stationary*–unique solution)
Let u, v comply with (1.23) and (1.24) hold. Let a and B satisfy resp. (1.20), (1.26). In addition to Ineq. 2.1 let the following upper bounds apply:

(2.6) $$\max_{\bar{D}} |v_{xx}|(\tau) \leq c_s ,$$

(2.7) $$\| a' \|_{0,\infty} \leq c_A .$$

Then

(2.8) $$\| B \|_{1,\infty} \leq c_v [1 + a_H + c_A + (c_s + 1)(2a_H + |x_1 - x_0|)c_v] \| V \|_{\mathscr{G}(\tau)}$$

where $\| V \|_{\mathscr{G}(\tau)} := \| V(\tau) \|_{2,\infty} + \| V_t(\tau) \|_{0,\infty}$.

Proof. Integration of Eq. 1.7 (see the proof of THM. 2.1) starts at any $\xi \in E_a$, instead of x_0, because $a(\xi)$ is known from Eq. 1.27. The norms $\| B \|_{0,\infty}$ and $\| B' \|_{0,\infty}$ are estimated in sequence. Ineq. 2.8 follows from adding up the upper bounds. □

2.2 Estimates for solutions to singular Cauchy problems

If $u_x(.)$, hence $v_x(.)$, vanishes somewhere, estimates can no longer yield uniform upper bounds. This is the subject of the next two Theorems, which are based on the uniqueness conditions of PROP. 1.4 and require the data and the solutions to be more regular. Another result will be given in § 2.4 below.

THEOREM 2.3 (stability related to HP. 1.32). Let $\bar{I} \subseteq \bar{D}$ denote a closed interval or a collection thereof. Let u, $v \in \mathscr{G}$ comply with (1.28) and (1.29). Assume there $\exists \tau \in \bar{T}$ and an exponent p, $1 \leq p < \infty$ s.t., u, v satisfy HP. 1.32 and the following conditions

(2.9) $$u_x(\tau), v_x(\tau) \in \mathcal{C}^0(\bar{I}),$$
(2.10) $$E_u(\tau), E_v(\tau) \subseteq \bar{I},$$
(2.11) $$\left(\frac{1}{u_x}\right)(\tau), \left(\frac{1}{v_x}\right)(\tau) \in L^p(D),$$
(2.12) $$\left\| \frac{1}{v_x} \right\|_{0,p}(\tau) \leq c_v .$$

Let the reference solution be continuous wherever either u or v may be stationary

(2.13) $$a \in \mathcal{A}_S := \mathcal{A}_{ad} \cap \mathcal{C}^0(\bar{I}) .$$

Finally assume the unique $b(v; f)$ satisfies (1.5). Then

(2.14) $$\| B \|_{0,p} \leq c_v (1 + 2a_H) \| V \|_{\mathscr{G}(\tau)} .$$

Proof. Integrate Eq. 1.7 starting from $\xi_v(\tau) \in E_v(\tau)$ and obtain

(2.15) $$(Bv_x)(\tau) =_{d.w.} \left(\int_{\xi_v}^{x} V_t \, dy - aV_x\right)(\tau) + c_r(\tau).$$

The dependence on τ will be omitted for the rest of the proof. The hypotheses imply that the r.h.s. is in $L^\infty(D)$ and continuous at ξ_v, hence

(2.16) $$c_r = (aV_x)(\xi_v^+).$$

As a consequence there exists an antiderivative, $r_{\xi_v^+}^{[-1]}$, which vanishes at ξ_v^+, in agreement with the property of the l.h.s.

(2.17) $$0 = (Bv_x)(\xi_v^+) = r_{\xi_v^+}^{[-1]}(\xi_v^+).$$

Elsewhere in $\xi_v < x < x_1$, $B =_{a.e.} -\frac{1}{v_x} r_{\xi_v^+}^{[-1]}$. A similar relationship is obtained in the interval $x_0 < x < \xi_v$ by reversing the x direction and integrating Eq. 1.7 from ξ_v towards x_0. Denote by $r_{\xi_v^-}^{[-1]}$ the corresponding r.h.s, which also vanishes at ξ_v because of the essential hypotheses (2.9) and (2.13). A unique antiderivative, named $r_{\xi_v}^{[-1]}$, has thus been determined s.t.,

(2.18) $$B =_{a.e.} -\frac{1}{v_x} r_{\xi_v}^{[-1]}.$$

No uniform estimate applies to Eq. 2.18, unlike the situation of THM. 2.1. From Ineq. 2.12, the L^p − norms of both sides of Eq. 2.18 are related by

(2.19) $$\| B \|_{0,p} \leq c_v \| r_{\xi_v}^{[-1]} \|_{0,\infty}(\tau).$$

Given (2.15) and (2.16), the norm on r.h.s. is now estimated from above by

(2.20) $$\| r_{\xi_v}^{[-1]} \|_{0,\infty} \leq \sqrt{|x_1 - x_0|} \, \| V_t \|_{0,2} + 2a_H \, \| V_x \|_{0,\infty},$$

which leads to Ineq. 2.14. □

An estimate can be established even if the second conductivity is not uniquely determined by the second potential alone, provided ambiguity is removed at some stage.

COROLLARY 2.1 (to THM. 2.3).

Let u and a satisfy the same HP. as in THM. 2.3. Let $v \in \mathcal{B}$ and $\exists \, \tau \in \overline{T}$ s.t., (2.9) and (2.10) continue to hold for v, whereas HP. 1.32, (2.11) and (2.12) are replaced by

(2.21) $$\text{meas}[E_v(\tau)] > 0,$$

(2.22) $$\left(\tfrac{1}{v}\right)_x(\tau) \in L^p(D \setminus E_v(\tau)) \equiv \mathcal{Y}(\tau),$$

(2.23) $$\left\|\tfrac{1}{v_x}\right\|_{\mathcal{Y}(\tau)} \leq c_v,$$

If $b(v; f)$ complies with (1.5) in $D \setminus E_v(\tau)$ and ambiguity in $E_v(\tau)$ is removed by

(2.24) $$b = a \text{ in } E_v(\tau),$$

then Ineq. 2.14 continues to hold.

Proof. Because of Ineq. 2.21, b is unique, and generally different from a, in $D \setminus E_v(\tau)$. Begin with the simplest case: E_v a (closed) interval. The counterparts of Eq. 2.15 through 2.17 and of Ineq. 2.19 are written separately for the two *disjoint* right and left subintervals, denoted by J_L and J_R, resp., into which $D \setminus E_v(\tau)$ at most decomposes. Namely

(2.25) $$\| B(i) \|_{L^p(J_i)} \leq c_v \| r_i^{[-1]} \|_{L^\infty(J_i)}(\tau), \quad i = L(EFT), R(IGHT)$$

where $r_R^{[-1]}$ vanishes at the left endpoint of J_R and $r_L^{[-1]}$ at the right endpoint of J_L. These two Inequalities are added term by term. On each side L^p- and, resp., L^∞- norms in $\{J_L \cup J_R\}$ are obtained. If the procedure were stopped here, the result would depend on a property of v, $\text{meas}[E_v(\tau)]$, which is not acceptable. Inside $E_v(\tau)$, b is arbitrary, whereas a is continuous and known. Eq. 2.24 however implies $B = 0$ in E_v: this interval adds no contribution to the estimate, even if $\tfrac{1}{v_x}$ is not defined here. As a consequence $\{J_L \cup J_R\}$ is replaced by D and Ineq. 2.14 follows. More complicated sets E_v of positive measure are similarly dealt with by the same decomposition argument. Ineq. 2.14 is eventually obtained. Finally note that passing to the limit $\| V \|_{\mathcal{G}(\tau)} \to 0$ implies $v_x(\tau) \to u_x(\tau)$ in $C^0(\bar{I})$, hence $\text{dist}(E_v(\tau), E_u(\tau)) \to 0$ i.e., $\text{meas}[E_v(\tau)] \to 0$. □

When uniqueness is due to HP. 1.33, the stability result is essentially the same, although the procedure is more complicated.

THEOREM 2.4 (stability related to HP. 1.33)

Assume that the potentials u, v are in $\mathcal{G}$ and satisfy (1.33). Let there exist $\bar{I} \subseteq \bar{D}$ s.t.,

(2.26) $$\bigcup_{t \in T} E_u(t), \bigcup_{t \in T} E_v(t) \subseteq \bar{I}.$$

Moreover, $\forall t \in \bar{T}$ and for some p as in THM. 2.3, assume

(2.27) $$u_x, v_x \in C^0(\bar{I}),$$

(2.28) $$\frac{1}{u_x}(\cdot, t) \in L^p(D \setminus E_u(t)), \quad \frac{1}{v_x}(\cdot, t) \in L^p(D \setminus E_v(t)),$$

and

(2.29) $$\max_{T} \left\| \frac{1}{v_x} \right\|_{L^p(D \setminus E_v(t))}(t) \leq c_v.$$

Let (2.13) hold and the unique $b(v; f)$ comply with (1.5). Then

(2.30) $$\| B \|_{0,p} \leq c_v (1 + 2a_H) \| V \|_{\mathcal{G}}.$$

The proof, which is not given in detail for the sake of conciseness, shares some features with that of THM. 2.3. A point $\xi_v^* \in \bigcap_{t \in T} E_u(t)$ exists s.t., right and left antiderivatives (see Eq. 2.17 and related comments) can be determined. The counterpart of Ineq. 2.19 is obtained, which eventually yields Ineq. 2.30.

The next result is the counterpart of COR. 2.1, which is a generalization of the arguments presented so far.

COROLLARY 2.2 (to THM. 2.3). *Let v single out because of*

(2.31) $$\text{meas}\left[\bigcap_{t \in T} E_v(t)\right] > 0,$$

and let all remaining HP. in THM. 2.4 be satisfied by u, v, and a. If

(2.32) $$b = a \text{ in } \bigcap_{t \in T} E_v(\tau),$$

then Ineq. 2.30 continues to hold.

REMARK 2.2. Even under the restrictions specified by REM. 2.1, no uniform estimate can be obtained when uniqueness comes from either HP. 1.32 or 1.33. Namely, it is wrong to apply the procedure of REM. 2.1 to solutions of singular Cauchy problems, because the datum " $v_x = 0$ at some point " prevents one from carrying out step *ii*).

2.3 Estimates for solutions to non – local problems

THEOREM 2.5. *Let $u, v \in \mathcal{G}$ and let the following hold at $t = \tau$: conditions (1.14),*

(2.33) $$\text{meas}[E_u(\tau)] = \text{meas}[E_v(\tau)] = 0$$

where these sets may be empty, and

(2.34) $$<au_x>(\tau) = <bv_x>(\tau) = k_\tau,$$

where a, b satisfy the usual requirements. If $\exists\ p, 1 \leq p \leq \infty$ and a constant c_v s.t.,

(2.35) $$\left\| \frac{1}{v_x} \right\|_{0,p}(\tau) \leq c_v,$$

then

$$\text{(2.36)} \qquad \|B\|_{0,p} \leq c_v (2 + a_H) \|V\|_{\mathcal{U}(\tau)} .$$

The *proof* relies on integrating Eq. 1.7, then averaging both sides over D. Eqs. 2.34 imply that the integration constant is $c_\tau = - <\int_{x_0}^{x} V_t \, dy >(\tau) - (aV_x)(x_0^+, \tau)$. If $p = \infty$, the uniform estimate follows as in THM. 2.1, otherwise an integral estimate is easily obtained.

2.4 Generalization of the estimates of integral type

The choice of $\mathcal{U}$ is by no means restrictive. Several ways of decreasing the regularity of the data can be devised. One could e.g., require the antiderivatives of u_t, v_t, f to be in some L^q, $q \geq 1$ or affect the spatial derivatives. For the sake of simplicity, only the latter case is described.

THEOREM 2.6. *Replace the $\mathcal{U}$ of (1.2) by*

$$\text{(2.37)} \qquad \mathcal{U} := \mathcal{C}^0(\overline{T}; W^{1,q}(D)) \cap \mathcal{C}^1(\overline{T}, L^2(D)), \quad q \geq \max\{2, \frac{p}{p-1}\},$$

where p is the same as in (2.11) or (2.28) and let all other respective hypotheses of THEOREMS *2.3, 2.4, 2.5,* COROLLARIES *2.1 and 2.2 hold. Then the L^p estimates for B are replaced by L^r estimates, where either*

$$\text{(2.38)} \qquad r = \frac{pq}{p+q} \quad \text{if } p > 1$$

or

$$\text{(2.39)} \qquad r = 1 \quad \text{if } p = 1 \text{ (and } q = \infty \text{)}.$$

E.g., Ineq. 2.36 for q finite becomes

$$\text{(2.40)} \qquad \|B\|_{0,r} \leq c_v (2 \sqrt[q-1]{|x_1 - x_0|} + a_H) \|V\|_{\mathcal{U}(\tau)}$$

Proof. The constraint on q simultaneously yields $V \in \mathcal{C}^0(\overline{D})$ and $r \geq 1$. Hoelder's inequality is applied to quotients like the r.h.s. of Eq. 2.18: their L^r-norms are estimated from above by products of L^p- and L^q- norms. If q is finite, Minkowsky's inequality is finally needed to replace $\| r_{\xi_v}^{[-1]} \|_{0,p}$ by the $\mathcal{U}(\tau)$- or the $\mathcal{U}$-norm, respectively. $\square$

Discussion and Conclusion

The uniqueness and stability of conductivity identified from potential and source term data have been assessed. These results have been classified according to the type of available information (local *vs.* non − local), the regularity of the Cauchy problem and time span.

The unique determination of distributional antiderivatives has relied on the continuity of the latter at endpoints. In all proofs derivatives of products like au_x, resp. Bv_x have never been expanded: this has kept the regularity requirements to a minimum.

Some examples have been provided: other relevant examples of non uniqueness and instability, arising because (1.32) or (2.1) do not hold, can be found in Ch. IV, § 2 of Banks and Kunisch [2].

About the *time span* of data, only THM. 2.4 and COR. 2.2 require that data be known everywhere in Q, as the related uniqueness condition does. For the sake of conciseness, the time – independent and the time – averaged situations are not discussed herewith. Intuitively, all results shown above have a straightforward time – independent counterpart, because the ODE structure of Eq. 1.4 and Eq. 1.7 is left unchanged. A related result by Marcellini [9] is however worth mentioning: in considering the homogeneous Dirichlet 2–point BVP for $(au')' = 1$, with $u \in H^2(D)$ and $a \in H^1(D)$, said Author was able obtain an L^2– stability estimate. From the discussion in § 1 and § 2 above, this is to be expected, in spite of the more regular a, because a singular Cauchy problem ($u' = 0$ somewhere inside D) for the defect equation is met.

An aspect, which is beyond the scope of this paper is the *discretization* of space, time and of the function spaces. In particular, time sampling has not been considered: stability estimates based on time sampled potentials and affecting smooth conductivity functions of two space variables have been given in a different context by Alessandrini and Vessella [1]. Elementary stability results for the algebraic counterpart of Eqs. 1.27 and 1.46 can be found in [10]. Indeed, experimental data are always processed in a finite dimensional setting. Nonetheless the uniqueness and stability results herewith ought to help in predicting the (asymptotic) performance of identification algorithms which rely upon either special features of the measured potential or upon prior knowledge about the solution.

Appendix A

Consider the space of test functions

(A.1) $\quad \mathcal{W}_{ad} := \{ w \mid w \in L^2(T; H_0^1(D)),\ w_t \in L^2(T; H^{-1}(D));\ w(\cdot, t_1) = 0 \}$

and a distribution $g \in L^2(T; H^{-1}(D))$. Introduce the one–parameter (ϵ) class of $C_0^\infty(\overline{D})$-functions defined by

(A.2) $\quad \omega_\epsilon(x) := \chi_{[-\epsilon, +\epsilon]}\, c_\epsilon \exp\left[\dfrac{\epsilon^2}{x^2 - \epsilon^2}\right]$,

where $\chi_{[\cdot,\cdot]}$ is the characteristic function of an interval and c_ϵ is the normalization constant.

Let $\zeta_0 \in D$ and choose $\epsilon \ni 0 < \epsilon < \min\{\zeta_0 - x_0\ ;\ x_1 - \zeta_0\}$. The function w_ϵ defined by

(A.3) $\quad w_\epsilon(x,t) := \displaystyle\int_{x_0}^{x}\left[w(\zeta,t) - \omega_\epsilon(\zeta - \zeta_0)\int_D w(\xi,t)\, d\xi \right]\, d\zeta,\quad w \in \mathcal{W}_{ad}$

is also in $\mathcal{W}_{ad}$.

A *spatial antiderivative* of g, denoted by $g^{[-1]}$, is defined by

(A.4) $\quad \langle g^{[-1]} \mid w_x \rangle = -\langle g \mid w \rangle,\ \forall w \in \mathcal{W}_{ad}$.

Let $c_1(.)$ be an arbitrary function in $L^2(T)$. A consequence of (A.4) is

(A.5) $$\langle g^{[-1]} \mid w \rangle = -\langle g \mid w_\xi \rangle + \langle c_1 \mid w \rangle, \forall w \in \mathcal{W}_{ad}.$$

This statement is similar to the Theorem in Ch. 1, § 1.2 of Vladimirov [11]: the only difference is needed to encompass the time dependence of some quantities. Of course every $g^{[-1]}$ of Eq. A.5 satisfies $g^{[-1]} \in L^2(T; L^1_{loc}(D))$. Since antidifferentiation is a set – valued map, the notation $\{ g^{[-1]} \} \subset L^2(T; L^1_{loc}(D))$ is used.

If there exists a representative of the class $\{ g^{[-1]} \}$, which is continuous at x_0^+, the antiderivative of g, which vanishes at x_0^+ is defined and is denoted by $g_0^{[-1]}$.

References

[1] Alessandrini G, Vessella S, *Error Estimates in an Identification Problem for a Parabolic Equation*, Boll. U.M.I. – Analisi Funzionale e Appl., Ser. VI, **IV–C** (1985), pp. 183 – 203.

[2] Banks H T, Kunisch K, *Estimation Techniques for Distributed Parameter Systems*, Birkhäuser: Basel, 1989.

[3] Baumeister J, Kunisch K, *Identifiability and Stability of a Two Parameter Estimation Problem*, Applicable Analysis, **40**, # 4 (1991), pp. 263 – 79.

[4] Cannon J R, *The One – Dimensional Heat Equation*, in Rota G. C., Ed., Encyclopaedia of Mathematics and Its Applications, vol. **23**, Addison Wesley: Reading, MA, 1984.

[5] Fattorini H O, *The Cauchy Problem*, in Rota G. C., Ed., Encyclopaedia of Mathematics and Its Applications, vol. **18**, Addison Wesley: Reading, MA, 1983.

[6] Haraux A, *Nonlinear Evolution Equations – Global Behaviour of Solutions*, Springer: Berlin, 1981.

[7] Isakov V, *Inverse Source Problems*, AMS: Providence, RI, 1990.

[8] Kitamura S, Nakagiri S, *Identifiability of Spatially – Varying and Constant Parameters in Distributed Systems of Parabolic Type*, SIAM J. Control and Optimiz., **15** (1977), pp. 785 – 802.

[9] Marcellini P, *Identificazione di un Coefficiente in un' Equazione Differenziale Ordinaria del Secondo Ordine*, Ricerche di Matematica dell' Universita' di Napoli, **31**, #2 (1982), pp. 223–43.

[10] Ponzini G, Crosta G, Giudici M, *Identification of Thermal Conductivities by Temperature Gradient Profiles: One–dimensional Steady Flow*, Geophysics, **54** (1989), pp. 643–53.

[11] Vladimirov V, *Distributions en Physique Mathématique*, MIR: Moscow, 1979.

Chapter 7
Estimation of Material Parameters in a Dynamic Nonlinear Plate Model with Norm Constraints[*]

L. W. White[†]

Abstract

The output-least-square estimation without regularization but with norm constraints of flexural rigidity with observation of deformations is considered for a dynamic von Karman system. Since such equations may have multiple solutions, we develop a theory for the estimation of coefficients in systems with solution sets that are compact in suitable spaces. For these problems in which the set of admissible parameters is norm bounded, we provide a convergence theory based on Galerkin approximations. A numerical estimation algorithm is developed and the results of numerical experiments are reported.

1 Introduction

Let Ω be a bounded domain in $\mathbb{R}^2$ with a Lipschitz boundary Γ. We set $Q = \Omega \times (0,T)$ and $\Sigma = \Gamma \times (0,T)$ where $T > 0$. Let $V = H_0^2(\Omega)$ and $H = L^2(\Omega)$, respectively. We denote standard normals on H and $V \subset H^2(\Omega)$ by $\|\cdot\|_0$ and $\|\cdot\|_2$, respectively. Also, we note that $\phi \mapsto \|\Delta\phi\|_0$ is a norm on V, cf. [2]. Furthermore, there exist positive constants k_1 and k_2 such that

$$k_1 \|\phi\|_2 \leq \|\Delta\phi\|_0 \leq k_2 \|\phi\|_2$$

for any $\phi \in V$, [2].

Introducing the von Karman bracket

$$[u,v] = u_{xx}v_{yy} + u_{yy}v_{xx} - 2u_{xy}v_{xy} ,$$

we consider the equations

(1.1)(i) $$u_{tt} + \Delta(a\Delta u) + \epsilon[\varphi, u] = f \text{ in } Q$$

and

(1.1)(ii) $$\Delta^2 \varphi = [u, u] \text{ in } Q$$

with boundary conditions

(1.2) $$u = \frac{\partial u}{\partial n} = 0 \text{ on } \Sigma$$

---

[*] This work was supported in part by AFOSR grant number AFOSR-91-0017.

[†] Department of Mathematics, University of Oklahoma, Norman, Oklahoma 73019

and initial conditions

(1.3)
$$u(0) = u_0$$
$$u_t(0) = u_1$$
in Ω.

The equations in (1.1) are refered to as von Karman equations and serve to model the motion of thin plates undergoing large deformations [3,9]. Global uniqueness of weak solutions for von Karman's equations is unknown, although the existence of unique of local classical solutions is proved in [14,16].

At the outset, we assume there is a positive constant ν such that

(1.4) $$a \in L^\infty(\Omega) \text{ and } a \geq \nu > 0,$$

that

(1.5) $$f \in L^2(Q),$$

and that

(1.6) $$u_0 \in V \text{ and } u_1 \in H.$$

The problem we consider here is to estimate the coefficient $a = a(x,y)$ from data obtained as measurements of the deformation u. We formulate this problem as an output-least-squares estimation problem in which we seek to determine a coefficient a_0 minimizing a fit-to-data criterion over an admissible parameter set. The coefficient "a" is called the flexural rigidity and depends on the thickness of the plate, the Young's modulus, and Poisson's ratio. Later, we make explicit assumptions on "a" as we specify admissible sets of parameters in the formulation of the estimation problem.

Since there my not be a unique solution to (1.1)–(1.6), we must take care in formulating the estimation problem. We utilize the fact that the set of solutions of (1.1)–(1.6) is weak $*$ compact in a suitable space to give a meaningful statement of the estimation problem. Data is usually obtained from point measurements and is processed in some manner. We assume only that the point measurements are interpolated in such a way that the data function z belongs to $L^2(Q)$. To solve the estimation problem numerically, we must accordingly deal with systems that may have multiple solutions. In [17] we consider the static problem which also does not possess a unique solution. There we present a theory that essentially treats the state equation as a constraint and penalizes the equation error in a weak sense. In the present case, however, we can take advantage of uniqueness results for initial value problems of the ordinary differential equations that are generated as the Galerkin approximating systems. This is not possible in the static case where the finite dimensional approximating systems may have multiple solutions as well. In [18] we consider a similar estimation problem with regularization but without norm constraints. In that case, however, it is necessary to assume stronger continuity properties than are needed here to obtain existence results.

In this paper we present a formulation of an estimation problem based on Galerkin approximating systems that supports a convergence theory. Hence, we link our results to an arbitrary but fixed Galerkin approximating scheme. We may then consider the estimation problem for the Galerkin systems and obtain approximation results for the associated estimation problem to the estimation problem for (1.1)–(1.6). Regularity results are established for optimal estimators in the Galerkin systems. These results are used to provide an approximation theory for the estimation problem. Finally, we provide a numerical algorithm for the estimation of parameters and report the results of several numerical experiments.

2 Preliminaries

In this section we cite several results that will be useful in the analysis of the estimation of the coefficient a in the model (1.1)–(1.6). We refer the reader to [3,9,18] for further properties and proofs.

The following is proved in [9] by a Galerkin argument.

PROPOSITION 2.1. *Under (1.4), (1.5), and (1.6) there exist solutions to the problem (1.1)–(1.3) satisfying*
$$u \in L^\infty(0,T;V), \quad u_t \in L^\infty(0,T,H)$$
and
$$\varphi \in L^\infty(0,T,V).$$
In fact,
$$u_{tt} \in L^\infty(0,T;V^*).$$

We briefly recall the Galerkin approximation for the purpose of obtaining some useful estimates. For ease of exposition, we take $u_0 = u_1 = 0$ since little is lost under this assumption. Let $S^N = \{\phi_i\}_{i=1}^N$ with $V^N = \text{span}\{\phi_i\}_{i=1}^N$ and $S = \{\phi_i\}_{i=1}^\infty$ be a basis of V. In this case we have $V^N \subset V^{N+1} \subset \cdots \subset V$. Setting $u^N(t) = \sum_{i=1}^N c_i(t)\phi_i$, it is required that

$$(2.1) \quad (u_{tt}^N(t), \phi_j) + \int_\Omega a \Delta u^N(t) \Delta \phi_j \, dx + \int_\Omega [B(u^N(t), u^N(t)), u^N(t)] \phi_j \, dx = (f(t), \phi_j)$$

for $j = 1, 2, \ldots, N$. Here we denote the solution of the equation
$$\Delta^2 \varphi = [u, v] \text{ in } \Omega$$
with boundary conditions
$$\varphi = \frac{\partial \varphi}{\partial n} = \text{ on } \Gamma$$
by $\varphi = B(u,v)$. From the theory of initial value problems for ordinary differential equations, it follows that for each N there exists an interval $(0, T_N)$ such that there is a unique solution $\{c_i(t)\}_{i=1}^N$ to the system (1.1) on $(0, T_N)$.

Multiplying equation (2.1) by the derivative $c_j'(t)$ and summing on j, we see that the Galerkin approximation u^N satisfies
$$\frac{d}{dt}(\|u_t^N(t)\|_H^2 + \int_\Omega a(\Delta u^N(t))^2 dx + \epsilon \|B(u^N(t), u^N(t))\|_V^2)$$
$$\leq 2(f(t), u_t^N(t))_H$$
$$\leq \|f(t)\|_H^2 + \|u_t^N(t)\|_H^2.$$

Integrating we find
$$\|u_t^N(t)\|_H^2 + \int_\Omega a(\Delta u^N(t))^2 dx + \epsilon \|B(u^N(t), u^N(t))\|_V^2$$

$$\le \int_0^t \|f(\tau)\|_H^2 d\tau + \int_0^t \|u_t^N(\tau)\|_H^2 d\tau.$$

Thus, by Gronwall's inequality we see that

(2.2)
$$\|u_t^N(t)\|_H^2 + \nu \int_\Omega (\Delta u^N(t))^2 dx + \epsilon \|B(u^N(t), u^N(t))\|_V^2$$
$$\le e^T \|f\|_{L^2(0,T;H)}^2.$$

From these estimates it follows that $T_N = T$ and there is a unique solution of (2.1) on $(0,T)$, [9].

Define the space
$$W = \{\psi \in L^1(0,T;V) : \psi_t \in L^1(0,T;H)\}$$

and note that its dual is given by
$$W^* = \{\psi \in L^\infty(0,T;V) : \psi_t \in L^\infty(0,T;H)\}.$$

Denote by $S_{W^*}(R)$ the closed ball of radius R in W^*, where in particular

(2.3)
$$R = e^{T/2} \|f\|_{L^2(0,T;H)} / \min(1,\nu)^{1/2}.$$

Then $S_{W^*}(R)$ is weak $*$ compact by Alaoglu's theorem. Noting that W^* is separable, we see that the weak $*$ topology on $S_{W^*}(R)$ is a metric topology [5]. Further, $S_{W^*}(R)$ is closed in the weak $*$ topology since the weak $*$ topology is Hausdorff. Hence, any sequence in $S_{W^*}(R)$ has a weak $*$ convergent subsequence whose limit belongs to $S_{W^*}(R)$. Thus, it follows that

$$u_{N_i} \to u \text{ weak } * \text{ in } W^*$$

and $u \in S_{W^*}(R)$. In addition, we also see from compactness of the embedding of W^* into $L^2(Q)$, see [15], that there exists a subsequence $\{u_{N_i}\}_{i=1}^\infty$ such that

$$u_{N_i} \to u \text{ in } L^2(Q).$$

Let $Q = H^2(\Omega)$ and define the subset

(2.4)
$$Q_{ad} = \{a \in Q : a \ge \nu \text{ and } \|a\|_Q \le K\}$$

where ν and K are such that $Q_{ad} \ne \varphi$. For the purpose of approximation in which we suppose that a specific set S of basis functions is given, we define the set $U_S(a)$ for $a \in Q_{ad}$ as follows

$U_S(a)$ = collection of weak $*$ cluster points in W^* of sequences of solutions $u^N(a_k)$ to the Galerkin systems for S where $\{a_k\}_{k=1}^\infty$ is any sequence in Q_{ad} converging weakly to a in Q.

Remark 2.2. An element $\tilde{u}$ belongs to $U_S(a)$ if and only if there exists a sequence $\{a_k\}_{k=1}^\infty \subset Q_{ad}$ such that $a_k \to a$ weakly in Q and a sequence of solutions $\{u^{N_k}(a_k)\}_{k=1}^\infty$ of Galerkin systems such that $u^{N_k}(a_k) \to \tilde{u}$ weak $*$ in W^*. The set $U_S(a)$ clearly depends on the set S of basis functions.

We state several results concerning compactness properties of $U_S(a)$ and refer the reader to [18] for their proof.

PROPOSITION 2.3. *If $a \in Q_{ad}$, then every member of the set $U_S(a)$ is a solution of (1.1)–(1.6) and $U_S(a) \subset S_{W^*}(R)$.*

PROPOSITION 2.4. *The set $U_S(a)$ is weak $*$ in W^*.*

PROPOSITION 2.5. *Let $\{a_n\}_{n=1}^\infty$ be a sequence in Q_{ad} such that $a_n \to a$ weakly in Q and let $u_n \in U_S(a_n)$. Then there exists a subsequence $\{u_{n_i}\}_{i=1}^\infty$ such that $u_{n_i} \to u$ weak $*$ in W^* and any such limit belongs to $U_S(a)$.*

3 The estimation problem

In this section we formulate the estimation problem as an output-least-squares minimization problem over the admissible set Q_{ad} that is contained in the Hilbert space Q. The space Q is chosen since, for $\Omega \subset \mathbb{R}^2$, it is the Sobolev space of smallest integer order that imbeds compactly in $L^\infty(\Omega)$. The problem is complicated by the fact that the parameter-to-state mapping $a \mapsto u(a)$ is not well-defined. Hence, we begin by giving meaning to the minimization problem when the parameter-to-state mapping is viewed as a set-valued mapping $a \mapsto U_S(a)$ from Q_{ad} into W^*. We focus on the set $U_S(a)$ since we wish to obtain approximation results based on the Galerkin approximations.

The output-least-squares problem where the parameter-to-state mapping $a \mapsto u(a)$ is well-defined is usually formulated as follows.

(3.1) \qquad Find $a_0 \in Q_{ad}$ such that $F(a_0) = \inf\{F(a) : a \in Q_{ad}\}$

where $F(a) = \|u(a) - z\|_{L^2(Q)}^2$.

However, since there may be multiple solutions to the problem (1.1)–(1.6), the mapping $a \mapsto F(a)$ may not be well-defined. To replace (3.1) by a suitable formulation, we note that W^* embeds compactly into $L^2(Q)$. Hence, the mapping taking W^* into $\mathbb{R}$ given by $u \mapsto \|u - z\|_{L^2(Q)}^2$ is continuous with respect to the weak $*$ topology on W^*. By proposition 2.4, $U_S(a)$ is compact in W^* under the weak $*$ topology. Setting

$$J(a) = \inf\{\|u - z\|_{L^2(Q)}^2 : u \in U_S(a)\}$$

there exists an element contained in $U_S(a)$ that we denote by $u(a)$ such that $J(a) = \|u(a) - z\|_{L^2(Q)}^2$. The estimation problem that we consider in place of (3.1) is given by the following.

(E) \qquad Find $a_0 \in Q_{ad}$ such that $J(a_0) = \inf\{J(a) : a \in Q_{ad}\}$.

Set $d = \inf\{J(a) : a \in Q_{ad}\}$.

THEOREM 3.1. *Given (1.1)–(1.6) and (2.4), there exists a solution to problem (E).*

Proof. Let $\{a_n\}_{n=1}^\infty$ be a minimizing sequence of (E). That is, the sequence $\{a_n\}_{n=1}^\infty$ satisfies

$$a_n \in Q_{ad} \text{ and } J(a_n) \to d.$$

Hence, the sequence of solutions $\{u(a_n)\}_{n=1}^\infty \subset U_S(a_n)$ is such that

$$\|u(a_n) - z\|_{L^2(Q)}^2 \to d.$$

The sequence $\{a_n\}_{n=1}^\infty$ is contained in Q_{ad}, and thus it is bounded as a set in Q. Hence, there is a subsequence $\{a_{n_i}\}_{i=1}^\infty$ such that

$$a_{n_i} \to a_0 \text{ weakly in } Q \text{ and strongly in } L^\infty(\Omega).$$

It follows that $a_0 \in Q_{ad}$. From Proposition 2.5 the subsequence $\{a_{n_i}\}_{i=1}^{\infty}$ may be chosen such that

$$u(a_{n_i}) \to u_0 \text{ weak} * \text{ in } W^* \text{ and strongly in } L^2(Q)$$

where $u_0 \in U_S(a_0)$. Therefore, we see that

$$d = \lim J(a_{n_i}) \geq \|u_0 - z\|_{L^2(Q)}^2.$$

In particular,

$$d \geq J(a_0).$$

Finally, since $a_0 \in Q_{ad}$, we conclude that $J(a_0) = d$, and a_0 is a solution of (E). □

We now turn to the estimation of the coefficient a in the Galerkin systems (2.1). In this case, since there exists a unique solution to the system of ordinary differential equations, the parameter-to-state mapping $a \mapsto u^N(a)$ defined by solving the system (2.1) is well-defined. Thus, we formulate the following problems

(E.N)   Find $a_N \in Q_{ad}$ such that $J^N(a_N) = \inf\{J^N(a) : a \in Q_{ad}\}$

where $J^N(a) = \|u^N(a) - z\|_{L^2(Q)}^2$.

Remark 3.2. The existence of a solution ot (E.N) may be proved similarly as for the problem (E).

We have the following limit result.

THEOREM 3.3. *Let $a_0 \in Q_{ad}$ be a solution to (E), that is, $J(a_0) = d$. Let $\{a_N\}_{N=1}^{\infty}$ be a sequence of solutions of (E.N). The sequence $\{a_N\}_{N=1}^{\infty}$ has a weak cluster point $\tilde{a} \in Q_{ad}$ and every weak cluster point is a solution of (E).*

Proof. The boundedness of the sequence $\{a_n\}_{N=1}^{\infty}$ implies there is a subsequence $\{a_{N_i}\}_{i=1}^{\infty}$ such that $a_{N_i} \to \tilde{a}$ weakly in Q, and $\tilde{a} \in Q_{ad}$. If $u(a_0) \in U_S(a_0)$, then there is a sequence $\{a_{0i}\}_{i=1}^{\infty} \subset Q_{ad}$ such that $a_{0i} \to a_0$ weakly in Q and a sequence $\{u^{M_i}(a_{0i})\}_{i=1}^{\infty}$ of Galerkin solutions such that $u^{M_i}(a_{0i}) \to u(a_0)$ weak $*$ in W^*. Further, if $\tilde{u}$ is a weak $*$ cluster point in W^* of the sequence $\{u^{M_i}(a_{N_i})\}_{i=1}^{\infty}$, then $\tilde{u} \in U_S(\tilde{a})$. Since a_{N_i} is optimal,

$$J^{M_i}(a_{0i}) \geq J^{M_i}(a_{N_i}).$$

Therefore, it follows that in the limit as $i \to \infty$

$$J(a_0) = \|u(a_0) - z\|_{L^2(Q)}^2 \geq \lim J^{M_i}(a_{N_i})$$
$$\geq \|\tilde{u} - z\|_{L^2(Q)}^2$$
$$\geq J(\tilde{a}) \geq d.$$

But, since $J(a_0) = d$, we conclude that $\tilde{a}$ is a solution of (E). □

Having demonstrated convergence properties of solutions of (E.N) to solutions of (E), we now turn to the approximation of solutions a_N of (E.N). To provide a theory of convergence, it is useful first to analyze the regularity properties of a_N. We introduce the following assumption for the purpose of applying classical elliptic regularity results [1].

(3.2) The set Ω has a C^2 boundary.

Our approach is to apply the Kuhn-Tucker theorem to (E.N). Recall that the admissible set Q_{ad} is defined by

$$Q_{ad} = \{a \in Q : a \geq \nu > 0, \|a\|_Q^2 \leq K\}$$

where $Q = H^2(\Omega)$ with inner product $(\cdot,\cdot)_Q$. We denote $L^2(\Omega)$ by $H = L^2(\Omega)$. By well-known arguments [7,10], we apply the Riesz representation theorem to see there is a continuous linear mapping $A: Q \mapsto Q^*$ such that $(\phi,\psi)_Q = \langle A\phi,\psi\rangle$ for any ϕ and ψ belonging to Q. Defining the subspace $D(A) = \{\phi \in Q : A\phi \in H\} \subset H$, we note that, as a mapping from H onto H, A is positive definite, self-adjoint and unbounded with compact inverse. We may define a scale of Hilbert spaces by $Q_\alpha = D(A^\alpha)$ for $\alpha \in \mathbb{R}$ with inner product $(\phi,\psi)_\alpha = (A^\alpha\phi, A^\alpha\psi)_H$ where we note that

$$Q_0 = H, \quad Q_{1/2} = Q, \quad Q_{(-1/2)} = Q^*, \quad \text{and} \quad Q_1 = D(A).$$

Also, $(Q_\alpha)^* = Q_{-\alpha}$. Moreover, under assumption (3.2), $D(A) = H^4(\Omega)$. Finally, there is a real number $\sigma \in (0, 1/2)$ such that if $\beta \geq \sigma$, then Q_β imbeds continuously onto $L^\infty(\Omega)$.

With the above background we define a mapping

$$G: Q \mapsto Y = Q_\beta \times \mathbb{R}$$

by

$$G(\phi) = (\nu - \phi, K - \|\phi\|_Q^2).$$

Since Q_β imbeds continuously into $L^\infty(\Omega)$, it follows that Y has a positive cone with nonempty interior. Hence, the constraints used in formulating Q_{ad} may be given in terms of the vector inequality

$$G(a) \leq \theta$$

where θ is the zero element in Y. Finally, to apply the Kuhn-Tucker theorem, it suffices that the constraints must satisfy a regularity condition [11], [12]. We refer the reader to the argument in [19] to see that any member of Q_{ad} satisfies the regular point condition.

By the Kuhn-Tucker theorem, there is a Lagrange multiplier $\lambda = (\lambda_1, \lambda_2) \in Y^* = Q_\beta \times \mathbb{R}$ such that with $u(a) = u^N(a)$

$$2(u(a_N) - z, Du(a_N)h)_{L^2(Q)} - \langle \lambda_1, h \rangle_{-\beta} + 2\lambda_2(a_N, h)_{1/2} = 0$$

for any $h \in Q_{1/2}$ where we are using that $Q = Q_{1/2}$. Here $Du(a_N)h$ is Fréchet derivative with increment h of the parameter-to-state mapping $a \mapsto u(a)$. In addition, $\lambda_1 \geq 0$ as a linear functional on Q_β and $\lambda_2 \geq 0$ with the property that

$$\langle \lambda_1, \nu - a \rangle_{-\beta} = 0$$

and

$$\lambda_2(\|a\|_Q^2 - K) = 0.$$

Setting $\nu = Du(a)(h)$, we find that v must satisfy the equations

$$(v_{tt}(t), \phi_j) + \int_\Omega a \Delta v(t) \Delta \phi_j \, dx$$

$$+ \int_\Omega \{2[B(v(t), u(t)), u(t)] + [B(u(t), u(t)), v(t)]\} \phi_j \, dx$$

$$= -\int_\Omega h \Delta u(t) \Delta \phi_j \, dx$$

for $j = 1, \ldots, N$. With initial conditions

$$v(0) = 0$$
$$v_t(0) = 0.$$

To calculate the Fréchet derivative of the functional $J(\cdot)$, we introduce the adjoint equation

$$(p_{tt}(t), \phi_j) + \int_\Omega a\Delta p(t)\Delta \phi_j \mathrm{dx}$$

$$+ \int_\Omega \{2[B(p(t), u(t)), u(t)] + [B(u(t), u(t)), p(t)]\}\phi_j \mathrm{dx}$$

$$= \int_\Omega (u(t) - z(t))\phi_j \mathrm{dx}$$

with terminal conditions given by

$$p(T) = 0$$
$$p_t(T) = 0.$$

Remark 3.4. Certainly, there exists a unique solution p to the adjoint system since it is a linear second order initial value problem.

By introducing the adjoint solution into the variational equation from the Kuhn-Tucker theorem, we obtain the expression

$$-2\int_0^T \int_\Omega (\Delta u \Delta p) h \mathrm{dxdt} - \langle \lambda_1, h \rangle + 2\lambda_2 (a_N, h)_{1/2} = 0.$$

Applying Fubini's theorem, we may interchange the integrals to obtain

$$-2\int_\Omega (\int_0^T (\Delta u \Delta p) \mathrm{dt}) h \mathrm{dx} - \langle \lambda_1, h \rangle + 2\lambda_2 (a_N, h)_{1/2} = 0.$$

From the regularity of the solutions u and p, we may conclude that the term

$$\mathcal{Z} = \int_0^T \Delta u \Delta p \mathrm{dt}$$

belongs to (at least) $L^2(\Omega) = \mathcal{Q}_0$. Hence, the quantity

$$\Lambda = \mathcal{Z} + \lambda_1/2$$

belongs to $\mathcal{Q}_{-\beta}$, and form the Riesz-Fréchet theorem there exists an element $\widetilde{\Lambda} \in \mathcal{Q}_\beta$ such that

$$\langle \Lambda, h \rangle = (\widetilde{\Lambda}, h)_\beta.$$

We may thus write the above equation as

$$\lambda_2 (a_N, H)_{1/2} = (\widetilde{\Lambda}, h)_\beta$$

for any $h \in \mathcal{Q}_{1/2}$. It follows that if $\lambda_2 \neq 0$ then $a_N \in D(A^{1-\beta})$ for $\beta \in (0, 1/2)$. We record this result.

PROPOSITION 3.5. *If $\lambda_2 > 0$ and if (3.2) holds, then any solution of (E.N) associated with λ_2 belongs to $H^{4-\sigma}(\Omega)$ where $\sigma = 4\beta$, $\beta \in (0, 1/2)$.*

Proof. By interpolation, we see that $D(A^{1-\beta}) = H^{4(1-\beta)}(\Omega)$ and the result follows form the above discussion. □

Given a family of subspaces $\{(Q^M\}_{M=1}^\infty$ of Q and an associated family of bounded linear operators $\mathcal{I}^M : Q \mapsto Q^M$ with the property that

$$\|\mathcal{I}^M \phi - \phi\|_Q \leq \widetilde{K}_1 \|\phi\|_Q$$

and

$$\|\mathcal{I}^M \phi - \phi\|_Q \leq \widetilde{K}_2(M) \|\phi\|_{H^4(\Omega)}$$

where $\widetilde{K}_2(M) \to 0$ as $M \to \infty$. From the interpolation theory of linear operators [10], we conclude that there is a function $\kappa(M)$ with $\kappa(M) \to 0$ as $M \to \infty$ for $\phi \in H^{4-\sigma}(\Omega)$ such that

$$\|\mathcal{I}^M \phi - \phi\|_Q \leq \kappa(M) \|\phi\|_{H^{4-\sigma}(\Omega)}.$$

If a solution "a" of (E.N) has a Lagrange multiplier $\lambda_2 > 0$, then we can utilize the above estimate to show that in the Q-norm approximations of a converge at a certain rate. The term $\|a\|_{H^{4-\sigma}(\Omega)}$ may be bounded in terms of f, z, λ_1, and λ_2

$$\|a\|_{H^{4-\sigma}(\Omega)} \leq C_1 = C_1(f, z, \lambda_1, \lambda_2).$$

Since Q imbeds continuously into $L^\infty(\Omega)$,

$$\|a\|_{L^\infty(\Omega)} \leq \widetilde{K}_2 \|a\|_Q,$$

we have

$$(\mathcal{I}^M a)(x) \geq \nu - |a(x) - (\mathcal{I}^M a)(x)|$$
$$\geq \nu - \widetilde{K} \|a - \mathcal{I}^M a\|_Q$$
$$\geq \nu - \widetilde{K} \kappa(M) C_1.$$

On the other hand, we see that

$$\|\mathcal{I}^M a\|_Q \leq \|a\|_Q + \|a - \mathcal{I}^M a\|_Q$$
$$\leq K^{1/2} + \kappa(M) C_1.$$

Setting $\nu_M = \nu - \widetilde{K}\kappa(M)C_1$ and $K_M = (K^{1/2} + \kappa(M)C_1)^2$, we define

$$Q_{ad}^M = \{a \in Q^M : a \geq \nu_M \text{ and } \|a\|_Q^2 \leq K_M\}.$$

It follows that for M sufficiently large and for a solution "a" of (E.N) with a positive Lagrange multiplier λ_2, $\mathcal{I}^M a \in Q_{ad}^M$.

Remark 3.6. In the case that $a > \nu$ then there are still elements a^M in Q^M such $a^M \to a$ in Q, cf. [6]. In this case Q_{ad}^M may be defined in terms of these elements. The important issue is to obtain a sequence of elements in Q_{ad}^M that converge strongly in Q to a solution a. We give the problem (E.N.M) as follows

(E.N.M) Find $a_N^M \in Q_{ad}^M$ such that $J^N(a_N^M) = \inf\{J^N(a) : a \in Q_{ad}^M\}$.

Having determined the above behavior, we may now obtain convergence properties of solutions to problems (E.N.M) over Q_{ad}^M in a manner similar to that as presented in [8,17,18]. Hence, we state the following.

THEOREM 3.7. *Let $\lambda_2 > 0$ and (3.2) hold and let $a_N^{M_k}$ be solutions of $(E.N.M_k)$. Then any weak cluster point $\tilde{a}$ of the sequence $\{a_N^{M_k}\}_{k=1}^{\infty}$ is a solution of $(E.N)$.*

The results of the Theorems may now be combined to obtain the following iterated limit result.

THEOREM 3.8. *Let the hypothesis of Theorems hold. The subsequentially we have*

$$w - \lim_{N \to \infty} (w - \lim_{k \to \infty} a_N^{M_k})a$$

where a is a solution of (E).

4. Numerical approximation.

We begin with the Galerkin system (1.1). Thus, let $\{\phi_k : k = 1, \ldots, N\}$ be a linearly independent subset of functions in V and set $V^N = \text{span}\{\phi_k\}_{k=1}^N$. Setting $u^N(t) = \sum_{k=1}^N c_k(t)\phi_k$, we recall that

(4.1)
$$(u_{tt}^N(t), \phi_j) + \int_\Omega (a\Delta u(t)\Delta \phi_j) dx$$
$$+ \int_\Omega \epsilon[B(u^N(t), u^N(t)), u^N(t)]\phi_j dx = (f(t), \phi_j)$$

for $j = 1, 2, \ldots, N$. We take nonhomogeneous initial conditions

(4.2)
$$u^N(0) = u_0^N$$
$$u_t^N(0) = u_1^N$$

where $u_0^N \to u_0$ in V and $u_1^N \to u_1$ in H.

To approximate the coefficient a, let $\{\psi_k : k = 1, \ldots, M\}$ be a linearly independent subset of functions in Q. We denote by $Q^M = \text{span}\{\psi_k : k = 1, \ldots, M\}$. We set $a^M = \sum_{i=1}^M a_i \psi_i$. Define the matrices

$$(G_0)_{ij} = \int_\Omega \phi_i \psi_j dx \text{ and } (G_2)_{ij} = \int_\Omega \Delta\phi_i \Delta\phi_j dx$$
$$(G^k)_{ij} = \int_\Omega \psi_k \Delta\phi_i \Delta\phi_j dx \text{ for } k = 1, \ldots, M$$

and

$$(H^k)_{ij} = \int_\Omega [\phi_i, \phi_j]\phi_k dx \text{ for } k = 1, \ldots, N.$$

Setting $\underline{c} = \text{col}(c_1, \ldots, c_N)$ and $\underline{d} = \text{col}(d_1, \ldots, d_N)$, we define

$$H(\underline{c}, \underline{d}) = \begin{bmatrix} \vdots \\ \underline{c}^* H^k \underline{d} \\ \vdots \end{bmatrix} = \begin{bmatrix} \vdots \\ \underline{d}^* H^k \underline{c} \\ \vdots \end{bmatrix} = \mathcal{H}(\underline{d})\underline{c}.$$

We denote the $N \times N$ matrix

$$G(\underline{a}) = \sum_{k=1}^M a_k G^k$$

where $\underline{a} = \text{col}(a_1, \ldots, a_m)$.

Finally, we define the N-vectors

$$\underline{c}(t) = \text{col}(c_1(t), \ldots, c_N(t)), \quad \underline{d}(t) = \text{col}(d_1(t), \ldots, d_N(t)),$$
$$\underline{\mu}_0 = \text{col}((u_0, \phi_1), \ldots, (u_0, \phi_N)), \quad \underline{\mu}_1 = \text{col}((u_1, \phi_1), \ldots, (u_1, \phi_N))$$

and

$$\underline{f}(t) = \mathrm{col}((f(\cdot,t),\phi_1),\ldots,(f(\cdot,t),\phi_N)).$$

The equations (4.1) and (4.2) become the system

$$G_0 \underline{c}_{tt} + G(\underline{a})\underline{c} + \epsilon H(\underline{c},\underline{d}) = \underline{f}$$

(4.3)
$$G_2 \underline{d} = H(\underline{c},\underline{c})$$

with initial conditions
$$G_0 \underline{c}(0) = \underline{\mu}_0$$
$$G_0 \underline{c}_t(0) = \underline{\mu}_1.$$

We use the following difference equations for temporal approximation

$$G_0 \frac{\underline{c}_{i+1} - 2\underline{c}_i + \underline{c}_{i-1}}{h^2} + G(\underline{a})\frac{\underline{c}_{i+1} + 2\underline{c}_i + \underline{c}_{i-1}}{4}$$
$$+ \epsilon H(\underline{c}_i,\underline{d}_i) = \underline{f}_i$$

where h denotes the time difference and the subscript i indicates the time $t_i = i\,h$. Thus, the following sequence is generated

$$G_0 \underline{c}_0 = \underline{\mu}_0$$
$$G_0 \underline{c}_1 = \underline{\mu}_0 + h\underline{\mu}_1$$
$$G_2 \underline{d}_i = H(\underline{c}_i,\underline{c}_i)$$
$$(G_0 + (\frac{h}{2})^2 G(\underline{a}))\underline{c}_{i+1} = 2(G_0 - \frac{h^2}{4}G(\underline{a}))\underline{c}_i$$
$$-(G_0 + (\frac{h}{2})^2 G(\underline{a}))\underline{c}_{i-1} + h^2(\underline{f}_i - \epsilon H(\underline{c}_i,\underline{d}_i)).$$

The derivative system is obtained by formally differentiating the equations in (4.3) with respect to a_ℓ for $\ell = 1,\ldots,M$, see [4]. Thus, we see that the partial derivatives of $\underline{c}$ and $\underline{d}$ with respect to a_ℓ, $D_\ell \underline{c}$ and $D_\ell \underline{d}$, satisfy the system

$$G_0(D_\ell \underline{c})'' + (G(\underline{a}) + \epsilon \mathcal{H}(\underline{d}))(D_\ell \underline{c})$$

(4.4)
$$+ \epsilon \mathcal{H}(\underline{c}) D_\ell \underline{d} = -G^\ell \underline{c}$$
$$G_2(D_\ell \underline{d}) = 2\mathcal{H}(\underline{c})(D_\ell \underline{c})$$

with initial conditions given by

$$(D_\ell \underline{c})(0) = (D_\ell \underline{c})'(0) = 0.$$

To specify the fit-to-data functional, we define the $M \times M$ matrix G^ψ by

$$(G^\psi)_{ij} = (\psi_i,\psi_j)_Q$$

for $i,j = 1,\ldots,M$, the N-vector valued function $t \mapsto \underline{\zeta}(t)$ by

$$\zeta_i(t) = (\phi_i, z(\cdot,t))_H$$

for $i = 1, \ldots, N$, and the real valued function

$$c_z(t) = \int_\Omega z^2(x,t)\mathrm{d}x.$$

With these definitions, the fit-to-data functional $J = J^N$ becomes

(4.5) $$J(\underline{a}) = \int_0^T (\underline{c}^*(t)G_0\underline{c}(t) - 2\underline{\zeta}^*(t)\underline{c}(t) + c_z(t))\mathrm{d}t.$$

Thus, J may be considered as a mapping from $\mathbb{R}^M$ onto $\mathbb{R}$. We may calculate the Fréchet partial derivative with respect to a_ℓ to obtain

$$D_\ell J(\underline{a}) = 2\int_0^Y (G_0\underline{c}(t) - \underline{\zeta}(t))^*(D_\ell \underline{c})(t)\mathrm{d}t$$

where $D_\ell \underline{c}$ is the solution of (4.4). To solve the estimation problem, we attempt to minimize the functional J. Our approach is to use a steepest descent method to obtain $\underline{a}$-vectors that decrease the size of J.

We present the results of several numerical experiments. Our computations are conducted for $\Omega = (0,1) \times (0,1)$ on which we specify a mesh obtained from uniform meshes on $(0,1)$, Δ_1, with 8 subintervals and, Δ_2, with 3 subintervals. The basis functions $\{\phi_i\}_{i=1}^N$ are obtained as tensor products of cubic B-splines [14] over the mesh Δ_1 that are adjusted to satisfy the clamped essential boundary conditions. Hence, for our examples, $N = 49$. On the other hand, for the approximation of the parameter, we use cubic B-splines over the mesh Δ_2 with no restriction on the boundary. The basis functions $\{\psi_i\}_{i=1}^M$ are obtained as tensor products of these functions. Hence, $M = 36$.

We consider the following test problems. Defining the function

$$u_T(x) = 16x^2(1-x)^2,$$

we specify the deformation as

$$u_{tst}(x,y,t) = u_T(x) * u_T(y) * \cos(t) \text{ in } \Omega \times (0,T).$$

Hence, the initial conditions for our examples are given by

$$u_0(x,y) = u_T(x) * u_T(y) \text{ in } \Omega$$

and

$$u_1(x,y) = 0 \text{ in } \Omega.$$

To carryout various experiments to recover the coefficient a, we specify a particular coefficient a_{0tst}. We then generate the resulting force vector $\underline{f}$ by using the equations (1.1) with basis functions ϕ_j. Data z for the problem is generated by

$$z(x,y,t) = u_{tst}(x,y,t) + e * \eta(x,y,t)$$

where $\eta(x,y,t)$ is a uniformly distributed random number between -1 and 1. The variable e is a constant that controls error. Hence, our data is given by u_{tst} plus a uniformly distributed error contribution. From the data z, the force vector $\underline{f}$, and the initial conditions u_0 and u_1, we wish to recover the coefficient a. We take $T = 0.5$.

We report the results of several experiments, see also [18]. Results are given in terms of the relative L^2-errors for a_0 at a particular iteration. The relative L^2-error is calculated as follows

$$\text{Relative } L^2 \text{ error} = \frac{(\int_\Omega (a_{0\text{calc}} - a_{0\text{tst}})^2 \,\mathrm{d}x\mathrm{d}y)^{1/2}}{\|a_{0\text{tst}}\|_{L^2(\Omega)}}.$$

We give several examples with smooth coefficients as well as a coefficient with a discontinuity.

Example 1. $\epsilon = 0.1$ and $e = 0.0$

$$a_{0\text{tst}}(x,y) = 1.0 + 0.25 * \cos(2\pi x) * \cos(2\pi y)$$

iteration	Relative L^2 error
0	0.462
1	0.299
5	0.143
10	0.0894
15	0.0684
35	0.0428

Example 2. $\epsilon = 0.1$ and $e = 0.25$

$$a_{0\text{tst}} = \begin{cases} 1.5, & x \geq 0.5 \text{ and } y \leq 0.5 \\ 1.0, & \text{otherwise} \end{cases}$$

iteration	Relative L^2 error
0	0.513
1	0.337
5	0.139
10	0.0691
15	0.0599
25	0.0443

Example 3. $\epsilon = 0.5$ and $e = 0.1$

$$a_{0\text{tst}} = \begin{cases} 1.5, & x \geq 0.5 \text{ and } y \leq 0.5 \\ 1.0, & \text{otherwise} \end{cases}$$

iteration	Relative L^2 error
0	0.875
1	0.594
5	0.290
10	0.160
15	0.119
25	0.0744

References

[1] S. Agmon, *Lectures on elliptic boundary value problems*, Van Nostrand, Princeton, 1965.
[2] P. G. Ciarlet, *The finite element method for elliptic problems*, North-Holland, New York, 1980.
[3] ———, *Les equations de von Karman*, Springer-Verlag, New York, 1980.
[4] J. Deiudonne, *Foundations of Modern Analysis*, Academic Press, New York, 1960.
[5] N. Dunford and J.T. Schwartz, *Linear operators Part I: General Theory*, Wiley, New York, 1988.
[6] R. Glowinski, *Numerical Solution of Nonlinear Variational Problems*, Springer-Verlag, New York, 1984.
[7] S. Hubert and S. Palencia, *Vibration and Coupling of Continuous Systems*, Springer-Verlag, New York, 1989.
[8] K. Kunisch and L. White, *Regularity properties in parameter estimation of diffusion coefficients in elliptic boundary value problems*, Appl. Anal., 21(1986), pp. 71–88.
[9] J. L. Lions, *Quelques methodes de resolution des problemes aux limites non lineaires*, Dunod, Paris, 1969.
[10] J. L. Lions and Magenes, *Non-homogeneous boundary value problems value problems and applications, Vol. 1*, Springer-Verlag, New York, 1969.
[11] D. G. Luenberger, *Optimization by vector space methods*, Wiley, New York, 1969.
[12] J. Maurer and J. Zowe, *First and second-order necessary and sufficient optimality conditions for infinite-dimensional programming problems*, Mathematical Programming, 16(1979), pp. 98–110.
[13] M. Schultz, *Spline analysis*, Prentice Hall, Englewood Cliffs, N.J., 1973.
[14] A. Stahel, *A remark on the equation of a vibrating plate*, Proc. of the Royal Society of Edinburgh, 106A(1987), pp. 307–314.
[15] R. Temam, *Navier-Stokes equations theory and numerical analysis*, North-Holland, 1979.
[16] W. von Wahl, *On nonlinear evolution equations in a Banach space and on nonlinear vibrations of the clamped plate*, Bayreuther mathematische Schriften, Heft 7, 1981.
[17] L. White, *Estimation of elastic parameters in a nonlinear elliptic model of a plate*, Appl. Math. and Comp., 42(1991), pp. 139–187.
[18] ———, *Estimation of material parameters in a dynamic nonlinear plate model*, J. Applied Math and Comp., to appear.
[19] ———, *Estimation of higher order damping terms in linear plate models*, J. Applied Math. and Comp., 33(1989), pp. 89–122.

Chapter 8
Global Stabilization of a von Kármán Plate Without Geometric Conditions[*]

M. E. Bradley[†]   I. Lasiecka[‡]

Abstract

We consider the problem of global exponential stabilization for a von Kármán plate by boundary feedback acting through bending moments and transverse shear forces. By use of results using microlocal analysis, we eliminate geometric conditions on the controlled portion of the boundary.

1 Introduction.

1.1 Statement of the Problem.

This paper is an abbreviated version of a paper accepted for publication in the *Journal of Mathematical Analysis and Applications*, where details of the proof may be found (see [3]).

Let Ω be a bounded open domain in R^2 with smooth boundary $\Gamma = \Gamma_0 \cup \Gamma_1$, where Γ_i are relatively open, $\bar{\Gamma}_0 \cap \bar{\Gamma}_1 = \emptyset$ and $\Gamma_0 \neq \emptyset$. We consider the von Kármán system in the variables $w(t,x)$ and $\chi(w(t,x))$:

$$(1.1) \quad \begin{cases} w_{tt} - \gamma^2 \Delta w_{tt} + \Delta^2 w \\ \quad + b(x)w_t = [w, \chi(w)] & \text{in } Q \\ w(0,\cdot) = w_0 \;;\; w_t(0,\cdot) = w_1 & \text{in } \Omega \\ w = \frac{\partial}{\partial \nu} w = 0 & \text{on } \Sigma_0 \\ \Delta w + (1-\mu)B_1 w = -\frac{\partial}{\partial \nu} w_t & \text{on } \Sigma_1 \\ \frac{\partial}{\partial \nu}\Delta w + (1-\mu)B_2 w \\ \quad -\gamma^2 \frac{\partial}{\partial \nu} w_{tt} = w_t - \frac{\partial^2}{\partial \tau^2} w_t & \text{on } \Sigma_1 \end{cases}$$

where $Q \equiv \Omega \times (0,T)$ and $\Sigma_i \equiv \Gamma_i \times (0,T)$, for $i = 0, 1$. Here, $b(x) \in L^\infty(\Omega)$ satisfies $b(x) > 0$ a.e. in Ω, $0 < \mu < 1/2$ is Poisson's ratio and the operators B_1 and B_2 are given by

$$B_1 w = 2n_1 n_2 w_{xy} - n_1^2 w_{yy} - n_2^2 w_{xx}$$
$$B_2 w = \frac{\partial}{\partial \tau}[(n_1^2 - n_2^2) w_{xy} + n_1 n_2 (w_{yy} - w_{xx})].$$

Also, $\chi(w)$ satisfies the system of equations

$$(1.2) \quad \left. \begin{array}{l} \Delta^2 \chi = -[w,w] \\ \chi = \frac{\partial}{\partial \nu}\chi = 0 \quad \text{on } \Sigma = \Gamma \times (0,\infty) \end{array} \right\}$$

---

[*]Adapted with permission from Journal of Mathematical Analysis and Applications, Vol. 174, 1993. Copyright © 1993 by Academic Press, Inc.

[†]Department of Mathematics, University of Louisville, Louisville, KY 40292.

[‡]Department of Applied Mathematics, University of Virginia, Charlottesville, VA 22903.

where

$$[\phi,\psi] = \frac{\partial^2 \phi}{\partial x^2}\frac{\partial^2 \psi}{\partial y^2} + \frac{\partial^2 \phi}{\partial y^2}\frac{\partial^2 \psi}{\partial x^2} - 2\frac{\partial^2 \phi}{\partial x \partial y}\frac{\partial^2 \psi}{\partial x \partial y}.$$

Define the bilinear form

(1.3)
$$a(w,v) = \int_\Omega \Delta w \Delta v d\Omega \\ + \int_\Omega (1-\mu)(2w_{xy}v_{xy} - w_{xx}v_{yy} - w_{yy}v_{xx}))d\Omega$$

and the energy functional is given by

(1.4)
$$E(t) = \tfrac{1}{2}\int_\Omega \{|w_t|^2 + \gamma^2|\nabla w_t|^2 + |\Delta\chi|^2\}d\Omega \\ + \tfrac{1}{2}a(w,w) \equiv E_1(t) + E_2(t),$$

where $E_2(t)$ is defined by

$$E_2(t) = \frac{1}{2}\int_\Omega |\Delta\chi|^2 d\Omega.$$

Our goal is to show that by implementing the stabilizing controls acting through forces w_t and bending moments $\frac{\partial}{\partial \nu}w_t$ and $\frac{\partial^2}{\partial \tau^2}w_t$ along a portion of the edge of the plate, that the energy (1.4) decays exponentially as time increases. These velocity feedbacks represent certain damping mechanisms introduced to the system which physically may be realized through the mechanical design of a damper or other form of friction.

1.2 Literature and Orientation.

The problem of stabilization and controllability for the von Kármán plate described above has attracted much attention in recent years. Indeed, the results on local controllability and stabilization can be found in [7, 8, 4]. As for the question of global decay rates (such as we consider here) we refer to [7]. There, it was proved that for the von Kármán model without rotational inertia or viscous damping (i.e. setting $\gamma = 0$ and $b = 0$) the energy of the resulting system is exponentially stable, provided that Ω is star-shaped. This stability was achieved by means of the boundary feedback acting on the whole boundary (i.e. $\Gamma_0 = \emptyset$). Notice that geometric conditions of star-shaped type were assumed in all existing works dealing with controllability and stabilization of the plate modeled by (1.1), even in the linear case, with the exception of [14] where some controllability results are obtained for linear plates by methods of geometric optics.

Our main contributions in this paper are the following: (i) We include rotational moments of inertia (i.e. $\gamma \neq 0$); (ii) we assume that only a portion of the boundary is available for control actions (i.e. $\Gamma_0 \neq \emptyset$); (iii) we do not impose any geometric conditions on the controlled portion of the boundary (i.e. on Γ_1).

It should be noted that the above mentioned features contribute to major technical difficulties in the problem and require techniques quite different from others found in the literature. Indeed, the basic approach developed in [7, 9, 10, 15] (references [9, 10] and [15] deal with a different plate model) is based on the construction of an appropriate Lyapunov function which does not appear to be applicable for our problem. This is due to the presence of certain boundary terms (related to the nonlinearity $\chi(w)$) and completely unstructured lower order terms appearing in the estimates for the Lyapunov function. These terms destroy the "good structure" of these estimates, making it impossible to prove a desired differential inequality for the Lyapunov function.

To cope with this difficulty, we propose a technique (see [11]) which is based on proving certain functional relations directly for the energy function. This allows us to "build in" an appropriately developed nonlinear compactness-uniqueness argument to "absorb" nonlinear boundary traces and undesirable lower order terms arising from energy estimates. In order to dispense with geometric conditions, we shall use "sharp" regularity results for the traces of the linear problem, which were proved in [12] by using microlocal analysis. (These estimates give rise to unstructured lower order terms, as is always the case when microlocal arguments are used.) Finally, a nonlinear semigroup argument leads to the desired decay rate for the solution to system (1.1). This method is fairly general and allows us to incorporate many other unstructured perturbations and can be applied when the usual Lyapunov technique fails (see [2] where the perturbed Kirchoff plate was considered). On the other hand, the limitation of the method is that we no longer have an exact estimate for the decay rate. (This information is lost in the compactness-uniqueness argument, which is proved by contradiction).

1.3 Statement of Results.

We begin by defining the space of finite energy, $\mathcal{H} \equiv H^2_{\Gamma_0}(\Omega) \times H^1_{\Gamma_0}(\Omega)$ where

$$H^2_{\Gamma_0}(\Omega) = \{w \in H^2(\Omega) : w = \tfrac{\partial}{\partial \nu} w = 0 \text{ on } \Gamma_0\}$$

and

$$H^1_{\Gamma_0}(\Omega) = \{w \in H^1(\Omega) : w = 0 \text{ on } \Gamma_0\}.$$

The following well-posedness and regularity result can be proved by combining the results of [2] and [5].

PROPOSITION 1.1. *Well-posedness(i) Given initial data* $(w_0, w_1) \in H^2_{\Gamma_0}(\Omega) \times H^1_{\Gamma_0}(\Omega) \equiv \mathcal{H}$, *there exists a unique solution* (w, w_t) *to (1.1) with*

(1.5)
$$\begin{aligned} w &\in C([0, \infty); H^2_{\Gamma_0}(\Omega)) \\ \text{and } w_t &\in C([0, \infty); H^1_{\Gamma_0}(\Omega)), \end{aligned}$$

where the solution depends continuously on the initial data. (ii) There exists a dense set[1] $D \subset \mathcal{H}$ *such that if* $(w_0, w_1) \in D$ *then the solution pair* (w, w_t) *for (1.1) satisfies*

(1.6)
$$\begin{aligned} w &\in C([0, T]; H^4(\Omega) \cap H^2_{\Gamma_0}(\Omega)) \\ w_t &\in C([0, T]; H^3(\Omega) \cap H^2_{\Gamma_0}(\Omega)) \\ \text{and } w_{tt} &\in C([0, T]; H^2_{\Gamma_0}(\Omega)) \end{aligned}$$

for any $0 < T < \infty$.

We now state the main result of this paper.

THEOREM 1.1. *Main Theorem Assume that the domain* $\Omega \subset R^2$ *has a sufficiently smooth boundary,* $\Gamma = \Gamma_0 \cup \Gamma_1$, *where* $\Gamma_0 \neq \emptyset$, *and that there exists* $x_0 \in R^2$ *such that*

(1.7)
$$\mathbf{h} \cdot \nu \equiv (x - x_0) \cdot \nu \leq 0 \text{ for } x \in \Gamma_0$$

Then for any initial data, $(w_0, w_1) \in \mathcal{H} \equiv H^2_{\Gamma_0}(\Omega) \times H^1_{\Gamma_0}(\Omega)$, *there exists a constant* C *and a constant* $\alpha = \alpha(\|w_0\|_{H^2_{\Gamma_0}(\Omega)}, \|w_1\|_{H^1_{\Gamma_0}(\Omega)})$ *such that the solution pair for (1.1) satisfies*

$$\|(w(t), w_t(t))\|_{\mathcal{H}} \leq C e^{-\alpha t} \{\|(w_0, w_1)\|_{\mathcal{H}}\}.$$

Here the constant C *depends on the size of the initial data.*

---

[1] $D \subset H^4(\Omega) \times H^3(\Omega)$ and elements of D satisfy appropriate compatibility conditions at the boundary.

Remark 1: The geometric condition (1.7) is used in developing the energy estimates related to multiplier $\mathbf{h} \cdot \nabla w$ (see Proposition 2.1).

Remark 2: It is interesting to note the role played by the "light" internal damping modeled by $b(x)w_t$. It is obvious that this damping alone (i.e. without the dissipation on the boundary) would not suffice to uniformly stabilize the plate. ("Strong" internal damping which would cause the energy to decay exponentially would be of the form $b(x)\Delta w_t$ where $b(x) \geq b_0 > 0$). On the other hand, the presence of this mild damping seems to be essential in proving the effectivness of the boundary dissipation. Thus, the combination of both the light interior damping and the boundary damping appears to provide the desired energy decay. We only note that the presence of light interior damping is physically motivated, since most vibrating materials possess some degree of interior damping.

Remark 3: If $\Gamma_0 = \emptyset$, it is enough to take $w_t - \frac{\partial^2}{\partial \tau^2}w_t + w$ in the last boundary condition in (1.1). The arguments are the same.

Remark 4: The techniques of this paper apply verbatim to much more general structures of viscous damping. For instance, the last term of the left hand side of equation (1.1) can be replaced by $B(w_t)$ where the operator $B \in \mathcal{L}(L^2(\Omega))$ and B is positive and injective.

Remark 5: One could also consider nonlinear monotone boundary feedbacks. The analysis in this case is essentially the same.

The remainder of this paper is organized as follows. We first set out the preliminary energy estimates which will be used in proving the main theorem. These will possess lower order terms which will then be absorbed by a compactness-uniqueness argument. We conclude with an argument from nonlinear semigroup theory, which will complete the proof of Theorem 1.1.

2 Proof of Theorem 1.1.

Our basic strategy in proving Theorem 1.1 is to use the method of multipliers to produce preliminary energy estimates which may then be used in conjunction with "sharp" trace regularity results to obtain the desired energy estimate, modulo traces of the nonlinear function $\chi(w)$ and lower order terms. These terms will be absorbed by an appropriately developed nonlinear compactness-uniqueness argument (Lemma 2.2). Finally, an argument from nonlinear semigroup theory will complete the proof.

2.1 Energy Estimates.

The first preliminary energy estimate we will prove is

PROPOSITION 2.1. *Let $(w_0, w_1) \in \mathcal{H} \equiv H^2_{\Gamma_0}(\Omega) \times H^1_{\Gamma_0}(\Omega)$, then the energy of system (1.1) as given by (1.4) satisfies the following estimate*

$$(2.1) \quad \int_0^T E(t)dt + E(0) + E(T)$$
$$\leq C_T \left\{ \int_{\Sigma_1} (w_t^2 + |\nabla w_t|^2)d\Sigma_1 + l.o.(w) \right.$$
$$+ \int_Q b(x)w_t^2 dQ + \int_\Sigma |\Delta \chi| d\Sigma$$
$$\left. + \int_{\Sigma_1} \left(\left|\frac{\partial^2}{\partial \tau^2}w\right|^2 + \left|\frac{\partial^2}{\partial \nu^2}w\right|^2 + \left|\frac{\partial^2}{\partial \nu \partial \tau}w\right|^2 \right) d\Sigma_1 \right\}$$

for all T sufficiently large where C_T depends on $E(0)$ in an increasing manner. Here, $l.o.(w)$ represents terms in w having order lower than the energy:

(2.2) $$l.o.(w) = \|w\|^2_{L^2([0,T];H^{3/2+\epsilon}(\Omega))} + \|w_t\|^2_{L^2(Q)}$$

where $0 < \varepsilon < 1/2$.

Our proof will rely on energy methods which require the regularity of Proposition (1.1)(ii). Since the set D of Proposition 1.1 is dense in $\mathcal{H} \equiv H^2_{\Gamma_0}(\Omega) \times H^1_{\Gamma_0}(\Omega)$, we will assume this additional regularity of our initial data for all of our computations. Then the final result will follow by a standard density argument.

Proof. Recalling (1.4) and by using the multiplier w_t, it is straightforward to calculate the identity

(2.3) $$E(t) + 2\int_0^t \int_{\Gamma_1} \{w_t^2 + |\nabla w_t|^2\} d\Gamma_1 \, dt \\ + 2\int_0^t \int_\Omega b(x) w_t^2 d\Omega \, dt = E(0).$$

This proves that energy is nonincreasing for the controlled system.

We now consider the multiplier w. We observe that

(2.4) $$\int_Q [w,\chi] w \, dQ = \int_Q [w,w]\chi \, dQ \\ = -\int_Q \Delta^2 \chi \cdot \chi \, dQ = -\int_Q (\Delta\chi)^2 dQ.$$

Now, multiplying (1.1) by w and integrating the result by parts (while using the boundary conditions in (1.1)), we obtain

$$\int_Q \{w_t^2 + \gamma^2 |\nabla w_t|^2\} dQ - \int_0^T a(w,w) dt - \int_Q (\Delta\chi)^2 dQ$$

$$= (w_t, w)_\Omega|_0^T + \gamma^2 (\nabla w_t, \nabla w)_\Omega|_0^T + \tfrac{1}{2}(b(x)w, w)_\Omega|_0^T$$

(2.5) $$+ (w, w)_{\Gamma_1}|_0^T + (\nabla w, \nabla w)_{\Gamma_1}|_0^T.$$

If we now multiply (1.1) by $\mathbf{h} \cdot \nabla w$ and integrate by parts, we obtain

$$\int_Q [w,\chi](\mathbf{h} \cdot \nabla w) dQ$$

$$= \int_Q \{w_{tt} - \gamma^2 \Delta w_{tt} + \Delta^2 w + b(x) w_t\}(\mathbf{h} \cdot \nabla w) dQ$$

$$= (w_t, \mathbf{h} \cdot \nabla w)_\Omega|_0^T + \gamma^2 (\nabla w_t, \mathbf{h} \cdot \nabla w)_\Omega|_0^T$$

$$+ \int_Q w_t^2 dQ + \int_0^T a(w,w) dt$$

$$+ \int_Q b(x) w_t (\mathbf{h} \cdot \nabla w) dQ$$

$$- \frac{1}{2}\int_{\Sigma_1} (w_t^2 + \gamma^2 |\nabla w_t|^2)(\mathbf{h} \cdot \nu) d\Sigma_1$$

$$- \frac{1}{2}\int_{\Sigma_0} (\mathbf{h} \cdot \nu)(\Delta w)^2 d\Sigma_0$$

$$+ \frac{1}{2}\int_{\Sigma_1} (\mathbf{h} \cdot \nu)(\Delta w)^2 d\Sigma_1$$

$$+ \int_{\Sigma_1} (\mathbf{h} \cdot \nu)\{(1-\mu)(w_{xy}^2 - w_{xx}w_{yy})\} d\Sigma_1$$

$$+ \int_{\Sigma_1} \left(w_t - \frac{\partial^2 w_t}{\partial \tau^2} \right)(h \cdot \nabla w) d\Sigma_1$$

$$+ \int_{\Sigma_1} \tfrac{\partial}{\partial \nu}(\mathbf{h} \cdot \nabla w)\tfrac{\partial}{\partial \nu} w_t d\Sigma_1.$$

From [7] (see page 115) we have

$$\int_Q [w, \chi](\mathbf{h} \cdot \nabla w) dQ$$

$$= -\frac{1}{2}\int_Q (\Delta \chi)^2 dQ - \frac{1}{2}\int_\Sigma \mathbf{h} \cdot \nu (\Delta \chi)^2 d\Sigma.$$

Consequently, we obtain

(2.6)
$$\int_Q w_t^2 \, dQ + \int_0^T a(w, w) dt + \frac{1}{2}\int_Q (\Delta \chi)^2 dQ$$
$$\leq -\frac{1}{2}\int_\Sigma (\mathbf{h} \cdot \nu)(\Delta \chi)^2 d\Sigma - \int_Q b(x) w_t (\mathbf{h} \cdot \nabla w) \, dQ$$
$$+ \left|(w_t, \mathbf{h} \cdot \nabla w)_\Omega \big|_0^T\right| + \gamma^2 \left|(\nabla w_t, \mathbf{h} \cdot \nabla w)_\Omega \big|_0^T\right|$$
$$+ C \int_{\Sigma_1} \{w_t^2 + |\nabla w_t|^2\} d\Sigma_1$$
$$+ C_1 \int_{\Sigma_1} \left\{|\nabla w|^2 + |\nabla (\mathbf{h} \cdot \nabla w)|^2\right\} d\Sigma_1$$
$$- \frac{1}{2}\int_{\Sigma_1} (\mathbf{h} \cdot \nu)(\Delta w)^2 d\Sigma_1$$
$$- \int_{\Sigma_1} (\mathbf{h} \cdot \nu)(1-\mu)(w_{xy}^2 - w_{xx}w_{yy}) d\Sigma_1.$$

Observe that the last two terms in (2.6) may be bounded by second order traces of the solution w and l.o.(w). We may also estimate

(2.7)
$$-\int_Q b w_t (\mathbf{h} \cdot \nabla w) dQ$$
$$\leq C_1 \int_Q b w_t^2 dQ + C_2 \int_Q |\nabla w|^2 dQ$$
$$\leq C_1 \int_Q b w_t^2 dQ + C_2 l.o.(w).$$

Using (2.6) and (2.7) together, we obtain the estimate

(2.8)
$$\int_Q w_t^2 dQ + \int_0^T a(w, w) dt + \frac{1}{2}\int_Q (\Delta \chi)^2 dQ$$
$$\leq C_1 \int_Q b w_t^2 dQ + C_2 \int_\Sigma (\Delta \chi)^2 d\Sigma$$
$$+ \left|(w_t, \mathbf{h} \cdot \nabla w)_\Omega\big|_0^T\right| + \gamma^2 \left|(\nabla w_t, \mathbf{h} \cdot \nabla w)_\Omega\big|_0^T\right|$$
$$+ C_3 \int_{\Sigma_1} \left\{w_t^2 + |\nabla w_t|^2\right\} d\Sigma_1 + C_5 l.o.(w)$$
$$+ C_4 \int_{\Sigma_1} \left(\left|\frac{\partial^2}{\partial \tau^2} w\right|^2 + \left|\frac{\partial^2}{\partial \nu^2} w\right|^2 + \left|\frac{\partial^2}{\partial \nu \partial \tau} w\right|^2 \right) d\Sigma_1.$$

From (2.5), (2.8), the compact Sobolev imbeddings and the trace theorem, we obtain

(2.9)
$$\int_0^T E(t)dt$$
$$\leq C(E(T) + E(0)) + C_1 \int_\Sigma (\Delta\chi)^2 d\Sigma + C_2 l.o.(w)$$
$$+ C_3 \int_{\Sigma_1} (w_t^2 + |\nabla w_t|^2) d\Sigma_1 + C_4 \int_Q b^2 w_t^2 dQ$$
$$+ C_5 \int_{\Sigma_1} \left(\left|\frac{\partial^2}{\partial \tau^2} w\right|^2 + \left|\frac{\partial^2}{\partial \nu^2} w\right|^2 + \left|\frac{\partial^2}{\partial \nu \partial \tau} w\right|^2 \right) d\Sigma_1.$$

Using Holder's inequality, we may estimate

(2.10)
$$\int_\Sigma (\Delta\chi)^2 d\Sigma \leq \left(\int_\Sigma |\Delta\chi|^3 d\Sigma \right)^{1/2} \left(\int_\Sigma |\Delta\chi| d\Sigma \right)^{1/2}$$
$$\leq \varepsilon \|\Delta\chi\|_{L^3(\Sigma)}^3 + \frac{1}{4\varepsilon} \int_\Sigma |\Delta\chi| d\Sigma.$$

By the Sobolev imbeddings (see [1], Theorem 5.22) and by the regularity of χ (see [13] p. 187 and [4]) we know that for $0 < \varepsilon < 2/3$

$$\|\Delta\chi(w)\|_{L^3(\Gamma)} \leq C_1 \|\Delta\chi(w)\|_{H^{1-\epsilon}(\Omega)} \leq C_2 \|w\|_{H^2_{\Gamma_0}(\Omega)}^2,$$

so that

$$\|\Delta\chi(w)\|_{L^3(\Sigma)}^3$$
$$\leq \int_0^T \|\Delta\chi(w)\|_{H^{1-\epsilon}(\Omega)}^3 dt$$
$$\leq \int_0^T \|\Delta\chi(w)\|_{H^{1-\epsilon}(\Omega)} \|\Delta\chi(w)\|_{H^{1-\epsilon}(\Omega)}^2 dt$$
$$\leq C \cdot E^2(0) \int_0^T \|\Delta\chi(w)\|_{H^{1-\epsilon}(\Omega)} dt$$
$$\leq C \cdot E^2(0) \int_0^T \|w(t)\|_{H^2_{\Gamma_0}(\Omega)}^2 dt$$

(2.11)
$$\leq C \cdot E^2(0) \int_0^T E(t) dt.$$

From (2.18)-(2.11), we obtain

(2.12)
$$\int_\Sigma |\Delta\chi(w)|^2 d\Sigma$$
$$\leq \varepsilon C \cdot E^2(0) \int_0^T E(t) dt + \frac{1}{4\varepsilon} \int_\Sigma |\Delta\chi(w)| d\Sigma.$$

Selecting $\varepsilon = \frac{1}{2C_1 C \cdot E^2(0)}$ and using (2.12) in (2.9) along with the estimate (2.3), we obtain the estimate

(2.13)
$$\int_0^T E(t) dt$$

$$\leq C \left\{ E(T) + \int_{\Sigma_1} (w_t^2 + |\nabla w_t|^2) d\Sigma_1 \right.$$
$$+ \int_Q bw_t^2 dQ + l.o.(w) + C(E(0)) \int_\Sigma |\Delta \chi| d\Sigma$$
$$\left. + \int_{\Sigma_1} \left(\left| \frac{\partial^2}{\partial \tau^2} w \right|^2 + \left| \frac{\partial^2}{\partial \nu^2} w \right|^2 + \left| \frac{\partial^2}{\partial \nu \partial \tau} w \right|^2 \right) d\Sigma_1 \right\},$$

where the constant $C(E(0))$ is increasing in $E(0)$.

By (2.3), we see that $\frac{d}{dt} E(t) \leq 0$, so that

(2.14) $$T E(T) \leq \int_0^T E(t) dt.$$

Consequently, (2.13) becomes

$$\{T - C\} E(T)$$
$$\leq C \left\{ \int_{\Sigma_1} (w_t^2 + |\nabla w_t|^2) d\Sigma_1 + \int_Q bw_t^2 dQ \right.$$
$$+ l.o.(w) + C(E(0)) \int_\Sigma |\Delta \chi| d\Sigma$$
$$\left. + \int_{\Sigma_1} \left(\left| \frac{\partial^2}{\partial \tau^2} w \right|^2 + \left| \frac{\partial^2}{\partial \nu^2} w \right|^2 + \left| \frac{\partial^2}{\partial \nu \partial \tau} w \right|^2 \right) d\Sigma_1 \right\}$$

Finally, selecting $T > T_0 = C$, we have the estimate for $E(T)$. Going back to (2.13), and using (2.3), recovers the full estimate of Proposition 2.1.

Our next step is to develop appropriate estimates for the traces of the solution, w, on the portion of the boundary, Γ_1. To accomplish this, we shall use the following result proved in [12].

PROPOSITION 2.2 ([12] THEOREM 1.1). *Let $p(t, x)$ be a solution to the following linear problem (in the sense of distributions)*

$$\begin{aligned} p_{tt} - \gamma^2 \Delta p_{tt} + \Delta^2 p &= f & \text{in } Q \\ p(0, \cdot) = p_0 \,;\, p_t(0, \cdot) &= p_1 & \text{in } \Omega \\ p = \tfrac{\partial}{\partial \nu} p &= 0 & \text{on } \Sigma_0 \\ \Delta p + (1 - \mu) B_1 p &= g_1 & \text{on } \Sigma_1 \\ \tfrac{\partial}{\partial \nu} \Delta p + (1 - \mu) B_2 p - \gamma^2 \tfrac{\partial}{\partial \nu} p_{tt} &= g_2 & \text{on } \Sigma_1 \end{aligned}$$

For every $T > \alpha > 0$ and $1/2 > \varepsilon > 0$ the following estimate holds

(2.15) $$\int_\alpha^{T-\alpha} \int_{\Gamma_1} \left(\left| \frac{\partial^2 p}{\partial \tau^2} \right|^2 + \left| \frac{\partial^2 p}{\partial \nu^2} \right|^2 + \left| \frac{\partial^2 p}{\partial \nu \partial \tau} \right|^2 \right) d\Gamma_1 \, dt$$
$$\leq C_{T,\alpha} \left\{ \|f\|^2_{L^2([0,T];H^{-\varepsilon}(\Omega))} + \|g_1\|^2_{L^2(\Sigma_1)} \right.$$
$$+ \|g_2\|^2_{L^2([0,T];H^{-1}(\Omega))} + \|p_t\|^2_{L^2(\Sigma_1)}$$
$$\left. + \|\nabla p_t\|^2_{L^2(\Sigma_1)} + \|p\|_{L^2([0,T];H^{3/2+\varepsilon}(\Omega))} \right\}.$$

PROPOSITION 2.3. *Let w satisfy (1.1). Then for any $\alpha > 0$ and $\varepsilon > 0$ we have*

$$\int_\alpha^{T-\alpha} \int_{\Gamma_1} \left(\left|\frac{\partial^2 w}{\partial \tau^2}\right|^2 + \left|\frac{\partial^2 w}{\partial \nu^2}\right|^2 + \left|\frac{\partial^2 w}{\partial \nu \partial \tau}\right|^2 \right) d\Gamma_1 \, dt$$

(2.16)
$$\leq C_{T,\alpha,\varepsilon} \left\{ \|w_t\|^2_{L^2(\Sigma_1)} + \|\nabla w_t\|^2_{L^2(\Sigma_1)} + E^2(0) \|\chi(w)\|_{L^1([0,T];H^{3-\varepsilon}(\Omega))} + l.o.(w) \right\}.$$

Proof. We apply the result of Proposition 2.2 to system (1.1) with

(2.17)
$$f \equiv -b(x)w_t + [\chi(w), w]$$
$$g_1 \equiv -\frac{\partial}{\partial \nu} w_t \qquad g_2 \equiv w_t - \frac{\partial^2}{\partial \tau^2} w_t.$$

It can be shown (see [4]) that for every $\varepsilon_0 > 0$ there exists $p > 1$ such that

(2.18)
$$\|[\chi(w), w]\|_{H^{-\varepsilon_0}(\Omega)} \leq C \|\chi(w)\|_{W^{2,p}(\Omega)} \|w\|_{H^2(\Omega)}.$$

Also, from [4], we have

(2.19)
$$\|\chi(w)\|_{H^{3-\varepsilon}(\Omega)} \leq C \|w\|^2_{H^2(\Omega)}.$$

By using the Sobolev imbeddings, we know that for all $p > 1$ there exists an $\varepsilon_1 > 0$ such that

(2.20)
$$H^{1-\varepsilon_1}(\Omega) \subset L^p(\Omega).$$

This, together with (2.18) and (2.19), implies

(2.21)
$$\|[\chi(w), w]\|_{H^{-\varepsilon_0}(\Omega)} \leq C \|w\|_{H^2(\Omega)} \|\chi(w)\|_{H^{3-\varepsilon_1}(\Omega)}$$
$$\leq C \|w\|^2_{H^2(\Omega)} \|\chi(w)\|^{1/2}_{H^{3-\varepsilon_1}(\Omega)}.$$

From (2.17), (2.21) and (2.3) we obtain

(2.22)
$$\int_0^T \|f(t)\|^2_{H^{-\varepsilon_0}(\Omega)} dt$$
$$\leq C \left\{ \int_0^T \|w_t(t)\|^2_{L^2(\Omega)} dt + \int_0^T \|w(t)\|^4_{H^2(\Omega)} \|\chi(w)\|_{H^{3-\varepsilon_1}(\Omega)} dt \right\}$$
$$\leq C \left\{ l.o.(w) + E^2(0) \int_0^T \|\chi(w)\|_{H^{3-\varepsilon_1}(\Omega)} dt \right\}.$$

On the other hand, from (2.17), we readily obtain

(2.23)
$$\|g_1\|_{L^2(\Sigma_1)} + \|g_2\|_{L^2([0,T];H^{-1}(\Gamma_1))}$$
$$\leq C \left\{ \left\|\frac{\partial}{\partial \nu} w_t\right\|_{L^2(\Sigma_1)} + \|w_t\|_{L^2(\Sigma_1)} + \left\|\frac{\partial}{\partial \tau} w_t\right\|_{L^2(\Sigma_1)} \right\}.$$

Here we have used Sobolev imbeddings and the fact that

$$\left\|\frac{\partial^2 w_t}{\partial \tau^2}\right\|_{H^{-1}(\Gamma)} \leq C \left\|\frac{\partial w_t}{\partial \tau}\right\|_{L^2(\Gamma)}.$$

After collecting the estimates in (2.23) and (2.22) and appropriately selecting $\varepsilon = \varepsilon_0$, we may apply the result of Proposition 2.2 and obtain the desired conclusion of Proposition 2.3.

We are now in a position to prove our main energy estimate.

LEMMA 2.1. *Let w satisfy the system (1.1) and let $0 < \alpha < T$ and let $\varepsilon > 0$ be arbitrary. Then*

$$E(0) + E(T) + \int_\alpha^{T-\alpha} E(t)dt$$

$$\leq C_{T,\alpha,\varepsilon} \left\{ \int_{\Sigma_1} (w_t^2 + |\nabla w_t|^2)d\Sigma_1 + \int_Q bw_t^2 dQ \right.$$

$$+ C(E(0)) \int_\Sigma |\Delta \chi(w)| d\Sigma$$

$$\left. + C(E(0)) \int_0^T \|\chi(w)\|_{H^{3-\varepsilon}(\Omega)} dt + l.o.(w) \right\}$$

Proof. Applying the result of Proposition 2.1 on the interval $[\alpha, T - \alpha]$ yields

(2.24)
$$\int_\alpha^{T-\alpha} E(t)dt + E(\alpha) + E(T-\alpha)$$

$$\leq C_T(E(\alpha)) \left\{ \int_{\Sigma_{1\alpha}} (w_t^2 + |\nabla w_t|^2)d\Sigma_{1\alpha} \right.$$

$$+ \int_\Sigma |\Delta \chi(w)| d\Sigma + \int_Q bw_t^2 dQ + l.o.(w)$$

$$\left. + \int_{\Sigma_{1\alpha}} \left(\left|\frac{\partial^2 w}{\partial \tau^2}\right|^2 + \left|\frac{\partial^2 w}{\partial \nu^2}\right|^2 + \left|\frac{\partial^2 w}{\partial \tau \partial \nu}\right|^2 \right) d\Sigma_{1\alpha} \right\}$$

where $\Sigma_{1\alpha} \equiv \Gamma_1 \times [\alpha, T - \alpha]$. By the dissipation of energy given in (2.3), we have $E(T) \leq E(\alpha) \leq E(0)$. Also, since $C_T(E(t))$ is an increasing function of $E(t)$, we have

(2.25) $$C_T(E(\alpha)) \leq C_T(E(0)).$$

Applying Proposition 2.3 to the last term on the right hand side of (2.24) and recalling (2.3) and (2.25) leads to the desired inequality in Lemma 2.1.

2.2 Compactness-Uniqueness Argument.

At this point, we have proven that the energy of system (1.1) is bounded by the feedback controls plus lower order and nonlinear terms. To obtain our desired goal, we must prove that the energy is bounded by the feedbacks alone. To accomplish this, we first observe that

(2.26) $$\int_\Sigma |\Delta \chi| d\Sigma \leq mes(\Gamma)^{1/2} \int_0^T \|\Delta \chi\|_{L^2(\Gamma)} dt$$

$$\leq C \int_0^T \|\Delta \chi\|_{H^{1/2+\varepsilon}(\Omega)} dt \leq C \int_0^T \|\chi\|_{H^{3-\varepsilon}(\Omega)} dt$$

Thus, it suffices to prove

LEMMA 2.2. *Compactness-Uniqueness.* Let (w, w_t) be a solution pair for (1.1). Then for any $\varepsilon > 0$,

$$\int_0^T \|\chi(w)\|_{H^{3-\varepsilon}(\Omega)} dt + l.o.(w) \tag{2.27}$$

$$\leq C(E(0)) \left\{ \int_{\Sigma_1} \{w_t^2 + |\nabla w_t|^2\} d\Sigma_1 + \int_Q b(x) w_t^2 dQ \right\}$$

where $C(E(0))$ is an increasing function of the initial energy, $E(0)$, and $l.o.(w)$ are as in (2.2).

Remark: Notice that the nonlinear term on the left hand side of (2.27) is *not* a lower order term. Indeed, its definition requires two spatial derivatives of the function w which is exactly at the level of the energy functional.

Proof. The proof is by contradiction. Suppose (2.27) does not hold. Then there exists a sequence of functions $\{w_n(t)\}$ in $\mathcal{H}$ which satisfies the system

$$\begin{aligned}
w_n'' - \gamma^2 \Delta w_n'' + \Delta^2 w_n + b w_n' & \\
= [w_n, \chi(w_n)] & \quad \text{in } Q \\
w_n(0, \cdot) = w_{n0} \; ; \; w_n'(0, \cdot) = w_{n1} & \quad \text{in } \Omega \\
w_n = \tfrac{\partial}{\partial \nu} w_n = 0 & \quad \text{on } \Sigma_0 \\
\Delta w_n + (1 - \mu) B_1 w_n = -\tfrac{\partial}{\partial \nu} w_n' & \quad \text{on } \Sigma_1 \\
\tfrac{\partial}{\partial \nu} \Delta w_n + (1 - \mu) B_2 w_n - \gamma^2 \tfrac{\partial}{\partial \nu} w_n'' & \\
= w_n' - \tfrac{\partial^2}{\partial \tau^2} w_n' & \quad \text{on } \Sigma_1.
\end{aligned} \tag{2.28}$$

Denoting the sequence

$$c_n \equiv \left\{ l.o.(w_n) + \int_0^T \|\chi(w_n)\|_{H^{3-\varepsilon}(\Omega)} dt \right\}^{1/2} \tag{2.29}$$

we see that

$$\lim_{n \to \infty} \frac{c_n}{\|w_n'\|_{H^1(\Sigma_1)}^2 + \int_Q b(w_n')^2 dQ} = \infty \tag{2.30}$$

where the initial energy (as prescribed by initial data (w_{n0}, w_{n1})) are uniformly bounded in n. (Note: for convenience, we denote the time derivatives by $'$.)

We now introduce the new variable

$$v_n \equiv \frac{w_n}{c_n}. \tag{2.31}$$

We observe that v_n satisfies the system

$$\begin{aligned}
v_n'' - \gamma^2 \Delta v_n'' + \Delta^2 v_n + b v_n' & \\
= [v_n, \chi(w_n)] & \quad \text{in } Q \\
v_n(0, \cdot) = v_{n0} \; ; \; v_n'(0, \cdot) = v_{n1} & \quad \text{in } \Omega \\
v_n = \tfrac{\partial}{\partial \nu} v_n = 0 & \quad \text{on } \Sigma_0 \\
\Delta v_n + (1 - \mu) B_1 v_n = -\tfrac{\partial}{\partial \nu} v_n' & \quad \text{on } \Sigma_1 \\
\tfrac{\partial}{\partial \nu} \Delta v_n + (1 - \mu) B_2 v_n - \gamma^2 \tfrac{\partial}{\partial \nu} v_n'' & \\
= v_n' - \tfrac{\partial^2}{\partial \tau^2} v_n' & \quad \text{on } \Sigma_1.
\end{aligned} \tag{2.32}$$

By using (2.30), we see that v_n satisfies

(2.33) $$\lim_{n\to\infty} \|v'_n\|^2_{H^1(\Sigma_1)} + \int_Q b(v'_n)^2 dQ = 0$$

and (by the quadratic dependence of χ on w_n)

(2.34) $$l.o.(v_n) + \int_0^T \|\chi(v_n)\|_{H^{3-\epsilon}(\Omega)} dt = \frac{l.o.(w_n) + \int_0^T \|\chi(w_n)\|_{H^{3-\epsilon}(\Omega)} dt}{l.o.(w_n) + \int_0^T \|\chi(w_n)\|_{H^{3-\epsilon}(\Omega)} dt} \equiv 1$$

By (2.33), we have the following convergence properties:

(2.35)
(i) $v'_n \to 0$ in $L^2(Q)$
(ii) $v'_n \to 0$ in $H^1(\Sigma_1)$.

In order to pass with a limit on (2.32), we need first to determine the convergence properties for v_n and for our nonlinear terms.

Using the energy estimates which were derived in the previous section and the well-posedness theorem, we have, by the compact Sobolev imbeddings and trace theory,

(2.36) $$v_n \xrightarrow{w} v \text{ in } L^2([0,T]; H^2(\Omega))$$
$$\text{and} \quad v'_n \xrightarrow{w} v' \text{ in } L^2([0,T]; H^1(\Omega))$$
$$\Longrightarrow \quad v_n \xrightarrow{w} v \text{ in } H^1([0,T] \times \Omega)$$
$$\Longrightarrow \quad v_n \to v \text{ in } L^2(\Sigma).$$

Also, since $E_n(0) \le M$, the well-posedness for (1.1) yields

(2.37) $$w_n \xrightarrow{w} w \text{ in } L^2([0,T]; H^2(\Omega))$$
$$\text{and} \quad w'_n \xrightarrow{w} w' \text{ in } L^2([0,T]; H^1(\Omega)),$$

so that $\{w_n\}$ has the same convergence properties as $\{v_n\}$.

We now examine the convergence properties of the von Kármán nonlinearity, $[v_n, \chi(w_n)]$. We wish to prove that

$$[v_n, \chi(w_n)] \to [v, \chi(w)] \quad \text{as } n \to \infty$$

in some meaningful sense on Q. It is straightforward to prove

PROPOSITION 2.4. *Let $w_n \xrightarrow{w} w$ in $H^2(\Omega)$. Then $\chi(w_n) \xrightarrow{w} \chi(w)$ in $H^2_0(\Omega)$.*

We now seek to obtain the convergence of $\chi(w_n)$ in the space-time cylinder, Q.

PROPOSITION 2.5. *Assume that*

$$\|w_n\|_{C([0,T]; H^2(\Omega))} + \|w'_n\|_{C([0,T]; H^1(\Omega))} \le C$$

and

$$w_n \xrightarrow{w} w \quad \text{in } L^2([0,T]; H^2(\Omega))$$
$$w'_n \xrightarrow{w} w' \quad \text{in } L^2([0,T]; H^1(\Omega)).$$

Then for every $\varepsilon > 0$,

$$\chi(w_n) \to \chi(w) \text{ in } C([0,T]; H^{3-\varepsilon}(\Omega)).$$

Proof. By the results in [4], we obtain that for any $\varepsilon > 0$,

(2.38) $$\|\chi(w_n)\|_{L^2([0,T];H^{3-\varepsilon/2}(\Omega))} \leq C_1.$$

We shall prove
(2.39) $$\|\chi(w_n)\|_{H^1([0,T];H^{2-\varepsilon/2}(\Omega))} \leq C.$$

To accomplish this, notice first that

(2.40) $$\frac{d}{dt}\chi(w_n) = -2\mathcal{A}^{-1}[w_n', w_n]$$

where $\mathcal{A}$ corresponds to the biharmonic operator with zero boundary conditions as in (1.2).

Now let $\phi \in H_0^{2+\varepsilon/2}(\Omega)$. Then

(2.41) $$\int_\Omega [w_n', w_n]\phi\, d\Omega = \int_\Omega [\phi, w_n] w_n'\, d\Omega.$$

By the assumptions imposed on w_n' and by the Sobolev imbeddings, we have

(2.42) $$\|w_n'\|_{C([0,T];L^p(\Omega))} \leq C \quad \text{for any } p > 1.$$

It can be easily verified that

(2.43) $$\|[\phi, w_n]\|_{C([0,T];L^{1+\frac{\varepsilon}{4-\varepsilon}}(\Omega))} \leq C.$$

From (2.41)-(2.43), taking $p = 4/\varepsilon$ in (2.42), we obtain

(2.44) $$\|[w_n', w_n]\|_{C([0,T];H^{-2-\varepsilon/2}(\Omega))} \leq C.$$

By the results of Grisvard in [6] and by duality, we have

(2.45) $$H^{-2-\varepsilon/2}(\Omega) \sim \left(\mathcal{D}(\mathcal{A}^{1/2+\varepsilon/8})\right)',$$

where this duality is with respect to the L^2 inner product.

Combining (2.40), (2.44) and (2.45), we infer that

$$\left\|\frac{d}{dt}\chi(w_n)\right\|_{C([0,T];\mathcal{D}(\mathcal{A}^{1/2-\varepsilon/8}))} \leq C.$$

This, in particular, implies (2.39).

Now we are in a position to apply the compactness result of Simon (see [16]) to conclude that (2.38) and (2.39) together imply

$$\chi(w_n) \to z \quad \text{strongly in } C([0,T]; H^{3-\varepsilon}(\Omega)).$$

But by the result of Proposition 2.4, together with our assumptions, $z = \chi(w)$, as desired.

We may now prove

PROPOSITION 2.6. *Suppose that* $v_n \xrightarrow{w} v$ *in* $H^2(\Omega)$ *and* w_n *satisfies the assumptions of Proposition 2.5. Then* $[v_n, \chi(w_n)] \to [v, \chi(w)]$ *in* $\mathcal{D}'(\Omega)$.

Proof. Let $\phi \in \mathcal{D}(\Omega)$. Then

$$\int_\Omega [v_n, \chi(w_n)]\phi \, d\Omega = \int_\Omega [\phi, \chi(w_n)] v_n \, d\Omega.$$

Now by the Proposition 2.4, we know that $\chi(w_n) \overset{w}{\to} \chi(w)$ in $H_0^2(\Omega)$, so that

$$\frac{\partial^2}{\partial x_i \partial x_j} \chi(w_n) \overset{w}{\to} \frac{\partial^2}{\partial x_i \partial x_j} \chi(w)$$

in $L^2(\Omega)$ for $i, j \in \{1, 2\}$. Since $v_n \to v$ in $L^\infty(\Omega)$, we have

$$\int_\Omega [\phi, \chi(w_n)] v_n \, d\Omega \to \int_\Omega [\phi, \chi(w)] v \, d\Omega,$$

which gives us the proof.

In passing with a limit on (2.32), we will consider two cases.

Case 1: $c_0 = \{l.o.(w) + \int_0^T \|\chi(w)\|_{H^{3-\epsilon}(\Omega)} dt\}^{1/2} \neq 0$. By the result of Proposition 2.5, (2.37) and (2.29), and by the compactness properties of $l.o.(w)$, we have $c_n \to c_0$, hence $v = w/c_0$. Using (2.35), Proposition 2.6 and passing with a limit on (2.32), we obtain the limit system

$$(2.46) \quad \begin{array}{ll} \Delta^2 v = [v, \chi(w)] = \frac{1}{c_0}[w, \chi(w)] & \text{in } Q \\ v = \frac{\partial}{\partial \nu} v = 0 & \text{on } \Sigma_0 \\ \Delta v + (1-\mu) B_1 v = 0 & \text{on } \Sigma_1 \\ \frac{\partial}{\partial \nu} \Delta v + (1-\mu) B_2 v = 0 & \text{on } \Sigma_1. \end{array}$$

Multiplying (2.46) by v and integrating by parts, we obtain

$$(2.47) \quad \begin{aligned} \int_\Omega [v, \chi(w)] v \, d\Omega &= \int_\Omega (\Delta v)^2 \, d\Omega + \int_\Gamma \left(\frac{\partial}{\partial \nu} \Delta v \cdot v - \Delta v \frac{\partial}{\partial \nu} v \right) d\Gamma \\ &= a(v, v). \end{aligned}$$

Next we examine the term

$$(2.48) \quad \begin{aligned} \int_\Omega [v, \chi(w)] v \, d\Omega &= \int_\Omega [v, v] \chi(w) \, d\Omega \\ &= -\int_\Omega \Delta^2 \chi(v) \chi(w) \, d\Omega \\ &= -\frac{1}{c_0^2} \int_\Omega (\Delta \chi(w))^2 \, d\Omega. \end{aligned}$$

Using (2.48) in (2.47), we obtain

$$0 = a(v, v) + \frac{1}{c_0^2} \int_\Omega (\Delta \chi(w))^2 \, d\Omega,$$

and conclude, by the positivity of $a(v, v)$ that

$$(2.49) \quad v \equiv 0 \quad \text{in } Q.$$

In order to complete the contradiction (see (2.34)), we must prove

$$\left(l.o.(v_n) + \int_0^T \|\chi(v_n)\|_{H^{3-\epsilon}(\Omega)} dt\right) \longrightarrow 0.$$

But this follows from Proposition 2.5, (2.36) and (2.49). This provides the contradiction, and hence the proof of Lemma 2.2 for Case 1.

Case 2: $c_0 = 0$, (i.e. $\chi(w) \equiv 0$ and $l.o.(w) = 0$.) In this case, we will again use the result of Proposition 2.5. Here we use the fact that $\chi(w_n) \to 0$ in $C([0,T]; H^{3-\epsilon}(\Omega))$ in combination with (2.36) and Proposition 2.6 in order to obtain that $[v_n(t), \chi(w_n)(t)] \to 0$ in $\mathcal{D}'(\Omega)$). By using (2.35) and passing with a limit on system (2.46), we obtain

$$(2.50) \quad \begin{array}{ll} \Delta^2 v = 0 & \text{in } Q \\ v = \frac{\partial}{\partial \nu} v = 0 & \text{on } \Sigma_0 \\ \Delta v + (1-\mu) B_1 v = 0 & \text{on } \Sigma_1 \\ \frac{\partial}{\partial \nu} \Delta v + (1-\mu) B_2 v = 0 & \text{on } \Sigma_1. \end{array} \Bigg\}$$

The same argument as in Case 1 yields $v \equiv 0$ and

$$\left(l.o.(v_n) + \int_0^T \|\chi(w_n)\|_{H^{3-\epsilon}(\Omega)} dt\right) \longrightarrow 0,$$

which gives us the contradiction.

2.3 Completion of Proof of Theorem 1.1.

By using (2.3) and (2.26) along with Lemmas 2.1 and 2.2, we have shown that, for T sufficiently large, the energy for system (1.1) satisfies

$$(2.51) \quad E(T) \leq C(E(0)) \left(\int_{\Sigma_1} (w_t^2 + |\nabla w_t|^2) d\Sigma_1 + \int_Q b w_t^2 dQ \right),$$

where the function $\rho \equiv C(E(0))$ is increasing in $E(0)$. From (2.3) we have that

$$\int_{\Sigma_1} (w_t^2 + |\nabla w_t|^2) d\Sigma_1 + \int_Q b w_t^2 dQ = 1/2(E(0) - E(T))$$

so that by using (2.51) we obtain

$$2E(T) \leq \rho \cdot (E(0) - E(T))$$

or equivalently,

$$(2.52) \quad E(T) \leq \frac{\rho(E(0))}{2 + 2\rho} \cdot E(0) \equiv \rho_0(E(0)) E(0).$$

Observe that ρ_0 is an increasing function of $E(0)$ and $\rho_0 < 1$. The proof of Theorem 1.1 now follows by using (2.51) and (2.52) in conjunction with results from nonlinear semigroup theory.

References

[1] R. Adams. *Sobolev Spaces*. Academic Press, New York, 1975.

[2] M. E. Bradley. Global exponential decay rates for a Kirchoff plate with boundary nonlinearities. To appear.

[3] M. E. Bradley and I. Lasiecka. Global decay rates for the solutions to a von Kármán plate without geometric conditions. to appear.

[4] M. E. Bradley and I. Lasiecka. Local exponential stabilization of a nonlinearly perturbed von Kármán plate. *Journal of Nonlinear Analysis: Techniques, Methods and Applications*, 18(4):333–343, 1992.

[5] A. Favini and I. Lasiecka. Well-posedness and regularity of second order abstract nonlinear evolution equations. To appear.

[6] P. Grisvard. *Elliptic Problems in Nonsmooth Domains*. Pitman, London, 1985.

[7] J. Lagnese. *Boundary Stabilization of Thin Plates*. Society for Industrial and Applied Mathematics, Philadelphia, 1989.

[8] J. Lagnese. Local controllability of dynamic von Karman plates. *Control and Cybernetics*, 1990.

[9] J. Lagnese. Uniform asymptotic energy estimates for solutions of the equations of dynamical plane elasticity with nonlinear dissipation at the boundary. *Nonlinear Analysis*, 16:35–54, 1991.

[10] J. Lagnese and G. Leugering. Uniform stabilization of a nonlinear beam by nonlinear feedback. *Journal of Differential Equations*, 91:355–388, 1991.

[11] I. Lasiecka. Global uniform decay rates for the solutions to wave equations with nonlinear boundary conditions. To appear.

[12] I. Lasiecka and R. Triggiani. Sharp trace estimates of solutions to Kirchoff and Euler-Bernoulli equations. In *Proceedings of the International Conference on Abstract Evolution Equations*, Bologne, July 1991.

[13] J. L. Lions and E. Magenes. *Non-Homogeneous Boundary Value Problems and Applications*, volume 1. Springer-Verlag, New York, 1972.

[14] W. Littman. Boundary control theory for beams and plates. In *Proceedings of the 24th Conference on Decision and Control*, pages 2007–2009, Ft. Lauderdale, 1985.

[15] J. Puél and M. Tucsnak. Boundary stabilization for the von Kármán equations. To appear.

[16] J. Simon. Compact sets in the space $L^p(0,T;B)$. *Annali di Mathematica Pura et Applicata*, 4:65–96, 1987.

Chapter 9
On the Standard Problem of H_∞-Optimal Control for Infinite Dimensional Systems

A. Bensoussan* P. Bernhard†

1 Introduction

We investigate in this paper the so called "standard problem" of H_∞ optimal control in an infinite dimensional setup, general enough to account for distributed parameter systems for instance.

Our objective is to extend to infinite dimensional systems the methods and results of [4], thus recovering in a simple way the results of [1] in the case of perfect information (state feedback control) and of [2] in the partial information case (output feedback control). The motivation for doing so is, on the one hand that the methods seem much simpler, and more importantly on the other hand that one can then extend them to other problems, such as the sampled data problem which is of paramount importance to control theoretists, and various information structures such as considered in [4].

Several difficulties arise in trying to carry out this program, mainly for the infinite horizon problem that we tackle here. One is connected with the conjugate point theory as developped in [6] and [4]. It is completely overcomed here, through a technique trivially extendable to finite time problems, and that yields a stronger result.

However, in the partial information case, while the certainty equivalence theorem of [4] can easily be extended, several technical difficulties arise related to the asymptotic behaviour of the Riccati equation. We propose here a solution based upon the use of duality, more in the line of [7], which seems significantly simpler than [2].

All the necessary tools in infinite dimensional system theory needed in the
sequel can be found in [5].

2 General Presentation of the Problem

2.1 Notation and Assumptions

Let X be a real separable Hilbert space, and A be the infinitesimal generator of a C_0-semigroup e^{At} in X. We denote by $D(A)$ the domain of the operator A. We say that the semigroup is exponentially stable if one has

$$||e^{At}|| \leq M e^{-\alpha t}, \alpha > 0.$$

---
*University Paris Dauphine and INRIA.
†INRIA Sophia Antipolis.

A linear unbounded operator A in X is said *exponentially stable* if it is the infinitesimal generator of a C_0-semigroup e^{At} in X which is exponentially stable. We recall the important result of DATKO [3], namely that A is exponentially stable iff for any h in X the solution of
$$x' = Ax, x(0) = h$$
belongs to $L^2(0, \infty; X)$. Let also U and W be real separable hilbert spaces, and B and D linear bounded operators from respectively U and W to X. Consider the dynamic system governed by the equation

(2. 1)
$$\begin{aligned} x' &= Ax + Bv + Dw \\ x(0) &= 0. \end{aligned}$$

In equation (2. 1) w stands for a disturbance and v stands for a control. We now make precise what is meant by *robust control*. Let Z be an addditional Hilbert space and $H \in \mathcal{L}(X; Z)$. Suppose the controller is interested in the cost function
$$K_0(v, w) = \int_0^\infty (|Hx|^2 + |v|^2)\, dt$$
(which may take the value $+\infty$) that he wishes to keep small in spite of the disturbances that are not known in advance. Clearly, if w is, in some sense, "large", it will be able to force a large value of K_0. A reasonable goal, in view of the linearity of the system, is to try to insure
$$K_0(v, w) \leq \gamma^2 \int_0^\infty |w|^2\, dt$$
for some given (positive) number γ, provided of course that the disturbance be square integrable, which will always be assumed in this paper. We therefore introduce the ratio
$$\rho(v, w) = \frac{K_0(v, w)}{\int_0^\infty |w|^2\, dt},$$
that the controller shall try to hold below a fixed value regardless of w.

The next question to address is that of the *admissible control strategies*. In this section, we assume that the controller has access to instantaneous perfect measurements of the state of the system 2. 1. We therefore want to allow controls of the form $v(t) = \mu(x(t))$ for a large class of μ's. We shall let $\mathcal{M}$ be the class of all applications from X into V that are such that the differential equation
$$x' = Ax + B\mu(x) + Dw, \qquad x(0) = h$$
has a solution over $(0, \infty)$ for all $h \in X$ and for all square integrable $w(t)$, and such that the control $v(t) = \mu(x(t))$ generated be square integrable. As a matter of fact, we could replace state feedbacks by a larger class of *causal* functions of state. The sequel is unchanged by such a generalization.

We shall use the unambiguous notation $\rho(\mu, w)$ to mean the value taken by $\rho(v, w)$ when $v = \mu(x)$. We can then define the property of interest in this section:

DEFINITION 2.1. *We say that the γ^2 robustness property (with full observation) holds for the equation (2. 1) and the cost function $K_0(v, w)$ if one has*
$$\inf_{\mu \in \mathcal{M}} \sup_w \rho(\mu, w) < \gamma^2.$$

2.2 Review of some results

Consider the pair of operators A, B, H as above.

DEFINITION 2.2. *We say that the pair A, B is H stabilizable if for any $h \in X$ there exists $v \in L^2(0, \infty; U)$ such that the solution x of the equation*

$$(2.\ 2) \qquad \begin{aligned} x' &= Ax + Bv \\ x(0) &= h \end{aligned}$$

satisfies $Hx \in L^2(0, \infty; Z)$. We say that the pair A, B is stabilizable if it is I stabilizable.

It is a classical result that the pair A,B is stabilizable iff there exists an operator $F \in \mathcal{L}(X, U)$ such that $A + BF$ is exponentially stable. We now give another

DEFINITION 2.3. *We say that the pair A, H is detectable if the pair A^*, H^* is stabilizable.*

It follows from the characterization of stabilizability that the pair A,H is detectable iff there exists an operator $G \in \mathcal{L}(Z, X)$ such that $A + GH$ is exponentially stable. We now state an important result in Control Theory, whose proof for infinite dimensional systems can be found in [5].

THEOREM 2.1. *We assume that A, B is H stabilizable and that A, H is detectable. Then there exists one and only one operator $\Gamma \in \mathcal{L}(X, X)$ with $\Gamma = \Gamma^* \geq 0$ satisfying $A - BB^*\Gamma$ is exponentially stable and*

$$(2.\ 3) \qquad \Gamma A + A^*\Gamma - \Gamma BB^*\Gamma + H^*H = 0.$$

The interpretation of the operator Γ is important. Consider the functional

$$K_h(v) = \int_0^\infty (|Hx|^2 + |v|^2)\, dt$$

where x is the solution of (2. 2). Then one has

$$(\Gamma h, h) = \min K_h(v).$$

The minimum in $K_h(v)$ is attained for $u = -B^*\Gamma y$ where y is the solution of

$$y' = (A - BB^*\Gamma)y \quad y(0) = h.$$

REMARK 2.1. *In equation 2. 3, one must interpret the operator*

$$\Gamma A + A^*\Gamma$$

according to the general theory of algebraic Riccati equations (see [5] for details). It is sufficient to notice that for any pair h, k in $D(A)$ the bilinear form

$$< (\Gamma A + A^*\Gamma)h, k > = (Ah, \Gamma k) + (\Gamma h, Ak)$$

makes sense.
Note also that if $h \in D(A)$ then $\Gamma h \in D(A^*)$.

3 γ^2 robustness property with full observation

3.1 Setting of the Result

We state the result due to [1].

THEOREM 3.1. *We assume that A, B is stabilizable and that A, H is detectable. Then the γ^2 robustness property with full observation holds for the equation (2. 1) and the cost function $K_0(v,w)$ iff there exists a $P \in \mathcal{L}(X,X)$ with $P = P^* \geq 0$ satisfying*

$$(3.\ 1) \qquad PA + A^*P - P(BB^* - \frac{1}{\gamma^2}DD^*)P + H^*H = 0$$

and $A - (BB^* - \frac{1}{\gamma^2}DD^*)P$ is exponentially stable.

The operator P has an interesting interpretation. Indeed let be

$$K_h(v,w) = \int_0^\infty (|Hx|^2 + |v|^2)\,dt$$

where x is the solution of

$$(3.\ 2) \qquad \begin{aligned} x' &= Ax + Bv + Dw \\ x(0) &= h. \end{aligned}$$

Let also

$$J_h(v,w) = K_h(v,w) - \gamma^2 \int_0^\infty |w|^2\,dt.$$

Then one has

$$(3.\ 3) \qquad (Ph, h) = \max_w \min_v J_h(v,w).$$

It is important to check the following result

LEMMA 3.1. *If P satisfies the properties stated in Theorem 3.1 then one has also $A - BB^*P$ is exponentially stable.*

Proof Considering the solution x of (3. 2), computing the derivative of $(Px(t), x(t))$ and integrating between 0 and T yields

$$(3.\ 4) \qquad \begin{aligned} (Px(T), x(T)) - (Ph, h) &+ \int_0^T |Hx|^2\,dt + \int_0^T |v|^2\,dt \\ -\gamma^2 \int_0^T |w|^2\,dt &= \int_0^T |v + B^*Px|^2\,dt - \gamma^2 \int_0^T |w - \frac{1}{\gamma^2}D^*Px|^2\,dt \end{aligned}$$

Choose $v = -B^*Px$ and $w = 0$, then we see that x is the solution of

$$x' = (A - BB^*P)x \ , \ x(0) = h$$

and from the equation (3. 4) it follows that

$$Hx \in L^2(0, \infty; Z), B^*Px \in L^2(0, \infty; U), D^*Px \in L^2(0, \infty; W).$$

Since we already know that $A - (BB^* - \frac{1}{\gamma^2}DD^*)P$ is exponentially stable, we can assert that $x \in L^2(0, \infty; X)$, hence the desired result. ♠

3.2 Proof of Main result

Sufficiency.
The control strategy that achieves the required result will be $\mu(x) = -B^*Px$. We consider the solution of
$$x' = (A - BB^*P)x + Dw \ , x(0) = 0.$$
Since $A - BB^*P$ is exponentially stable and $w \in L^2(0, \infty; W)$ the solution x belongs to $L^2(0, \infty; X)$, as well as the control v generated. Thus this μ is admissible. We apply the relation (3. 4) with $v = -B^*Px$ and $h = 0$, letting T tend to ∞. This yields
$$\int_0^\infty |Hx|^2 \, dt + \int_0^\infty |B^*Px|^2 \, dt = \gamma^2 \int_0^\infty |w|^2 \, dt - \gamma^2 \int_0^\infty |w - \frac{1}{\gamma^2}D^*Px|^2 \, dt.$$
Therefore
$$\rho(-B^*Px, w) = \gamma^2 - \gamma^2 \frac{\int_0^\infty |w - \frac{1}{\gamma^2}D^*Px|^2 \, dt}{\int_0^\infty |w|^2 \, dt}.$$
It follows that
$$\sup_w \rho(-B^*Px, w) \leq \gamma^2 - \gamma^2 \inf_w \frac{\int_0^\infty |w - \frac{1}{\gamma^2}D^*Px|^2 \, dt}{\int_0^\infty |w|^2 \, dt}.$$
Now x can be viewed as the solution of
$$x' = \left(A - (BB^* - \frac{1}{\gamma^2}DD^*)P\right)x + D(w - \frac{1}{\gamma^2}D^*Px) \ , x(0) = 0.$$
Since $A - (BB^* - \frac{1}{\gamma^2}DD^*)P$ is exponentially stable we have
$$\int_0^\infty |x|^2 \, dt \leq c_0 \int_0^\infty |w - \frac{1}{\gamma^2}D^*Px|^2 \, dt.$$
Hence immediately
$$\int_0^\infty |w|^2 \, dt \leq c_1 \int_0^\infty |w - \frac{1}{\gamma^2}D^*Px|^2 \, dt.$$
Therefore we can assert that
$$\text{(3. 5)} \qquad \inf_w \frac{\int_0^\infty |w - \frac{1}{\gamma^2}D^*Px|^2 \, dt}{\int_0^\infty |w|^2 \, dt} > 0.$$
We deduce
$$\sup_w \rho(-B^*Px, w) \leq \gamma^2.$$

Necessity.
We shall prove a result stronger than that of the theorem (3.1). Namely the following

PROPOSITION 3.1. *If*
$$\sup_w \inf_v \rho(v, w) < \gamma^2,$$
then the conditions of theorem (3.1) hold.

This is indeed stronger than the theorem, since clearly, one has

$$\inf_v \rho(v,w) \leq \rho(\mu,w)$$

for any admissible μ, and therefore

$$\sup_w \inf_v \rho(v,w) \leq \sup_w \rho(\mu,w), \quad \forall \mu \in \mathcal{M}$$

and thus

$$\sup_w \inf_v \rho(v,w) \leq \inf_{\mu \in \mathcal{M}} \sup_w \rho(\mu,w).$$

We now prove the proposition. We consider a control problem where the system is defined by (3. 2) in which w is given, the control being v. We minimize the cost $K_h(v,w)$. Note that the assumptions of Theorem 2.1 are satisfied, thus the equation

(3. 6) $$\Gamma A + A^*\Gamma - \Gamma BB^*\Gamma + H^*H = 0.$$

has a unique solution $\Gamma = \Gamma^* \geq 0$, such that the operator $A - BB^*\Gamma$ is exponentially stable. Consider next the linear equation

(3. 7) $$r' + (A^* - \Gamma BB^*)r + \Gamma Dw = 0$$

where the solution $r(.)$ belongs to $L^2(0,\infty;X)$. Note that in (3. 7) the initial condition is given at ∞ and not at 0. The equation (3. 7) has a unique solution. The optimal feedback is then descibed by

$$\hat{u} = -B^*(\Gamma \hat{x} + r)$$

where $\hat{x}$ the optimal state is the solution of

$$\frac{d}{dt}\hat{x} = (A - BB^*\Gamma)\hat{x} - BB^*r + Dw \,;\, \hat{x}(0) = h.$$

We can also express the value of $J_h(\hat{u},w)$ as follows

(3. 8) $$J_h(\hat{u},w) = -\int_0^\infty |B^*r|^2 \, dt \\ -\gamma^2 \int_0^\infty |w|^2 \, dt + 2\int_0^\infty (r, Dw) \, dt + 2(h, r(0)) + (\Gamma h, h).$$

The calculations leading to the expression (3. 8) are standard and not detailed here. We now look at the problem of maximizing the expression $J_h(\hat{u},w)$ for $w \in L^2(0,\infty;W)$. Note that although we have an LQ problem, the concavity is not a priori verified. This is where we use the assumption of γ^2 robustness property. Set $\Phi_h(w) = J_h(\hat{u},w)$, then we note the relation

$$\Phi_h(w) = \Phi_0(w) + 2(h,r(0)) + (\Gamma h, h).$$

By the assumption of γ^2 robustness there exists a positive $\delta < \gamma$ such that $\sup_w \inf_v \rho(v,w) \leq \gamma^2 - \delta^2$. It follows that

$$J_0(\hat{u},w) \leq -\delta^2 \int_0^\infty |w|^2 \, dt$$

for any w. Therefore, we can assert that

$$\Phi_0(w) \leq -\delta^2 \int_0^\infty |w|^2 \, dt. \tag{3.9}$$

Note also the relation

$$\Phi_h(\theta w_1 + (1-\theta)w_2) = \theta \Phi_h(w_1) + (1-\theta)\Phi_h(w_2) - 2\theta(1-\theta)\Phi_0(w_1 - w_2).$$

This relation shows immediately that $\Phi_h(w)$ is strictly concave. From (3.9) it is a coercive functional. Therefore we can apply to the control problem (3.7) and cost function $\Phi_h(w)$ the standard theory of existence, uniqueness of an optimal control. Moreover, the theory of necessary and sufficient conditions of optimality hold. Calling $\hat{w}$ and $\hat{r}$ the optimal control and state, we have the relations

$$\begin{aligned} \frac{d}{dt}\hat{r} + (A^* - \Gamma BB^*)\hat{r} + \Gamma D\hat{w} &= 0, \\ \frac{d}{dt}p &= (A - BB^*\Gamma)p - BB^*\hat{r} + D\hat{w}, \end{aligned} \tag{3.10}$$

$$\begin{aligned} p(0) &= h, \\ \gamma^2 \hat{w} - D^*\hat{r} - D^*\Gamma p &= 0. \end{aligned} \tag{3.11}$$

The classical decoupling argument applies to the system (3.10) and (3.11), therefore there exists $\Sigma = \Sigma^*$, such that $\hat{r} = \Sigma p$. The operator Σ is a solution of the Riccati equation

$$\Sigma\left(A - (BB^* - \frac{1}{\gamma^2}DD^*)\Gamma\right) + \left(A^* - \Gamma(BB^* - \frac{1}{\gamma^2}DD^*)\right)\Sigma$$
$$-\Sigma(BB^* - \frac{1}{\gamma^2}DD^*)\Sigma + \frac{\Gamma DD^*\Gamma}{\gamma^2} = 0.$$

Set $P = \Gamma + \Sigma$, then P is self adjoint and is a solution of (3.1), as easily seen. Note that p is the solution of

$$\begin{aligned} \frac{d}{dt}p &= \left(A - (BB^* - \frac{1}{\gamma^2}DD^*)P\right)p \\ p(0) &= h. \end{aligned}$$

Since $p \in L^2(0,\infty;X)$, we deduce that $A - (BB^* - \frac{1}{\gamma^2}DD^*)P$ is exponentially stable. Moreover

$$\max \Phi_h(w) = \Phi_h(\hat{w}) = (Ph, h).$$

Therefore, we have

$$\max_w \min_v J_h(v,w) = (Ph, h).$$

It follows among other things that

$$(Ph, h) \geq \min_v J_h(v, 0) \geq 0.$$

The proof of the necessity part has been completed. This concludes the end of the proof.
♠

It should be pointed out that we have avoided the classical conjugate point argument, such as in [4] chapt. 8 for instance, or the arguments of [8] in the stationary case. It is completely clear that the present proof extends to the finite horizon non stationary case.

4 γ^2 robustness property with partial observation

4.1 Presentation of the problem

Consider again the system

(4. 1)
$$\begin{aligned} x' &= Ax + Bv + Dw \\ x(0) &= 0. \end{aligned}$$

and the cost function

$$K_0(v,w) = \int_0^\infty (|Hx|^2 + |v|^2)\, dt.$$

In the present situation, the controller has access only to a partial observation described as follows

(4. 2) $$y = Cx + \eta$$

where $C \in \mathcal{L}(X;Y)$, Y being a new Hilbert space, called the space of observations. Next η is another disturbance, modelling the measurement error. The controller can use only a causal functional on the observation y. A natural class of controllers is the following one

$$v = Lp,$$
$$p' = (A+M)p + Ny, \qquad p(0) = 0.$$

In other words the controller is characterized by three maps $L \in \mathcal{L}(X;U)$, $M \in \mathcal{L}(X;X)$, $N \in \mathcal{L}(Y;X)$. We call it a *feedback controller*. We have not combined $A+M$ into a single operator since A is unbounded, whereas M is bounded. If we use such a controller in the equation 4. 1, then we get a coupled system as follows

(4. 3)
$$\begin{aligned} x' &= Ax + BLp + Dw, \\ p' &= (A+M)p + NCx + N\eta, \\ x(0) &= 0, \\ p(0) &= 0. \end{aligned}$$

The coupled system introduces an operator

$$\mathcal{A} = \begin{pmatrix} A & BL \\ NC & A+M \end{pmatrix}.$$

The operator $\mathcal{A}$ is the operator related to the feedback controller L, M, N. We shall consider only controllers whose corresponding operator is exponentially stable. For such a controller the cost

$$K_0(Lp, w) = \int_0^\infty (|Hx|^2 + |Lp|^2)\, dt$$

is finite. We can then define the ratio corresponding to the feedback controller L, M, N

$$\rho(L, M, N) = \sup_{w, \eta} \frac{K_0(Lp, w)}{\int_0^\infty (|w|^2 + |\eta|^2)\, dt}.$$

We state the

DEFINITION 4.1. *We say that the γ^2 robustness property (with partial observation) holds for the equation (4. 1), the observation (4. 2) and the cost function $K_0(v, w)$ if there exists a feedback controller L, M, N such that the corresponding operator $\mathcal{A}$ is exponentially stable and if one has*

$$\rho(L, M, N) < \gamma^2.$$

Our objective is naturally to give a necessary and sufficient condition of γ^2 robustness property (with partial observation) which leads to a computable feedback. We shall see that this property is equivalent to 3 systems and corresponding costs enjoying the γ^2 robustness property (with full observation). From section 3 corresponding Riccati equations can be introduced. In fact, the solution of one of them is expressible in terms of the others. Moreover one of the system and cost is (4. 1) with cost $K_0(v,w)$. Therefore γ^2 robustness property (with partial observation) implies γ^2 robustness property (with full observation).

4.2 Statement of the main result

We state the result due to [2].

THEOREM 4.1. *We assume that A, D is stabilizable and that A, H is detectable. Then the γ^2 robustness property with partial observation holds for the equation (4. 1), the observation (4. 2) and the cost function $K_0(v,w)$ iff there exist solutions of the Riccati equations*

(4. 4)
$$PA + A^*P - P(BB^* - \frac{1}{\gamma^2}DD^*)P + H^*H = 0$$
$$P \text{ symmetric}, \geq 0$$
$$A - (BB^* - \frac{1}{\gamma^2}DD^*)P \text{ is exponentially stable}$$

(4. 5)
$$\Sigma A^* + A\Sigma - \Sigma(C^*C - \frac{1}{\gamma^2}H^*H)\Sigma + DD^* = 0$$
$$\Sigma \text{ symmetric}, \geq 0$$
$$A^* - (C^*C - \frac{1}{\gamma^2}H^*H)\Sigma \text{ is exponentially stable}$$

(4. 6)
$$I - \frac{1}{\gamma^2}P\Sigma \text{ is invertible}; \Sigma\left(I - \frac{1}{\gamma^2}P\Sigma\right)^{-1} \geq 0$$

If the conditions (4. 4), (4. 5), (4. 6) hold, then the feedback controller

$$L = -B^*P, M = -(BB^* - \frac{1}{\gamma^2}DD^*)P - \Pi C^*C, N = \Pi C^*$$

where

$$\Pi = \Sigma\left(I - \frac{1}{\gamma^2}P\Sigma\right)^{-1}$$

satisfies the conditions of the definition 4.1.

REMARK 4.1. *The Riccati equation (4. 4) is the same as the one characterizing the γ^2 robustness property with full observation. Note that the pair A, B is stabilizable, as a consequence of the fact that $\mathcal{A}$ is exponentially stable. From formula (3. 3) the solution is unique. A similar property holds for (4. 5) since it characterizes the γ^2 robustness property with full observation of a different system and cost. This will be made precise in the proof.*

4.3 Duality Considerations

We shall prove here some norm equalities which will be instrumental in the proof of Theorem 4.1 and are interesting in themselves. Consider Hilbert spaces Ξ, Φ, Ψ, identified with their duals. Consider operators $\mathcal{A}$, $\mathcal{B}$, $\mathcal{C}$, which are linear, $\mathcal{A}$ is unbounded in Ξ and is the infinitesimal generator of a C_0 semigroup in Ξ. Next

$$\mathcal{B} \in \mathcal{L}(\Phi; \Xi), \mathcal{C} \in \mathcal{L}(\Xi; \Psi).$$

To the triple $\mathcal{A}, \mathcal{B}, \mathcal{C}$ we associate a linear system

(4.7) $$\xi' = \mathcal{A}\xi + \mathcal{B}\phi\ ; \xi(0) = 0$$

and a corresponding observation $\mathcal{C}\xi$. We define next a "dual" triple $\mathcal{A}^*, \mathcal{C}^*, \mathcal{B}^*$ to which corresponds the system

(4.8) $$\zeta' = \mathcal{A}^*\zeta + \mathcal{C}^*\psi;\ \zeta(0) = 0,$$

and the corresponding observation $\mathcal{B}^*\zeta$. In equations (4.7) and (4.8) the quantities ϕ and ψ are inputs (or controls). We now state the following

PROPOSITION 4.1. *Assume that $\mathcal{A}$ is exponentially stable. Then one has the relation*

$$\sup_\phi \frac{\int_0^\infty |\mathcal{C}\xi|^2\,dt}{\int_0^\infty |\phi|^2\,dt} = \sup_\psi \frac{\int_0^\infty |\mathcal{B}^*\zeta|^2\,dt}{\int_0^\infty |\psi|^2\,dt}.$$

We begin with some preliminary results.

LEMMA 4.1. *Let $\phi \in L^2(-\infty;+\infty;\Phi)$, then there exists one and only one function*

$$\xi \in L^2(-\infty;+\infty;\Xi) \cap C^0(-\infty;+\infty;\Xi)$$

solution of the equation

(4.9) $$\xi' = \mathcal{A}\xi + \mathcal{B}\phi.$$

Proof

The solution of (4.9) is interpreted as

(4.10) $$\xi(t) = \int_{-\infty}^t e^{\mathcal{A}(t-\tau)} \mathcal{B}\phi(\tau)\,d\tau$$

for any t. Using the estimate $\|e^{\mathcal{A}t}\| \leq M e^{-\alpha t}$, for some convenient constants M and $\alpha > 0$, since $\mathcal{A}$ is exponentially stable, we deduce the inequality

$$|\xi(t)| \leq M\|\mathcal{B}\|\beta(t)$$

where

$$\beta(t) = \int_{-\infty}^t e^{-\alpha(t-\tau)} |\phi(\tau)|\,d\tau.$$

¿From Schwartz' inequality, one checks that

$$|\beta(t)| \leq \frac{1}{\sqrt{2\alpha}} \left(\int_{-\infty}^{+\infty} |\phi(s)|^2\,ds \right)^{1/2}$$

and using the differential equation satisfied by β and

$$0 \leq \frac{1}{2}|\beta(\tau)|^2 = \int_{-\infty}^\tau \beta\beta'\,dt$$

one gets the other estimation

$$\int_{-\infty}^{+\infty} |\beta(t)|^2\,dt \leq \frac{1}{\alpha^2} \int_{-\infty}^{+\infty} |\phi(t)|^2\,dt.$$

Therefore the expression (4. 10) is a solution. The solution is unique. Indeed, it is sufficient to prove that when $\phi = 0$, then the solution is necessarily 0. But if $\phi = 0$, we can write

$$\xi(t) = e^{\mathcal{A}(t-s)}\xi(s)$$

hence

$$|\xi(t)| \le e^{-\alpha(t-s)}|\xi(s)|.$$

For fixed t let $s \to -\infty$, using the fact that $|\xi(s)|$ remains bounded, we conclude that $|\xi(t)| = 0$. This proves the uniqueness of the solution. ♠

Let

$$\alpha = \sup_\phi \frac{\int_0^\infty |\mathcal{C}\xi|^2\, dt}{\int_0^\infty |\phi|^2\, dt}$$

where we refer to the system (4. 7). Similarly, set

$$\beta = \sup_\phi \frac{\int_{-\infty}^\infty |\mathcal{C}\xi|^2\, dt}{\int_{-\infty}^\infty |\phi|^2\, dt}$$

referring now to the system (4. 9). Then we have

LEMMA 4.2. *The two numbers α and β are equal.*

Proof

It is first clear that

$$\alpha \le \beta.$$

This is due to the fact that, when in (4. 9) we restrict ourselves to inputs ϕ which are equal to 0 for $t < 0$, then (4. 9) reduces immediately to (4. 7). It will be convenient to introduce the linear map Θ from $L^2(-\infty; +\infty; \Phi)$ to $L^2(-\infty; +\infty; \Psi)$ defined by

$$\Theta(\phi) = \mathcal{C}\xi$$

where ξ is the solution of (4. 9) corresponding to the input ϕ. Clearly

$$\beta = \|\Theta\|^2.$$

Now to any $\phi \in L^2(-\infty; +\infty; \Phi)$ we associate ϕ_T defined by

$$\phi_T(t) = \begin{cases} \phi(t) & \text{if } t > -T \\ 0 & \text{otherwise} \end{cases}$$

Let us denote by ξ_T the solution of (4. 9) corresponding to the input ϕ_T. Setting

$$\tilde{\xi}_T(t) = \xi_T(-T + t) \ ; \tilde{\phi}_T(t) = \phi_T(-T + t)$$

we see immediately that $\tilde{\xi}_T$ is the solution of (4. 7) corresponding to the input $\tilde{\phi}_T$. From the definition of α we can write

$$\int_0^\infty |\mathcal{C}\tilde{\xi}_T(t)|^2\, dt \le \alpha \int_0^\infty |\tilde{\phi}_T(t)|^2\, dt.$$

But

$$\int_0^\infty |\mathcal{C}\tilde{\xi}_T(t)|^2 \, dt = \int_{-\infty}^\infty |\mathcal{C}\xi_T(t)|^2 \, dt$$

and

$$\int_0^\infty |\tilde{\phi}_T(t)|^2 \, dt = \int_{-\infty}^\infty |\phi_T(t)|^2 \, dt$$
$$\leq \int_{-\infty}^\infty |\phi(t)|^2 \, dt$$

Therefore we have proved that

$$|\Theta(\phi_T)|^2 \leq \alpha \int_{-\infty}^\infty |\phi(t)|^2 \, dt..$$

Letting $T \to \infty$ and noting that ϕ_T tends to ϕ in $L^2(-\infty; +\infty; \Phi)$, we obtain

$$|\Theta(\phi)|^2 \leq \alpha \int_{-\infty}^\infty |\phi(t)|^2 \, dt.$$

Since ϕ is arbitrary, it follows from the definition of β that

$$\beta \leq \alpha.$$

The proof has been completed. ♠

We can now proceed with the

Proof of Proposition 4.1 In a way similar to (4.9) we consider the dual system on $-\infty, +\infty$

(4.11) $$\zeta' = \mathcal{A}^*\zeta + \mathcal{C}^*\psi$$

where $\psi \in L^2(-\infty; +\infty; \Psi)$ and the solution ζ belongs to $L^2(-\infty; +\infty; \Xi) \cap C^0(-\infty; +\infty; \Xi)$. Define a linear map from $L^2(-\infty; +\infty; \Psi)$ to $L^2(-\infty; +\infty; \Phi)$ by setting

$$\Upsilon(\psi) = \mathcal{B}^*\zeta.$$

In view of Lemma 4.1 the desired result will be demonstrated if we prove

(4.12) $$\|\Theta\| = \|\Upsilon\|$$

To any ψ associate $\bar{\psi}$ by setting $\bar{\psi}(t) = \psi(-t)$. The key point is to verify that

(4.13) $$\Upsilon(\psi)(t) = \Theta^*(\bar{\psi})(-t).$$

This property implies the result (4.12). Now (4.13) is easily deduced from the explicit formula (4.10) and the corresponding one for ζ the solution of (4.11). Details are left to the reader. The proof has been completed. ♠

5 Proof of Theorem 4.1

5.1 Necessary Conditions

Proof of (4.4)
In fact, the assumptions of Theorem 3.1 are satisfied, since the property of γ^2 robustness

with full observation holds and that the pair A, B is stabilizable(see Remark 4.1). Therefore There exists a unique P solution of (4. 4).

Proof of (4. 5)

We shall use the duality considerations of paragraph 4.3, see Proposition 4.1. Let $\Xi = X \times X$, $\Phi = W \times Y$, $\Psi = Z \times U$. Let next

$$\mathcal{A} = \begin{pmatrix} A & BL \\ NC & A+M \end{pmatrix}$$

and

$$\mathcal{B} = \begin{pmatrix} D & 0 \\ 0 & N \end{pmatrix} \quad \mathcal{C} = \begin{pmatrix} H & 0 \\ 0 & L \end{pmatrix}.$$

Setting $\xi = (x, p)$, $\phi = (w, \eta)$, then we see immediately that

$$\rho(L, M, N) = \alpha$$

where the number α has been defined in the proof of Lemma 4.2. We can then make use of Proposition 4.1. The dual system of (4. 3) is (apply (4. 8) and set $\zeta = m, q$, $\psi = \lambda, \mu$)

(5. 1)
$$\begin{aligned} m' &= A^*m + C^*N^*q + H^*\lambda \\ q' &= L^*B^*m + (A^* + M^*)q + L^*\mu \\ m(0) &= 0 \\ q(0) &= 0. \end{aligned}$$

Using then Proposition 4.1 we can assert that

$$\rho(L, M, N) = \sup_{\lambda, \mu} \frac{\int_0^\infty (|D^*m|^2 + |N^*q|^2)\, dt}{\int_0^\infty (|\lambda|^2 + |\mu|^2)\, dt}.$$

Therefore from the assumption, we have the property

$$\sup_{\lambda, \mu} \frac{\int_0^\infty (|D^*m|^2 + |N^*q|^2)\, dt}{\int_0^\infty (|\lambda|^2 + |\mu|^2)\, dt} < \gamma^2.$$

In particular, restricting to $\mu = 0$, we have

(5. 2)
$$\begin{aligned} m' &= A^*m + C^*N^*q + H^*\lambda \\ q' &= L^*B^*m + (A^* + M^*)q \\ m(0) &= 0 \\ q(0) &= 0. \end{aligned}$$

and

(5. 3)
$$\sup_{\lambda} \frac{\int_0^\infty (|D^*m|^2 + |N^*q|^2)\, dt}{\int_0^\infty |\lambda|^2\, dt} < \gamma^2.$$

Consider the dynamic system

(5. 4)
$$\begin{aligned} m' &= A^*m + C^*v + H^*\lambda \\ m(0) &= 0 \end{aligned}$$

where v is the control and λ is the perturbation. We first note that the pair A^*, C^* is stabilizable, as a consequence of the fact that the operator $\mathcal{A}$, hence its dual $\mathcal{A}^*$ is exponentially stable. Consider also the cost function

$$\mathcal{K}_0(v, \lambda) = \int_0^\infty (|D^*m|^2 + |v|^2)\, dt.$$

¿From the assumptions A^*, D^* is detectable. Moreover, from (5. 3) we can assert that the γ^2 robustness property holds for the system (5. 4) and the cost function $\mathcal{K}_0(v, \lambda)$. Therefore, relying on Theorem 3.1 we obtain the existence and uniqueness of the solution Σ of (4. 5).

Proof of (4. 6)

The proof will be decomposed in several steps. We begin with
-a:The matrix operator

$$\mathcal{A}_P = \begin{pmatrix} A + \frac{1}{\gamma^{*2}}DD^*P & BL \\ NC & A + M \end{pmatrix}$$

is exponentially stable. For that purpose consider the dynamic system

(5. 5)
$$\begin{array}{rcl}
\bar{x}' & = & \left(A + \frac{1}{\gamma^{*2}}DD^*P\right)\bar{x} + BL\bar{p} \\
\bar{p}' & = & (A+M)\bar{p} + NC\bar{x} \\
\bar{x}(0) & = & h \\
\bar{p}(0) & = & k.
\end{array}$$

Then we must prove that

(5. 6) $$\bar{x} \in L^2(-\infty; +\infty; X)\ ; \bar{p} \in L^2(-\infty; +\infty; X)$$

Consider then the system

(5. 7)
$$\begin{array}{rcl}
x' & = & Ax + BLp + Dw \\
p' & = & (A+M)p + NCx \\
x(0) & = & h \\
p(0) & = & k.
\end{array}$$

where $w \in L^2(-\infty; +\infty; W)$ and the system corresponding to $w = 0$

(5. 8)
$$\begin{array}{rcl}
x_1' & = & Ax_1 + BLp_1 \\
p_1' & = & (A+M)p_1 + NCx_1 \\
x_1(0) & = & h \\
p_1(0) & = & k.
\end{array}$$

Considering the differences $x - x_1$ and $p - p_1$ we can make use of the assumption

$$\rho(L, M, N) < \gamma^2$$

to assert that there exists a number $\delta < \gamma$ such that

(5. 9)
$$\int_0^\infty (|H(x - x_1)|^2 + |L(p - p_1)|^2)\, dt - \gamma^2 \int_0^\infty |w|^2\, dt \leq -\delta^2 \int_0^\infty |w|^2\, dt$$

Furthermore, since $\mathcal{A}$ is exponentially stable

$$\int_0^\infty (|x_1|^2 + |p_1|^2)\, dt \le C(|h|^2 + |k|^2)$$

where C is a constant independant of h, k. Combining this estimate and (5.9), it is easy to deduce the following

(5.10)
$$\int_0^\infty (|Hx|^2 + |Lp|^2)\, dt - \gamma^2 \int_0^\infty |w|^2\, dt \le -\delta_0^2 \int_0^\infty |w|^2\, dt + C_0(|h|^2 + |k|^2)$$

where $\delta_0 < \delta$ and C_0 is an appropriate constant. Let us set

$$\mathcal{J}_{h,k}(w) = \int_0^\infty (|Hx|^2 + |Lp|^2)\, dt - \gamma^2 \int_0^\infty |w|^2\, dt.$$

Note that in the functional $\mathcal{J}_{h,k}(w)$ the triple L, M, N is fixed and the control is w. The estimate (5.10) shows easily that the functional $\mathcal{J}_{h,k}(w)$ is strictly concave and tends to $-\infty$ as $w \to \infty$. Therefore for any pair h, k there exists an optimal $\hat{w}_{h,k}$ which maximizes $\mathcal{J}_{h,k}(w)$ with respect to w. We denote by $\hat{x}_{h,k}, \hat{p}_{h,k}$ the corresponding optimal state. Let T be arbitrary and set $\bar{x}_T = \bar{x}(T), \bar{p}_T = \bar{p}(T)$ where $\bar{x}, \bar{p}$ is the solution of (5.5). We now define

$$\tilde{x}_T(t) = \begin{vmatrix} \bar{x}(t) & \text{if } t \le T \\ \hat{x}_{\bar{x}_T, \bar{p}_T}(t-T) & \text{if } t > T \end{vmatrix} \qquad \tilde{p}_T(t) = \begin{vmatrix} \bar{p}(t) & \text{if } t \le T \\ \hat{p}_{\bar{x}_T, \bar{p}_T}(t-T) & \text{if } t > T \end{vmatrix}$$

and

$$\tilde{w}_T(t) = \begin{vmatrix} \frac{1}{\gamma^2} D^* P \bar{x}(t) & \text{if } t < T \\ \hat{w}_{\bar{x}_T, \bar{p}_T}(t-T) & \text{if } t > T \end{vmatrix}$$

By construction $\tilde{x}_T, \tilde{p}_T$ is the solution of (5.7) corresponding to $\tilde{w}_T$. Therefore from (5.10) we can write the estimate

(5.11)
$$\int_0^\infty (|H\tilde{x}_T|^2 + |L\tilde{p}_T|^2)\, dt - \gamma^2 \int_0^\infty |\tilde{w}_T|^2\, dt \le -\delta_0^2 \int_0^\infty |\tilde{w}_T|^2\, dt + C_0(|h|^2 + |k|^2)$$

Now we have

$$\int_T^\infty \left(|H\tilde{x}_T|^2 + |L\tilde{p}_T|^2 - \gamma^2|\tilde{w}_T|^2\right) dt = \int_0^\infty \left(|H\hat{x}_{\bar{x}_T,\bar{p}_T}|^2 + |L\hat{p}_{\bar{x}_T,\bar{p}_T}|^2 - \gamma^2|\hat{w}_{\bar{x}_T,\bar{p}_T}|^2\right) dt$$

which is by construction

$$\max_w \left[\int_0^\infty \left(|H x_{\bar{x}_T,\bar{p}_T}|^2 + |L p_{\bar{x}_T,\bar{p}_T}|^2 - \gamma^2|w|^2\right) dt \right]$$

where we have denoted by $x_{\bar{x}_T,\bar{p}_T}, p_{\bar{x}_T,\bar{p}_T}$ the solution of (5.7) corresponding to initial conditions $h = \bar{x}_T, k = \bar{p}_T$. Clearly this quantity is larger or equal to

$$\max_w \min_v J_{\bar{x}_T}(v, w) = (P\bar{x}_T, \bar{x}_T).$$

Therefore we have proved that

$$\int_T^\infty \left(|H\tilde{x}_T|^2 + |L\tilde{p}_T|^2 - \gamma^2 |\tilde{w}_T|^2 \right) dt \geq (P\bar{x}_T, \bar{x}_T)$$

But from the Riccati equation (4. 4) and the first equation (5. 5) we have

$$(P\bar{x}_T, \bar{x}_T) = (Ph, h) + \int_0^T \left(|B^*P\bar{x} + L\bar{p}|^2 + \frac{1}{\gamma^2}|D^*P\bar{x}|^2 - |L\bar{p}|^2 - |H\bar{x}|^2 \right) dt.$$

Combining the two last relations we deduce

(5. 12) $$\int_0^\infty \left(|H\tilde{x}_T|^2 + |L\tilde{p}_T|^2 - \gamma^2|\tilde{w}_T|^2 \right) dt \geq (Ph, h) + \int_0^T |B^*P\bar{x} + L\bar{p}|^2 dt$$

Finally from (5. 11) and (5. 12) we get the estimate

$$(Ph, h) + \int_0^T |B^*P\bar{x} + L\bar{p}|^2 dt \leq -\delta_0^2 \int_0^\infty |\tilde{w}_T|^2 dt + C_0(|h|^2 + |k|^2)$$

Recalling the definition of $\tilde{w}_T$, we deduce in particular

$$\delta_0^2 \int_0^T |\frac{1}{\gamma^2}D^*P\bar{x}|^2 dt \leq C_0(|h|^2 + |k|^2).$$

Since T is arbitrary, we have proved that $D^*P\bar{x} \in L^2(-\infty; +\infty; W)$. Since $\mathcal{A}$ is exponentially stable, this suffices (see (5. 5)) to prove (5. 6).
-b:Consider the system

(5. 13) $$\begin{array}{rcl} x' & = & \left(A + \frac{1}{\gamma^2}DD^*P\right)x + BLp + Dw \\ p' & = & (A+M)p + NCx + N\eta \\ x(0) & = & 0 \\ p(0) & = & 0. \end{array}$$

then we shall prove

(5. 14) $$\sup_{w,\eta} \frac{\int_0^\infty |B^*Px + Lp|^2 dt}{\int_0^\infty (|w|^2 + |\eta|^2) dt} < \gamma^2.$$

Computing $\frac{d}{dt}(Px, x)$ and integrating between 0 and T, then letting $T \to \infty$ we obtain the relation

$$\int_0^\infty |B^*Px + Lp|^2 dt - \gamma^2 \int_0^\infty (|w|^2 + |\eta|^2) dt =$$
$$\int_0^\infty (|Hx|^2 + |Lp|^2) dt - \gamma^2 \int_0^\infty \left(|w + \frac{1}{\gamma}D^*Px|^2 + |\eta|^2 \right) dt$$

Next, as above using the basic assumption $\rho(L, M, N) < \gamma^2$, we can write

$$\int_0^\infty (|Hx|^2 + |Lp|^2) dt \leq (\gamma^2 - \delta^2) \int_0^\infty \left(|w + \frac{1}{\gamma^2}D^*Px|^2 + |\eta|^2 \right) dt.$$

Therefore combining the two above relations yields

$$(5.15) \quad \frac{\int_0^\infty |B^*Px + Lp|^2 \, dt}{\int_0^\infty (|w|^2 + |\eta|^2) \, dt} \leq \gamma^2 - \delta^2 \frac{\int_0^\infty \left(|w + \frac{1}{\gamma^2} D^*Px|^2 + |\eta|^2\right) dt}{\int_0^\infty (|w|^2 + |\eta|^2) \, dt}.$$

Now let us check that

$$(5.16) \quad \frac{\int_0^\infty \left(|w + \frac{1}{\gamma^2} D^*Px|^2 + |\eta|^2\right) dt}{\int_0^\infty (|w|^2 + |\eta|^2) \, dt} \geq c_0$$

where c_0 is a positive constant. Indeed consider

$$(5.17) \quad \begin{array}{rcl} x_0' & = & Ax_0 + BLp_0 + Dw_0 \\ p_0' & = & (A + M)p_0 + NCx_0 + N\eta \\ x_0(0) & = & 0 \\ p_0(0) & = & 0. \end{array}$$

Then from the exponential stability of $\mathcal{A}$ we have among other things

$$\int_0^\infty \left(|w_0 - \frac{1}{\gamma^2} D^*Px_0|^2 + |\eta|^2\right) dt \leq c_1 \int_0^\infty (|w_0|^2 + |\eta|^2) \, dt.$$

Applying this estimate with

$$w_0 = w + \frac{1}{\gamma^2} D^*Px \quad x_0 = x \quad p_0 = p$$

where w, x, p correspond to (5.13) we deduce immediately (5.16) with $c_0 = 1/c_1$. Note that we can take c_0 as small as we wish, yet strictly positive. In particular $\gamma^2 - \delta^2 c_0 > 0$. Using (5.16) in (5.15) we easily deduce the desired property (5.14).

-c:duality considerations We again use the duality considerations of paragraph 4.3, see Proposition 4.1. Let $\Xi = X \times X$, $\Phi = W \times Y$, and this time $\Psi = U$. Let then

$$\mathcal{A} = \begin{pmatrix} A + \frac{1}{\gamma^{*2}} DD^*P & BL \\ NC & A + M \end{pmatrix}$$

and

$$\mathcal{B} = \begin{pmatrix} D & 0 \\ 0 & N \end{pmatrix} \quad \mathcal{C} = \begin{pmatrix} B^*P & L \end{pmatrix}.$$

The dual system is (apply (4.8) and set $\zeta = m, q, \psi = \mu$)

$$(5.18) \quad \begin{array}{rcl} m' & = & (A^* + \frac{1}{\gamma^{*2}} PDD^*)m + C^*N^*q + PB\mu \\ q' & = & L^*B^*m + (A^* + M^*)q + L^*\mu \\ m(0) & = & 0 \\ q(0) & = & 0. \end{array}$$

Using Proposition 4.1 and (5.14) we can assert that

$$(5.19) \quad \sup_\mu \frac{\int_0^\infty (|D^*m|^2 + |N^*q|^2) \, dt}{\int_0^\infty |\mu|^2 \, dt} < \gamma^2.$$

Consider the dynamic system

(5. 20)
$$\begin{aligned} m' &= (A^* + \tfrac{1}{\gamma^{*2}}PDD^*)m + C^*v + PB\mu \\ m(0) &= 0 \end{aligned}$$

where v is the control and μ is the perturbation. We observe that the pair $A^* + \tfrac{1}{\gamma^{*2}}PDD^*, C^*$ is stabilizable, as a consequence of the fact that the operator $\mathcal{A}_P$, hence its dual $\mathcal{A}_P^*$ is exponentially stable (see part -a of the present proof and beware of the fact $\mathcal{A}_P$ has been designated now by $\mathcal{A}$ by consistency with the generic notation used when dealing with duality considerations). Consider also the cost function

$$\mathcal{K}_0(v,\mu) = \int_0^\infty (|D^*m|^2 + |v|^2)\,dt.$$

¿From the assumption that A^*, D^* is detectable it follows that the pair $A^* + \tfrac{1}{\gamma^{*2}}PDD^*, D^*$ is detectable. Using (5. 19) we can assert that the γ^2 robustness property holds for the system (5. 20) and the cost function $\mathcal{K}_0(v,\mu)$. Therefore, we may rely on Theorem 3.1 to obtain the existence and uniqueness of a self adjoint operator $\Pi \in \mathcal{L}(X,X) \geq 0$, solution of the Riccati equation

(5. 21)$$\Pi(A^* + \tfrac{1}{\gamma^2}PDD^*) + (A + \tfrac{1}{\gamma^2}DD^*P)\Pi - \Pi(C^*C - \tfrac{1}{\gamma^2}PBB^*P)\Pi + DD^* = 0$$

and $A^* + \tfrac{1}{\gamma^2}PDD^* - (C^*C - \tfrac{1}{\gamma^2}PBB^*P)\Pi$ is exponentially stable. Note also that as in Lemma 3.1 we have also $A^* + \tfrac{1}{\gamma^2}PDD^* - C^*C\Pi$ is exponentially stable.
- d:algebraic manipulations We check here that

(5. 22)
$$\Sigma = \Pi(I - \tfrac{1}{\gamma^2}P\Sigma)$$

If (5. 22) is true then clearly (using also the symmetry of Σ)

$$(I - \tfrac{1}{\gamma^2}P\Sigma)(I + \tfrac{1}{\gamma^2}P\Pi) = (I + \tfrac{1}{\gamma^2}P\Pi)(I - \tfrac{1}{\gamma^2}P\Sigma) = I$$

which proves that $I - \tfrac{1}{\gamma^2}P\Sigma$ is invertible. Moreover

$$\Pi = \Sigma(I - \tfrac{1}{\gamma^2}P\Sigma)^{-1} \geq 0$$

and the proof of (4. 6) will then be complete. To prove (5. 22) we proceed by algebraic manipulations combining the Riccati equations of P, Σ, Π namely (4. 4),(4. 5) and (5. 21). To simplify notation write

$$\Lambda = -\Sigma + \Pi(I - \tfrac{1}{\gamma^2}P\Sigma)$$

$$A_1 = A^* - (C^*C - \tfrac{1}{\gamma^2}H^*H)\Sigma$$

$$A_2 = A^* + \tfrac{1}{\gamma^2}PDD^* - (C^*C - \tfrac{1}{\gamma^2}PBB^*P)\Pi$$

Then we can check after an easy calculus

(5. 23)
$$\Lambda A_1 + A_2^*\Lambda = 0.$$

Note that A_1, A_2 are exponentially stable. Then the relation (5. 23) implies $\Lambda = 0$. Indeed consider for $h \in D(A_1), k \in D(A_2^*)$ the solutions x_1, x_2 of

$$x_1' = A_1 x_1 \quad ; x_1(0) = h$$

$$x_2' = A_2 x_2 \quad ; x_2(0) = k$$

then thanks to (5. 23) we check immediately that

$$\frac{d}{dt}(\Lambda x_1, x_2) = 0$$

hence $(\Lambda x_1, x_2)$ is constant. Since it vanishes at infinity by virtue of the exponential stability, it is also 0 at $t = 0$. Hence $(\Lambda h, k) = 0$ which extends by density to all values of h, k. Therefore $\Lambda = 0$ and (5. 22) has been proven.

5.2 Sufficient Conditions

So we assume that (4. 4),(4. 5),(4. 6) hold. We set

$$\Pi = \Sigma(I - \frac{1}{\gamma^2} P\Sigma)^{-1}.$$

Note that Π is symmetric. This follows from the relation

$$\Sigma(I - \frac{1}{\gamma^2} P\Sigma) = (I - \frac{1}{\gamma^2} \Sigma P)\Sigma.$$

By assumption we know that $\Pi \geq 0$. Moreover using the notation A_1, A_2, Λ as above, we have the relation (5. 23) since $\Lambda = 0$. After using the equations of Σ and P we deduce easily that the left hand side of (5. 21) multiplied to the right by $I - \frac{1}{\gamma^2} P\Sigma$ vanishes. ¿From the invertibility of $I - \frac{1}{\gamma^2} P\Sigma$ we deduce that Π is a solution of the left hand side of (5. 21). Details to make precise this formal calculation are left to the reader. Note also that

$$A_2 = (I - \frac{1}{\gamma^2} P\Sigma) A_1 (I - \frac{1}{\gamma^2} P\Sigma)^{-1}.$$

Consider the equation

$$x' = A_2 x \quad ; x(0) = h.$$

Then setting

$$x_1 = (I - \frac{1}{\gamma^2} P\Sigma)^{-1} x$$

we have

$$x_1' = A_1 x_1 \quad ; x_1(0) = (I - \frac{1}{\gamma^2} P\Sigma)^{-1} h.$$

Since we know that A_1 is exponentially stable we deduce that $x_1 \in L^2(-\infty; +\infty; X)$ hence also $x \in L^2(-\infty; +\infty; X)$. Therefore we get that A_2 is exponentially stable. Hence the operator Π satisfies all the properties stated in the part -c of the proof of (4. 6) in the necessary conditions, paragraph 5.1.

We define next L, M, N as in the statement of Theorem 4.1, definition of the feedback controller and we shall prove that this feedback controller satisfies the conditions of γ^2

robustness property(with partial observation), as stated Definition 4.1. We associate to the triple L, M, N the matrix operator $\mathcal{A}$ as in the proof of (4. 4) in paragraph 5.1. With the present choice of L, M, N it amounts to

$$\mathcal{A} = \begin{pmatrix} A & -BB^*P \\ \Pi C^*C & A - (BB^* - \frac{1}{\gamma^2}DD^*)P - \Pi C^*C \end{pmatrix}$$

We must prove that
(5. 24) $\qquad\qquad\qquad \mathcal{A}$ is exponentially stable.

We decompose the proof in several steps.
-a: The matrix operator

$$\mathcal{A}_P = \begin{pmatrix} A + \frac{1}{\gamma^{*2}}DD^*P & -BB^*P \\ \Pi C^*C & A - (BB^* - \frac{1}{\gamma^2}DD^*)P - \Pi C^*C \end{pmatrix}$$

is exponentially stable. Consider indeed the dynamic system

(5. 25)
$$\begin{aligned} x' &= (A + \tfrac{1}{\gamma^{*2}}DD^*P)x - BB^*Pp \\ p' &= \left(A - (BB^* - \tfrac{1}{\gamma^2}DD^*)P - \Pi C^*C\right)p + \Pi C^*Cx \\ x(0) &= h \\ p(0) &= k. \end{aligned}$$

Setting
$$\xi = x - p$$
we see that ξ is the solution of

(5. 26)
$$\begin{aligned} \xi' &= (A + \tfrac{1}{\gamma^2}DD^*P - \Pi C^*C)\xi \\ \xi(0) &= h - k. \end{aligned}$$

Since $A + \frac{1}{\gamma^2}DD^*P - \Pi C^*C$ is exponentially stable we get that $\xi \in L^2(-\infty; +\infty; X)$. Next x appears as the solution of

(5. 27)
$$\begin{aligned} x' &= (A + \tfrac{1}{\gamma^2}DD^*P - BB^*P)x + BB^*P\xi \\ x(0) &= h. \end{aligned}$$

Since $A + \frac{1}{\gamma^2}DD^*P - BB^*P$ is exponentially stable we deduce that $x \in L^2(-\infty; +\infty; X)$ and thus also $p \in L^2(-\infty; +\infty; X)$. This completes the proof of -a.
- b: consider the system

(5. 28)
$$\begin{aligned} x' &= (A + \tfrac{1}{\gamma^{*2}}DD^*P)x - BB^*Pp + Dw \\ p' &= \left(A - (BB^* - \tfrac{1}{\gamma^2}DD^*)P - \Pi C^*C\right)p + \Pi C^*Cx + \Pi C^*\eta \\ x(0) &= h \\ p(0) &= k \end{aligned}$$

then one has the estimate

(5. 29)
$$\begin{aligned} \int_0^\infty (|Hx|^2 + |B^*Pp|^2)\,dt &- \gamma^2 \int_0^\infty \left(|w + \tfrac{1}{\gamma^{*2}}D^*Px|^2 + |\eta|^2\right) dt \\ &< -\delta_0^2 \int_0^\infty (|w|^2 + |\eta|^2)\,dt + C_0(|h|^2 + |k|^2) \end{aligned}$$

where δ_0 and C_0 are appropriate positive constants. To prove the estimate (5. 29) we shall exploit the Riccati equation

(5. 21) whose solution is Π. Consider the system

$$
\begin{aligned}
(5.\ 30) \qquad m' &= (A^* + \tfrac{1}{\gamma^{*2}}PDD^*)m + C^*v + PB\mu \\
m(0) &= 0
\end{aligned}
$$

where v is the control and μ is the perturbation. The pair $A^* + \tfrac{1}{\gamma^{*2}}PDD^*, C^*$ is stabilizable. Consider the cost function

$$\mathcal{K}_0(v,\mu) = \int_0^\infty (|D^*m|^2 + |v|^2)\, dt.$$

The pair $A^* + \tfrac{1}{\gamma^{*2}}PDD^*, D^*$ is detectable. The existence of Π implies that the γ^2 robustness property with full observation holds for the system (5. 30) and the cost function $\mathcal{K}_0(v,\mu)$. In fact consider

$$
\begin{aligned}
(5.\ 31) \qquad m' &= (A^* + \tfrac{1}{\gamma^{*2}}PDD^* - C^*C\Pi)m + PB\mu \\
m(0) &= 0
\end{aligned}
$$

then one has the property

$$\sup_\mu \frac{\int_0^\infty (|D^*m|^2 + |C\Pi m|^2)\, dt}{\int_0^\infty |\mu|^2\, dt} < \gamma^2.$$

Using duality considerations we introduce the dual system

$$
\begin{aligned}
(5.\ 32) \qquad \xi' &= (A + \tfrac{1}{\gamma^2}DD^*P - \Pi C^*C)\xi + Dw + \Pi C^*\eta \\
\xi(0) &= 0.
\end{aligned}
$$

Then we can assert the estimate

$$(5.\ 33) \qquad \sup_{w,\eta} \frac{\int_0^\infty |B^*P\xi|^2\, dt}{\int_0^\infty (|w|^2 + |\eta|^2)\, dt} < \gamma^2.$$

In particular, we can find $\delta < \gamma$ such that

$$(5.\ 34) \qquad \frac{\int_0^\infty |B^*P\xi|^2\, dt}{\int_0^\infty (|w|^2 + |\eta|^2)\, dt} \leq \gamma^2 - \delta^2.$$

Consider now the system (5. 28) for initial values $h = 0; k = 0$. We denote by x_0, p_0 the corresponding solution. We see that the solution ξ of (5. 32) is equal to $x_0 - p_0$. We shall also use the following relation

$$\int_0^\infty \left(|Hx_0|^2 + |B^*Pp_0|^2 - |B^*P\xi|^2 + \gamma^2(|w|^2 - |w + \tfrac{1}{\gamma^2}D^*Px_0|^2)\right) dt = 0$$

which is obtained by computing $\tfrac{d}{dt}(Px_0, x_0)$ and integrating between 0 and ∞. In this equality, we make use of the estimate (5. 34) to obtain

(5. 35)
$$\int_0^\infty \left(|Hx_0|^2 + |B^*Pp_0|^2 - \gamma^2(|\eta|^2 + |w + \tfrac{1}{\gamma^2}D^*Px_0|^2)\right) dt \leq -\delta^2 \int_0^\infty (|w|^2 + |\eta|^2)\, dt.$$

Introduce $x_1 = x - x_0, p_1 = p - p_0$ which depend only on h, k and not on w, η. We replace in (5.29) x by $x_0 + x_1$ and p by $p_0 + p_1$. Using inequalities like $|Hx|^2 \leq (1+\epsilon)|Hx_0|^2 + (1+\frac{1}{\epsilon})|Hx_1|^2$ where ϵ is arbitrarily small and making use of (5.35) we easily deduce the desired estimate (5.29).

- c:Proof of (5.24) Consider the system

(5.36)
$$\begin{aligned} \bar{x}' &= A\bar{x} - BB^*P\bar{p} \\ \bar{p}' &= \left(A - (BB^* - \frac{1}{\gamma^2}DD^*)P - \Pi C^*C\right)\bar{p} + \Pi C^*C\bar{x} \\ \bar{x}(0) &= h \\ \bar{p}(0) &= k. \end{aligned}$$

then we shall prove
(5.37) $$\bar{x}, \bar{p} \in L^2(-\infty; +\infty; X).$$

Let us denote by $x_{h,k}, p_{h,k}$ the solution of

(5.38)
$$\begin{aligned} x' &= (A + \frac{1}{\gamma^{*2}}DD^*P)x - BB^*Pp \\ p' &= \left(A - (BB^* - \frac{1}{\gamma^2}DD^*)P - \Pi C^*C\right)p + \Pi C^*Cx \\ x(0) &= h \\ p(0) &= k \end{aligned}$$

and for any T define the system

$$x_T(t) = \begin{vmatrix} \bar{x}(t) & \text{if } t \leq T \\ x_{\bar{x}_T, \bar{p}_T}(t-T) & \text{if } t > T \end{vmatrix} \quad p_T(t) = \begin{vmatrix} \bar{p}(t) & \text{if } t \leq T \\ p_{\bar{x}_T, \bar{p}_T}(t-T) & \text{if } t > T \end{vmatrix}$$

and

$$w_T(t) = \begin{vmatrix} -\frac{1}{\gamma^2}D^*P\bar{x}(t) & \text{if } t < T \\ 0 & \text{if } t > T \end{vmatrix}$$

Clearly the pair $x_T(.), p_T(.)$ is the solution of the system (5.28) corresponding to the perturbation $w(.) = w_T(.), \eta = 0$. Hence we may apply the estimate (5.29) to assert that

(5.39)
$$\int_0^\infty (|Hx_T|^2 + |B^*Pp_T|^2)\,dt - \gamma^2 \int_0^\infty |w_T + \frac{1}{\gamma^{*2}}D^*Px_T|^2\,dt$$
$$< -\delta_0^2 \int_0^\infty |w_T|^2\,dt + C_0(|h|^2 + |k|^2)$$

Now from (5.38) it follows the relation

(5.40)
$$\int_0^\infty (|Hx_{h,k}|^2 + |B^*Pp_{h,k}|^2)\,dt - \gamma^2 \int_0^\infty |\frac{1}{\gamma^{*2}}D^*Px_{h,k}|^2\,dt$$
$$= (Ph, h) + \int_O^\infty |B^*P(x_{h,k} - p_{h,k})|^2\,dt$$

We apply (5.40) with $h = \bar{x}_T, k = \bar{p}_T$ and use the fact

$$\int_T^\infty \left(|Hx_T|^2 + |B^*Pp_T|^2 - \gamma^2|w_T + \frac{1}{\gamma^{*2}}D^*Px_T|^2\right)dt$$
$$= \int_0^\infty \left(|Hx_{\bar{x}_T,\bar{p}_T}|^2 + |B^*Pp_{\bar{x}_T,\bar{p}_T}|^2 - \gamma^2|\frac{1}{\gamma^{*2}}D^*Px_{\bar{x}_T,\bar{p}_T}|^2\right)dt..$$

Hence we have

$$\int_T^\infty (|Hx_T|^2 + |B^*Pp_T|^2) \, dt - \gamma^2 \int_T^\infty |w_T + \frac{1}{\gamma^2} D^*Px_T|^2 \, dt$$
(5. 41)
$$= (P\bar{x}_T, \bar{x}_T) + \int_O^\infty |B^*P(x_{\bar{x}_T,\bar{p}_T} - p_{\bar{x}_T,\bar{p}_T})|^2 \, dt$$

Therefore we deduce from (5. 39) and (5. 41) the estimate

$$\int_0^\infty (|Hx_T|^2 + |B^*Pp_T|^2) \, dt - \gamma^2 \int_0^\infty |w_T + \frac{1}{\gamma^{*2}} D^*Px_T|^2 \, dt = \int_0^T (|Hx_T|^2 + |B^*Pp_T|^2) \, dt + (P\bar{x}_T, \bar{x}_T)$$
(5. 42)
$$+ \int_O^\infty |B^*P(x_{\bar{x}_T,\bar{p}_T} - p_{\bar{x}_T,\bar{p}_T})|^2 \, dt < -\delta_0^2 \int_0^\infty |w_T|^2 \, dt + C_0(|h|^2 + |k|^2)$$

Among other things it follows from (5. 42) that

$$\int_0^T |D^*P\bar{x}|^2 \, dt \leq C_1(|h|^2 + |k|^2).$$

Letting T tend to ∞ we get $D^*P\bar{x} \in L^2(-\infty; +\infty; W)$. Since the pair $\bar{x}, \bar{p}$ appears as the solution of (5. 28) with values

$$w = -\frac{1}{\gamma^{*2}} D^*P\bar{x} \quad , \eta = 0$$

it follows that $\bar{x}, \bar{p} \in L^2(-\infty; +\infty; X)$ and (5. 37) hence (5. 24) is proven.

To complete the proof of the γ^2 robustness property (with partial observation) for the triple L, M, N consider the system

(5. 43)
$$\begin{aligned} x' &= Ax - BB^*Pp + Dw \\ p' &= \left(A - (BB^* - \frac{1}{\gamma^2}DD^*)P - \Pi C^*C\right)p + \Pi C^*Cx + \Pi C^*\eta \\ x(0) &= 0 \\ p(0) &= 0 \end{aligned}$$

we must prove

(5. 44)
$$\sup_{w,\eta} \frac{\int_0^\infty (|Hx|^2 + |B^*Px|^2) \, dt}{\int_0^\infty (|w|^2 + |\eta|^2) \, dt} < \gamma^2.$$

Apply the estimate (5. 29) to (5. 43) with $w = w - \frac{1}{\gamma^2} D^*Px$ and $h = 0, k = 0$ to obtain

(5. 45)
$$\int_0^\infty \left(|Hx|^2 + |B^*Pp|^2 - \gamma^2(|w|^2 + |\eta|^2)\right) dt \leq -\delta_0^2 \int_0^\infty \left(|w - \frac{1}{\gamma^2} D^*Px|^2 + |\eta|^2\right) dt$$

hence

$$\sup_{w,\eta} \frac{\int_0^\infty (|Hx|^2 + |B^*Px|^2) \, dt}{\int_0^\infty (|w|^2 + |\eta|^2) \, dt} \leq$$

$$\gamma^2 - \delta_0^2 \inf_{w,\eta} \frac{\int_0^\infty \left(|w - \frac{1}{\gamma^2} D^*Px|^2 + |\eta|^2\right) dt}{\int_0^\infty (|w|^2 + |\eta|^2) \, dt}$$

Using the exponential stability of $\mathcal{A}_P$ we can check as done previously that

$$\inf_{w,\eta} \frac{\int_0^\infty \left(|w - \frac{1}{\gamma^2}D^*Px|^2 + |\eta|^2\right) dt}{\int_0^\infty (|w|^2 + |\eta|^2) dt} > 0.$$

Therefore (5. 44) has been proven.
The proof of Theorem 4.1 has been completed.

References

[1] Bert van Keulen, Marc Peters and Ruth Curtain, H_∞ Control with state feedback :The infinite dimensional case, W-9015, University of Gronigen.
[2] Bert van Keulen, H_∞ Control with measurement feedback for linear infinite-dimensional systems,W-9103,University of Gronigen.
[3] Richard Datko ,Extending a theorem of A.M. Liapunov to Hilbert space, *Journal of Mathematical analysis and applications*,vol.32,pp. 610-616,1970.
[4] Tamer Başar, Pierre Bernhard, H_∞ *Optimal Control and Related Minimax Design Problems. A Dynamic Game Approach*, Birkhäuser, Boston, 1991.
[5] A. Bensoussan, G. Da Prato, M. Delfour, S. Mitter, *Infinite Dimensional System Theory*, Birkhäuser, Boston, 1992.
[6] Pierre Bernhard, Linear quadratic two-person zero-sum differential games: necessary and sufficient conditions, *Journal of Optimization Theory and Applications* vol 27, pp.51–69, 1979.
[7] John Doyle, Keith Glover, Pramod Khargonekar, and Bruce Francis, State-space solutions to standard $\mathcal{H}_2$ and $\mathcal{H}_\infty$ control problems. *IEEE Transactions on Automatic Control*, vol AC-34, pp.831–847, 1989.
[8] E. F. Mageirou, Values and strategies for infinite duration linear quadratic games, *IEEE Transactions on Automatic Control*, vol AC-21, pp.547–550, 1976.

Chapter 10
Pertubation of Well-Posed Systems by State Feedback[1]

K. A. Morris[*]

Abstract

Well-posed systems (A, B, C) remain well-posed after state-feedback is introduced. In general, however, the output operator of the controlled system is not C.

1 Introduction

Consider the system
$$\dot{x}(t) = Ax(t) + Bu(t), \qquad x(0) = x_o. \tag{1}$$

The output of the free system ($u = 0$) is described by an operator C so that, formally,
$$y(t) = Cx(t). \tag{2}$$

The operator A generates a strongly continuous semigroup of operators $S(t)$ on a Hilbert space $\mathcal{X}$, so A is closed with domain dom(A) dense in $\mathcal{X}$. For the case where boundary control is used, the operator B is not generally a bounded operator into $\mathcal{X}$. Similarly, if point sensing is used, the operator C will often be unbounded. The class of *well-posed* systems [9], defined in the next section, is a general framework for systems with unbounded control and/or observation. Pritchard-Salamon systems [7] and systems with bounded control and sensing are included in this class.

Suppose we have a well-posed system (A, B, C). Does this system remain well-posed after state feedback, $u = -Kx$, or more realistically, $u = -Kx + v$, is introduced? More precisely, we consider the well-posedness of the system (1,2) with $u = -Kx+v$. A number of papers have examined the problem of when the well-posedness of (A, B, C) implies that $(A - BK, B, C)$ is well-posed. For example, in [9] well-posed systems are defined, and the problem is studied fairly generally. Curtain [4] studies control of a special class of systems, Pritchard-Salamon systems, while [2] examines a slightly wider class of systems.

In this paper the well-posedness of the controlled system is examined, and a complete answer to the problem of perturbation of well-posed systems by state feedback for the case of bounded feedback is obtained. It is also shown that $(A - BK, B, C)$ is not, in general, a state-space realisation of this controlled system.

---

[1]This research was partially supported by the Fields Institute, which is funded by grants from Ontario Ministry of Colleges and Universities and the Natural Sciences and Engineering Research Council of Canada, and by a grant from the Natural Sciences and Engineering Research Council of Canada.

[*]Department of Applied Mathematics, University of Waterloo, Waterloo, Ontario N2L 3G1.

2 Well-Posed Systems

Throughout this paper it will be assumed that the input space U and output space Y are separable Hilbert spaces. The state space $\mathcal{X}$ is a Hilbert space. We begin with a definition of what we mean by *well-posed*.

Definition 2.1 *We say that the system defined by the triple (A, B, C) is* well-posed *if the following holds for any $T \geq 0$:*

- *(s0)* There exist three Hilbert spaces $\mathcal{W} \hookrightarrow \mathcal{X} \hookrightarrow \mathcal{V}$ and A generates a C_0-semigroup on all three. The semigroup on $\mathcal{V}$ restricts to the semigroup on $\mathcal{X}$, which in turn restricts to the semigroup on $\mathcal{W}$ and we denote all three by the same symbol $S(t)$.

- *(s1)* $B \in \mathcal{L}(U, \mathcal{V})$ and for some $D, \bar{D} = L_2(0, T; U)$ the controllability map

$$\mathcal{B}(T)u := \int_0^T S(T - s) B u(s) ds$$

is well-defined and bounded from $D \subset L_2(0, T; U)$ to $\mathcal{X}$.

- *(s2)* $C \in \mathcal{L}(\mathcal{W}, Y)$ and for $x \in \mathcal{W}$ the observability map

$$\mathcal{C}(T)x := CS(\cdot)x$$

is bounded from $\mathcal{W} \subset \mathcal{X}$ to $L_2(0, T; Y)$.

- *(s3)* For $u \in D_1, \bar{D}_1 = L_2(0, T; U)$, there exists a suitable input/output map $\mathcal{G}(T) : u \mapsto y$ compatible with (1) and (2) such that for some $k_T > 0$,

$$\|y\|_{L_2(0,T;Y)} \leq k_T \|u\|_{L_2(0,T;U)}.$$

Statement (s1) implies that $\mathcal{B}(T)$ has a unique extension to a bounded operator from $L_2(0, T; U)$ to $\mathcal{X}$, which is also denoted by $\mathcal{B}(T)$. Similarly, (s2) implies that $\mathcal{C}(T)$ can be extended to a bounded operator from $\mathcal{X}$ to $L_2(0, T; Y)$.

Assumptions (s0)–(s3) ensure that the control system is consistent with a general definition of a stationary dynamical system in state-space form [13]. This formulation is sufficiently general to include many cases of point sensing and boundary control.

Weiss [11] has shown that if the observability map is bounded (s2), C has an extension to a bounded operator from $[\text{dom}(A)]$ to Y, where $[\text{dom}(A)]$ denotes the Hilbert space $\text{dom}(A)$ taken with the graph norm. Thus we may take $\mathcal{W} = [\text{dom}(A)]$. Similarly, if the controllability map satisfies (s1), B has an extension to a bounded operator from U to $[\text{dom}(A^*)]'$ and so we may take $\mathcal{V} = [\text{dom}(A^*)]'$ [10]. Here Z' denotes the conjugate dual of Z.

Formally, the system is described by (1) and (2). However, since B and C are unbounded, $x(t)$ will not always be in the domain of C, and so (2) is not an appropriate description of the input/output map. Since, for any $\mu \in \rho(A)$,

$$x(t) = (\mu - A)^{-1}(\mu x(t) - \dot{x}(t)) + (\mu - A)^{-1} B u(t), \tag{3}$$

this suggests

$$y(t) := C(\mu - A)^{-1}(\mu x(t) - \dot{x}(t)) + G(\mu) u(t), \tag{4}$$

for some $G(\mu) \in \mathcal{L}(U, Y)$, as a logical extension of (2).

Salamon [9] showed that (s0)–(s3) plus the compatibility condition,

$$G(s) - G(\mu) = C(\mu - s)(\mu - A)^{-1}(s - A)^{-1} B, \qquad s, \mu \in \rho(A), \tag{5}$$

were sufficient to ensure that (4) is independent of the choice of μ and is well-defined. However, it is shown in [1, 3] that the compatibility condition is redundant. Conditions (s0)–(s3) are sufficient to ensure that (4) is well-defined. As for the case where B and C are bounded, the "generalised transfer function" $G(s)$ is precisely the Laplace transform of the input-output map.

In many cases of interest one considers *bounded* control systems (A, B, C), i.e., cases where $B \in \mathcal{L}(U, \mathcal{X})$ and $C \in \mathcal{L}(\mathcal{X}, Y)$, which are more restrictive (but still useful) hypotheses than the boundedness statements of (s1) and (s2). In these cases, the input/output map from $L_2(0, T; U)$ to $L_2(0, T; Y)$ is just a convolution of a function with the input and the transfer function $G(s)$ has the familiar form $CR(s; A)B$.

Reference to a well-posed system (A, B, C) will henceforth be understood to refer to the equations (1),(4).

3 Abstract Linear Systems

A well-posed system can also be defined without explicitly defining the operators A, B, C. The following definition, of an *abstract linear system* is given in ([12], Defn. 1.1) .

Let Z be any Hilbert space, and consider some function $z(t) \in Z$ for almost all t. Define the projection operator
$$P_T z := \begin{cases} z(t) & t < T \\ 0 & t \geq T \end{cases}$$

We say that $z \in L_{ps}(0, \infty; Z)$ if $P_T z \in L_p(0, \infty; Z)$ for all $T < \infty$.

We define the τ-concatenation of u and v, for $u, v \in L_{2e}(0, \infty; Z)$ as follows:
$$u \underset{\tau}{\diamond} v := \begin{cases} u(r) & 0 < r < \tau \\ v(r - \tau) & r \geq \tau \end{cases}$$

Definition 3.1 *Let $U, \mathcal{X}$ and Y be Hilbert spaces, and let $\mathcal{U} := L_2(0, \infty; U), \mathcal{Y} := L_2(0, \infty; Y)$. An abstract linear system on $\mathcal{X}, \mathcal{U}, \mathcal{Y}$ is a quadruple $\Sigma = (S, \mathcal{B}, \mathcal{C}, \mathcal{G})$ where for $t, \tau \geq 0$,*

(a0) $S(t)$, $t \geq 0$ *is a strongly continuous semigroup of bounded linear operators on $\mathcal{X}$.*

(a1) $\mathcal{B}(t)$, $t \geq 0$ *is a family of bounded operators from $\mathcal{U}$ to $\mathcal{X}$ with $\mathcal{B}(0) = 0$ such that for $u, v \in \mathcal{U}$,*
$$\mathcal{B}(t + \tau)(u \underset{\tau}{\diamond} v) = S(t)\mathcal{B}(\tau)u + \mathcal{B}(t)v.$$

(a2) $\mathcal{C}(t)$, $t \geq 0$ *is a family of bounded operators from $\mathcal{X}$ to $\mathcal{Y}$ with $\mathcal{C}(0) = 0$ such that for any $x \in \mathcal{X}$,*
$$\mathcal{C}(t + \tau)x = \mathcal{C}(\tau)x \underset{\tau}{\diamond} \mathcal{C}(t)S(\tau)x.$$

(a3) $\mathcal{G}(t)$, $t \geq 0$ *is a family of bounded linear operators from $\mathcal{U}$ to $\mathcal{Y}$ such that for $u, v \in \mathcal{U}$,*
$$\mathcal{G}(t + \tau)(u \underset{\tau}{\diamond} v) = \mathcal{G}(\tau)u \underset{\tau}{\diamond} [\, \mathcal{C}(t)\mathcal{B}(\tau)u + \mathcal{G}(t)v \,] .$$

U is the *input space* of Σ, $\mathcal{X}$ is the *state-space*, and Y is the *output space*. The operators $\mathcal{B}$ are called *input maps*, the operators $\mathcal{C}$ are called *output maps* and the operators $\mathcal{G}$ are the *input/output maps* of Σ.

Setting $v = 0$ and $t = 0$ in the above definitions the following formulae expressing causality can be derived:
$$\mathcal{B}(\tau)P_\tau = \mathcal{B}(\tau), \qquad \mathcal{G}(\tau)P_\tau = \mathcal{G}(\tau). \tag{6}$$

Also, for $0 \leq \tau \leq T$,
$$P_\tau \mathcal{C}(T) = \mathcal{C}(\tau), \qquad P_\tau \mathcal{G}(T) = \mathcal{G}(\tau). \tag{7}$$

The linear space $L_{2e}(0,\infty;U)$ with the family of seminorms $p_n(u) = \|P_n u\|_{L_2(0,T;U)}$, $n \in N$ is a topological space. Define $L_{2e}(0,\infty;Y)$ similiarly. The operators $\mathcal{B}(T), \mathcal{G}(T)$ can be extended to $L_{2e}(0,\infty;U)$ by continuity. Also, the following limits exist in $L_{2e}(0,\infty;Y)$ for any $x \in \mathcal{X}$ and any $u \in L_{2e}(0,\infty;U)$:

$$\mathcal{C}x := \lim_{t \to \infty} \mathcal{C}(t)x, \qquad \mathcal{G}u := \lim_{t \to \infty} \mathcal{G}(t)u.$$

The formulae (7) extend to

$$P_\tau \mathcal{C} = \mathcal{C}(\tau), \qquad P_\tau \mathcal{G} = \mathcal{G}(\tau),$$

and also (s2), (s3) extend to

$$\mathcal{C}x = \mathcal{C}x \underset{\tau}{\diamond} \mathcal{C}S(\tau)x, \qquad \mathcal{G}(u \underset{\tau}{\diamond} v) = \mathcal{G}u \underset{\tau}{\diamond} (\mathcal{C}\mathcal{B}(\tau)u + \mathcal{G}v).$$

Further details may be found in [12]. We will henceforth use the notation $\mathcal{C}(t)x$ to indicate a function defined on $L_2(0,t;Y)$ and $(\mathcal{C}x)(r) = (\mathcal{C}(t)x)(r)$ to indicate the value of that function at time $r \leq t$. Similar notation will be used for input/output maps.

It is clear that every well-posed system is associated with an abstract linear system. The converse is also true. Every semigroup $S(t)$ possesses a unique infinitesimal generator, A. In Weiss [10], an *abstract linear control system* was defined as a pair $(S,\mathcal{B})$ satisfying conditions (a0) and (a1) above. As mentioned above, there is a unique operator $B \in \mathcal{L}(U,\mathcal{V})$, $\mathcal{V} = [\text{dom}(A^*)]'$, associated with every abstract linear control system $(S,\mathcal{B})$ such that for every piecewise constant $u(t), t \geq 0$,

$$\mathcal{B}(t)u = \int_0^t S(t-r)Bu(r)dr.$$

This operator is called the *control operator* of the system and can be calculated for any $u \in U$ as

$$Bu := \lim_{\tau \to 0} \frac{\mathcal{B}(\tau)u}{\tau} \tag{8}$$

where the limit is taken with respect to the norm of $\mathcal{V}$.

In Weiss [11] an abstract observation system was defined as a pair $(S,\mathcal{C})$ satisfying (a0) and (a2) above. It was shown in this reference that there is a unique operator associated with every abstraction observation system. This operator $C \in \mathcal{L}(\mathcal{W},Y)$, where $\mathcal{W} = [\text{dom}(A)]$, is called the *observation operator* of the system, and satisfies for all $x \in \text{dom}(A), t > 0$,

$$\mathcal{C}(t)x = CS(t)x.$$

It is possible to give a formula for $\mathcal{C}(t)$ valid for any $x \in \mathcal{X}$, by calculating the *Lebesgue extension* of C:

$$C_L z := \lim_{\tau \to 0} \frac{1}{\tau} \int_0^\tau \mathcal{C}(r)z\, dr \tag{9}$$

with

$$\text{dom}(C_L) = \{z \in \mathcal{X} \mid \text{the above limit exists}\}.$$

Then, $\mathcal{C}(t)x = C_L S(t)x$, for $x \in \mathcal{X}$. The triple (A,B,C_L), with state space $\mathcal{X}, \mathcal{W} = [\text{dom}(A)], \mathcal{V} = [\text{dom}(A^*)]'$, obtained in this manner is a well-posed system. Thus, every abstract linear system can be associated with a well-posed system, and vice versa.

For an important class of systems the input/output map and the transfer function have a form similar to that for systems with bounded B and C.

Definition 3.2 *Let Σ be an abstract linear system, with input space U and output space Y. We say that Σ is a regular system if for any $u \in U$, the corresponding step response $y_u(\cdot)$ has a Lebesgue point at 0. That is, the following limit exists in Y:*

$$Eu = \lim_{\tau \to 0} \frac{1}{\tau} \int_0^\tau y_u(r)dr.$$

The operator E defined above is called the feedthrough operator *of the system Σ.*

Since $\frac{1}{\tau}\int_0^\tau y_u(r)dr \in \mathcal{L}(U,Y)$ for every $\tau > 0$, the Uniform Boundedness Principle implies that, if a system is regular, the feedthrough operator is bounded.

Theorem 3.1 *([12], Prop. 4.3) Let $\Sigma = (S, \mathcal{B}, \mathcal{C}, \mathcal{G})$ be an abstract linear system, with input space U, state space $\mathcal{X}$ and output space Y. Let A be the generator of S, let B be the control operator of Σ and let C be the observation operator. Denote by C_L the Lebesgue extension of C. The following conditions are equivalent:*

1. *Σ is regular,*

2. *for some $s \in \rho(A)$ and any $v \in U$, $(sI - A)^{-1}Bv \in \text{dom}(C_L)$.*

Theorem 3.2 *([12], Thm 4.5, Prop. 4.7) With the notation of the previous theorem, suppose that Σ is regular, and let E be its feedthrough operator. Then, for almost every $t \geq 0$*

$$\int_0^t S(t-r)Bu(r)dr \in \text{dom}(C_L),$$

$$(\mathcal{G}u)(t) = C_L \int_0^t S(t-r)Bu(r)dr + Eu(t)$$

where $u \in L_{2e}(0, \infty; U)$. The transfer function G of the system can be written

$$G(s) = C_L(sI - A)^{-1}B + E$$

and it is an analytic $\mathcal{L}(U,Y)$-valued function of s on $\rho(A)$. For real $s, u \in U$,

$$Eu = \lim_{s \to \infty} G(s)u. \tag{10}$$

The notation (A, B, C, E) shall henceforth be understood to imply that the quadruple (A, B, C, E) is a realisation of a regular system. In particular, it implies that for some $s \in \rho(A)$ and any $v \in U$, $(sI - A)^{-1}Bv \in \text{dom}(C)$.

The class of regular systems includes all systems where either the control operator B or the observation operator C are bounded, and Pritchard-Salamon systems [7]. In [2], the authors make two assumptions in addition to well-posedness. These assumptions imply that the class of systems considered therein is a subset of the class of regular systems. Furthermore, most well-posed systems with boundary control and/or point sensing are regular. This is shown below.

Lemma 3.3 *Let $\Sigma = (S, \mathcal{B}, \mathcal{C}, \mathcal{G})$ be an abstract system with input space U, output space Y and state-space $\mathcal{X}$. Suppose the input operator B associated with $(S, \mathcal{B})$ is injective and such that $\text{range}(B) \bigcap \mathcal{X} = \{0\}$. Then, Σ is regular.*

Proof: Let (A, B, C) be a realisation of Σ. Define the Hilbert space

$$\mathcal{Z} := \{z \in \mathcal{X} | Az + Bu \in \mathcal{X}, \text{for some } u \in U\},$$

$$\|z\|_{\mathcal{Z}}^2 = \|z\|_{\mathcal{X}}^2 + \|u\|_U^2 + \|Az + Bu\|_{\mathcal{X}}^2,$$

where u is the unique element of U such that $Az + Bu \in \mathcal{Z}$ [9].

In ([9], Sect. 2.2), it is shown that the assumptions imply that C has an extension to $K \in \mathcal{L}(\mathcal{Z}, Y)$ and furthermore, that for $\mu \in \rho(A), (\mu - A)^{-1}B \in \mathcal{L}(U, \mathcal{Z})$. Hence, by Theorem (3.1), Σ is regular. □

Any abstract system arising from a system with boundary control yields an input operator satisfying the above theorem. Thus, boundary control systems are regular. This statement is made more precise by the following theorem.

Theorem 3.4 Let $U, Y, Z, \mathcal{X}$ be Hilbert spaces with $Z \hookrightarrow \mathcal{X}$. For $\Delta \in \mathcal{L}(Z, \mathcal{X})$, $\Gamma \in \mathcal{L}(Z, U)$ and $K \in \mathcal{L}(Z, Y)$, consider the following <u>boundary control system:</u>

$$\dot{x}(t) = \Delta(t), \quad x(0) = x_o,$$
$$\Gamma x(t) = u(t),$$
$$y(t) = Kx(t).$$

It is assumed that Γ is onto, $\ker \Gamma$ is dense in $\mathcal{X}$, and that there exists $\mu \in R$ such that $\mu - \Delta$ is onto and $\ker(\mu - \Delta) \cap \ker \Gamma = \{0\}$.

If the above system is an abstract linear system with input space U, output space Y and state space $\mathcal{X}$, it is also regular.

Proof: In ([9], Prop. 2.8) it is shown that this system can be redefined as a semigroup system (A, B, C) where B is injective and $\text{range}(B) \cap \mathcal{X} = \{0\}$. This system is well-posed if and only if the original boundary control system is an abstract linear system ([9], Prop. 2.11). □

The following example illustrates these ideas.

Example:

$$\dot{x}(t) = \frac{\partial^2 x}{\partial r^2}, \quad x(\cdot, 0) = x_o,$$
$$x(0, t) = 0, \quad \frac{\partial x}{\partial r}(1, t) = u(t) - x(1, t),$$
$$y(t) = x(1, t).$$

This system is a model for control of the temperature in a rod where the temperature at $r = 0$ is fixed and the heat flux at $r = 1$ is proportional to the difference between the control $u(t)$ and the rod temperature at $r = 1$. Measurements of temperature are made at $r = 1$.

Using the notation of the previous theorem, $\mathcal{X} = L_2(0, 1)$, $Z := \{x \in H^2(0, 1) | x(0) = 0\}$, $U = Y = R$. For $x \in Z$,

$$\Delta x := \frac{\partial^2 x}{\partial r^2}, \quad \Gamma x := \frac{\partial x}{\partial r}(1) + x(1), \quad Kx := \frac{\partial x}{\partial r}(1).$$

This can easily be shown to be a boundary control system as defined above. Using ([9], Section 2) to transform this to a semigroup control system, we obtain

$$Ax := \Delta x, \quad \text{dom}(A) = \{x \in Z | \Gamma x = 0\}.$$

Note that for this example, A is self-adjoint, so $\mathcal{W} = [\text{dom}(A)]$ and $\mathcal{V} = \mathcal{W}'$. Indicating the canonical injection of $\mathcal{W}$ into Z by i,

$$C := Ki.$$

Define $B \in \mathcal{L}(U, \mathcal{V})$ as follows. Given $u \in U$, choose $x \in Z$ so that $\Gamma x = u$. Then, for any $z \in \text{dom}(A)$,

$$(Bu, z) := (\Delta x, z) - (x, Az)$$
$$= \frac{\partial x}{\partial r}(1) z(1) - x(1) \frac{\partial z}{\partial r}(1)$$
$$= u z(1).$$

Thus, $B = \delta(r - 1)$.

The spectral expansion of A can be used to show that (A, B, C) is well-posed, as in [3].

Taking Laplace transforms of the partial differential equation problem, and applying the boundary conditions, we obtain that the system transfer function is

$$G(s) := \frac{\hat{y}(s)}{\hat{u}(s)} = \frac{\sqrt{s}\sinh\sqrt{s}}{\sqrt{s}\cosh\sqrt{s} + \sinh\sqrt{s}}.$$

Since this is a boundary control system, it is regular. For $u \in U$, real s, we obtain (10)

$$Eu = \lim_{s \to \infty} G(s)u = u$$

and so $E = 1$.

4 Perturbation of Well-Posed Systems by State Feedback

Suppose we have a well-posed system (A, B, C) (1, 4) on $\mathcal{X}$ with input space U and output space Y. Does this system remain well-posed after state feedback, $u = -Kx$, or more realistically, $u = -Kx + v$, is introduced? More precisely, we consider the well-posedness of the system

$$\begin{align}
\dot{x}(t) &= Ax(t) + Bu(t), \quad x(0) = x_0, \tag{11}\\
y(t) &= C(\mu - A)^{-1}(\mu x(t) - \dot{x}(t)) + D_\mu u(t), \tag{12}\\
u &= -Kx + v, \tag{13}
\end{align}$$

with $x_o \in \mathcal{X}$ and exogenous input $v \in L_2(0, \infty; U)$. The case $K \in \mathcal{L}(\mathcal{X}, U)$ is considered.

Theorem 4.1 *Consider a well-posed system (A, B, C) with state space $\mathcal{X}$, input space U, output space Y and some operator $K \in \mathcal{L}(\mathcal{X}, U)$. The controlled system (11-13) is an abstract linear system.*

Proof: We associate (A, B, C) with an abstract linear system as follows. Let $S(t)$ be the semigroup generated by A, $\mathcal{B}(t)$ the controllability operator defined in (s1), $\mathcal{C}(t)$ the observability operator defined in (s2) and $\mathcal{G}(t)$ the input/output map defined in (s3). Considering (11-13), define $S_K(t)$ to be the map from x_o to $x(t)$ with $v = 0$, $\mathcal{B}_K(t)$ to be the map from v to x where $x(0) = 0$, $\mathcal{C}_K(t)$ the map from x_0 to y with $v = 0$ and $\mathcal{G}_N(t)$ the map from v to y with $x(0) = 0$. It is required to prove that these maps are well-defined and that the quadruple $(S_K, \mathcal{B}_K, \mathcal{C}_K, \mathcal{G}_N)$ is an abstract linear system on $\mathcal{X}$.

(1) It will first be shown that S_K is the C_o-semigroup generated by $A - BK$ on $\mathcal{X}$. This is shown in [8][Thm. 1.3.7]. However, a proof similar to that of ([5], Theorem 2.31) is given here, as details will be used below. Define for any $x_o \in \mathcal{X}$, $t \in [0, t_f]$,

$$T_o(t)x_o := S(t)x_o, \qquad T_n(t)x_o := \mathcal{B}(t)[-KT_{n-1}(\cdot)x_o], \qquad n = 1, 2\ldots$$

Let M, ω be such that

$$\|S(t)\| \leq Me^{\omega t}$$

and let β be the norm of $\mathcal{B}(t_f)$ as an operator from $L_2(0, t_f; U)$ to $\mathcal{X}$.

Assume that

$$\|T_n(t)x_o\| \leq (\beta\|K\|)^n M \max(1, e^{\omega t}) \left(\frac{t^n}{n!}\right)^{\frac{1}{2}} \|x_o\|. \tag{14}$$

Then,

$$\begin{align}
\|T_{n+1}(t)x_o\| &\leq \beta \left[\int_0^t \|KT_n(\sigma)x_o\|^2 d\sigma\right]^{\frac{1}{2}}\\
&\leq \beta\|K\|(\beta\|K\|)^n M \max(1, e^{\omega t}) \left[\int_0^t \frac{\sigma^n}{n!} d\sigma\right]^{\frac{1}{2}} \|x_o\|\\
&\leq (\beta\|K\|)^{n+1} M \max(1, e^{\omega t}) \left[\frac{t^{n+1}}{(n+1)!}\right]^{\frac{1}{2}} \|x_o\|.
\end{align}$$

Since (14) is clearly true for $n = 0$, it follows by an induction argument that (14) is valid for all $n \geq 0$. Defining $a := \beta^2 \|K\|^2$ and

$$S_K(t) := \sum_{n=0}^{\infty} T_n(t),$$

$$\|S_K(t)\| \leq M \max(1, e^{\omega t}) \sum_{n=0}^{\infty} \frac{(at)^{\frac{n}{2}}}{(n!)^{\frac{1}{2}}}.$$

Using $(n!)^{\frac{1}{2}} \geq (\frac{n}{2}!)$, it can be shown that for $t \leq t_f$,

$$\|S_K(t)\| \leq M \max(1, e^{\omega t})(1 + \sqrt{at}) \exp(at). \tag{15}$$

Since t_f was arbitrary, $S_K(t) \in \mathcal{L}(\mathcal{X}, \mathcal{X})$ for every $t \geq 0$. It is easy to show that $S_K(t)$ satisfies

$$S_K(t)x_o = S(t)x_o + \mathcal{B}(t)[-KS_K(\cdot)x_o]. \tag{16}$$

The solution to (16) is unique. To show this, assume that there is another solution, $T(t)$. Defining $g(t) := \|S_K(t)x_o - T(t)x_o\|$,

$$g(t) \leq \beta \|K\| \left[\int_0^t g^2(r) dr \right]^{\frac{1}{2}}.$$

Hence, by Gronwall's inequality, $g^2(\cdot) = 0$. It is clear that $S_K(0) = I$. Using (16), and the fact that $\mathcal{B}(t)$ has property (a1), it follows that for $x_o \in \mathcal{X}, t, s \geq 0$,

$$\mathcal{B}(t+s)[-KS_K(\cdot)x_o] = S(t)\mathcal{B}(s)[-KS_K(\cdot)x_o] + \mathcal{B}(t)[-KS_K(\cdot+s)x_o],$$

and so

$$S_K(t+s)x_o = S(t)S(s)x_o + S(t)\mathcal{B}(s)[-KS_K(\cdot)x_o] + \mathcal{B}(t)[-KS_K(\cdot+s)x_o].$$

Also,

$$S_K(t)S_K(s)x_o = S(t)S(s)x_o + S(t)\mathcal{B}(s)[-KS_K(\cdot)x_o] + \mathcal{B}(t)[-KS_K(\cdot)S_K(s)x_o].$$

Therefore,

$$S_K(t+s)x_o - S_K(t)S_K(s)x_o = \mathcal{B}(t)[-KS_K(\cdot+s)x_o + KS_K(\cdot)S_K(s)x_o].$$

Defining now $g(t) := \|S_K(t+s)x_o - S_K(t)S_K(s)x_o\|$, and $\beta := \|\mathcal{B}(t_f)\|, t \leq t_f$,

$$g(t) \leq \beta \|K\| \left[\int_0^t g^2(r) dr \right]^{\frac{1}{2}}.$$

Hence, by Gronwall's inequality, and the fact that $g(0) = 0$, $g^2(\cdot) = 0$. It follows that S_K has the semigroup property.

To prove that $S_K(t)$ is strongly continuous for all $t \geq 0$, choose $h, t \geq 0$ and define $\tau = t + h$. Using (a1),

$$\begin{aligned}
\|S_K(t)x_o - S_K(\tau)x_o\| &\leq \|S(t)x_o - S(\tau)x_o\| + \|\mathcal{B}(\tau)[KS_K(\cdot)x_o] - \mathcal{B}(t)[KS_K(\cdot)x_o]\| \\
&\leq \|S(t)x_o - S(\tau)x_o\| + \|S(h)\mathcal{B}(t)[KS_K(\cdot)x_o] - \mathcal{B}(t)[KS_K(\cdot)x_o]\| \ldots \\
&\quad + \|\mathcal{B}(h)[KS_K(t+\cdot)x_o]\|.
\end{aligned} \tag{17}$$

Since S is strongly continuous, the first two terms can be made arbitrarily small by choosing h sufficiently small. Now, define $\beta = \|\mathcal{B}(t_f)\|$ where $t_f \geq h$. Since $\|\mathcal{B}(h)\|$ is a non-decreasing function of h, we have, again defining $a := \beta^2\|K\|^2$, and using (15),

$$\|\mathcal{B}(h)[KS_K(t+\cdot)x_o]\| \leq \beta\|K\|\left[\int_0^h \|S_K(t+r)x_o\|^2 dr\right]^{\frac{1}{2}}$$

$$\leq Ma^{\frac{1}{2}}\max(1,e^{\omega\tau})(1+\sqrt{a\tau})\left[\int_0^h \exp(2a(t+r))dr\right]^{\frac{1}{2}}\|x_o\|$$

$$\leq Ma^{\frac{1}{2}}\max(1,e^{\omega\tau})(1+\sqrt{a\tau})e^{at}\left[\frac{\exp(2ah)-1}{2a}\right]^{\frac{1}{2}}.$$

Thus, the last term in (17) can also be made arbitrarily small. It follows that $S_K(t)$ is a strongly right continuous function of t.

Now consider $\tau = t - h, t \geq h$, to show continuity on the left.

$$\|S_K(t-h)x_o - S_K(t)x_o\| \leq \|S(t-h)x_o - S(t)x_o\| + \|\mathcal{B}(t-h)[KS_K(\cdot+h) - KS_K(\cdot)x_o]\|...$$
$$+\|S(t-h)\mathcal{B}(h)[-KS_K(\cdot)x_o]\|. \qquad (18)$$

The first term converges to zero as h approaches 0 because of the strong continuity of $S(t)$. Also, since

$$\left[\int_0^h \|S_K(r)x_o\|^2 dr\right]^{\frac{1}{2}} \leq \|x_o\|M\max(1,e^{\omega h})(1+\sqrt{ah})\left(\frac{e^{2ah}-1}{2a}\right)^{\frac{1}{2}},$$

the third term also converges to zero. To show that the second term converges to zero, define

$$f_h(s) = \begin{cases} \|S_K(s+h)x_o - S_K(s)x_o\|^2 & 0 \leq s \leq t-h \\ 0 & t-h < s \leq t \end{cases}.$$

The functions f_h are a sequence of real-valued, integrable functions on $[0,t]$, and for each $s \in [0,t]$,

$$\lim_{h \to 0} f_h(s) = 0.$$

It follows from Lebesgue's Convergence Theorem that $\int_0^t f_h(s)ds$ converges to zero, and hence, the second term in (18) converges to zero. The left continuity of $S_K(t)$ follows.

Thus, $S_K(t)$ is a strongly continuous semigroup of bounded operators on $\mathcal{X}$, and its infinitesimal generator, A_K, is a closed operator densely defined on $\mathcal{X}$. In [8][Thm. 1.3.7] it is shown that $A_K = A - BK$.

(2) Consider now the controlled system (11-13) with $x_o = 0$ and $v \in L_2(0, \infty; U)$:

$$\dot{x}(t) = A_K x(t) + Bv(t).$$

Considering the extensions of $A, S(t), K$ to operators on $\mathcal{V}$, we see that BK is a bounded extension of A so that S_K can be extended to a semigroup on $\mathcal{V}$. Thus,

$$\mathcal{B}_K(t)v := \int_0^t S_K(t-r)Bv(r)dr$$

is well-defined as an integral in $\mathcal{V}$ for $v \in L_2(0, \infty; U)$. The approach in ([2], Lemma 2.1) can be extended to show that this operator satisfies (s1). That is, for all $T \geq 0$, $\mathcal{B}_K(T)$ has a unique extension to a bounded operator from $L_2(0, \infty; U)$ to $\mathcal{X}$. Define for $t \geq 0, u \in L_2(0, \infty; U)$, and non-negative integers n,

$$\mathcal{B}_n(t)u(\cdot) := \int_0^t T_n(t-s)Bu(s)ds.$$

This integral is well-defined as an integral with values in $\mathcal{V}$.

Assume that for u in some dense subset D of $L_2(0,\infty;U)$, $\mathcal{B}_n(t)u \in \mathcal{X}$, for arbitrary $t \geq 0$. If this is the case, then

$$\begin{aligned}\mathcal{B}(t)[-K\mathcal{B}_n(\cdot)u] &= -\int_0^t S(t-r)BK\int_0^r T_n(r-s)Bu(s)dsdr \\ &= -\int_0^t \int_s^t S(t-r)BKT_n(r-s)Bu(s)drds \\ &= -\int_0^t \int_0^{t-s} S(t-s-\sigma)BKT_n(\sigma)d\sigma Bu(s)ds \\ &= \int_0^t T_{n+1}(t-s)Bu(s)ds \\ &= \mathcal{B}_{n+1}(t)u.\end{aligned}$$

Since the assumption is true for $n=0$, it follows by induction that $\mathcal{R}(\mathcal{B}_n(t)) \subset \mathcal{X}$ for arbitrary t, n. Now, for some $\beta > 0$,

$$\|\mathcal{B}_0(t)u\| \leq \beta\|u\|_{L_2(0,t;U)},$$

and so

$$\|\mathcal{B}_1(t)u\| \leq \beta(\beta\|K\|)\, t^{\frac{1}{2}}\|u\|_{L_2(0,t;U)}$$

where $\|\mathcal{B}_n(t)u\|$ indicates the norm on $\mathcal{X}$. Assume that

$$\|\mathcal{B}_n(t)u\| \leq \beta(\beta\|K\|)^n \left(\frac{t^n}{n!}\right)^{\frac{1}{2}} \|u\|_{L_2(0,t;U)}. \tag{19}$$

Then,

$$\begin{aligned}\|\mathcal{B}_{n+1}(t)u\| &\leq \beta\left(\int_0^t \|K\int_0^r T_n(r-s)Bu(s)ds\|^2 dr\right)^{\frac{1}{2}} \\ &\leq \beta\|K\|\left(\int_0^t \|\int_0^r T_n(r-s)Bu(s)ds\|^2 dr\right)^{\frac{1}{2}} \\ &\leq \beta(\beta\|K\|)^{n+1}\left(\frac{t^{n+1}}{(n+1)!}\right)^{\frac{1}{2}}\|u\|_{L_2(0,t;U)}.\end{aligned}$$

Since (19) is true for $n=1$, it follows by induction that (19) is a bound for this sequence of operators. Defining again $a = \beta^2\|K\|^2$,

$$\begin{aligned}\|\sum_{n=0}^\infty \mathcal{B}_i(t)u(\cdot)\| &\leq \beta\sum_{n=0}^\infty \frac{(at)^{\frac{n}{2}}}{\sqrt{n!}}\|u\|_{L_2(0,t;U)} \\ &\leq \beta[(1+\sqrt{at})\exp(at)]\|u\|_{L_2(0,t;U)}.\end{aligned}$$

Then,

$$\int_0^t S_K(t-r)Bv(r)dr = \int_0^t \sum_{n=0}^\infty T_i(t-s)Bv(s)ds = \sum_{n=0}^\infty \int_0^t T_i(t-s)Bv(s)ds = \sum_{n=0}^\infty \mathcal{B}_i(t)v.$$

It follows that $\mathcal{B}_K(t)$ is a bounded operator from $L_2(0,\infty;U)$ to $\mathcal{X}$ for $t \geq 0$. Thus, $(S_K, \mathcal{B}_K)$ is an abstract control system with input operator B.

(3) The output of (11-13) with zero input v and initial state x_o is identical to that of (1,4) with initial state x_o and input $u(t) = -KS_K(t)x_o$. Thus, the output y of the controlled system with $v = 0$ is described by

$$C_K(t)x_o := C(t)x_o + \mathcal{G}(t)\left[-KS_K(\cdot)x_o\right]. \tag{20}$$

It is clear that $C_K(t) \in \mathcal{L}(\mathcal{X}, L_2(0,t;Y))$ for any $t \geq 0$. Using the fact that $C(t)$ satisfies (s2) and $\mathcal{G}(t)$ satisfies (s3), we have for $t, \tau \geq 0$,

$$\begin{aligned}
C_K(t+\tau)x_o &= C(\tau)x_o + \mathcal{G}(\tau)\left[-KS_K(\cdot)x_o\right] \underset{\tau}{\diamond} C(t)S(\tau)x_o .... \\
&\quad + C(t)B(\tau)\left[-KS_K(\cdot)x_o\right] + \mathcal{G}(t)\left[-KS_K(\tau+\cdot)x_o\right] \\
&= C_K(\tau)x_o \underset{\tau}{\diamond} C(t)S_K(\tau)x_o + \mathcal{G}(t)\left[-KS_K(\tau+\cdot)x_o\right] \\
&= C_K(\tau)x_o \underset{\tau}{\diamond} C_K(t)S_K(\tau)x_o.
\end{aligned}$$

Therefore, (S_K, C_K) is an abstract observation system.

(4) Consider (11)–(13) with $x(0) = 0$. Since $K \in \mathcal{L}(\mathcal{X}, U)$, it is now clear that $(A_K, B, -K, I)$ is a regular system and also that

$$(\mathcal{G}_D v)(\tau) := v(\tau) - K\mathcal{B}_K(\tau)v \tag{21}$$

is the input/output map of this system. The function $\mathcal{G}_D(t)v$ is the difference between the uncontrolled input v and the controlled input $-Kx(\cdot)$ where $x(\cdot)$ is the state of (11). In other words, letting u be as defined in (13), $u = \mathcal{G}_D v$. Define now for $v \in L_2(0, \infty; U)$,

$$\mathcal{G}_N(t)v := \mathcal{G}(t)\left[\mathcal{G}_D(t)v\right]. \tag{22}$$

The above operator is the map from the uncontrolled input v to the measured output y in (11-13). Since $\mathcal{G}(t)$ and $\mathcal{G}_D(t)$ are both bounded operators for any $t \geq 0$, is follows that $\mathcal{G}_N(t)$ is a bounded operator from $L_2(0, \infty; U)$ to $L_2(0, \infty; Y)$ for any $t \geq 0$.

It only remains to show that $\mathcal{G}_N$ is an input/output map satisfying (a3) and consistent with the abstract control system $(S_K, \mathcal{B}_K)$ and abstract observation system (S_K, C_K). Let C_D indicate the abstract observation operator associated with $(A_K, B, -K, I)$. Then, for $t, \tau \geq 0, u, v \in L_2(0, \infty; U)$,

$$\mathcal{G}_D(t+\tau)(u \underset{\tau}{\diamond} v) = g \underset{\tau}{\diamond} h$$

where

$$g := \mathcal{G}_D(\tau)u, \qquad h = C_D(t)\mathcal{B}_K(\tau)u + \mathcal{G}_D(t)v.$$

Now,

$$\begin{aligned}
\mathcal{B}_K(t)v &= \mathcal{B}(t)v + \sum_{n=1}^{\infty} \mathcal{B}_n(t)v \\
&= \mathcal{B}(t)v + \sum_{n=0}^{\infty} \mathcal{B}(t)[-K\mathcal{B}_n(\cdot)v] \\
&= \mathcal{B}(t)v + \mathcal{B}(t)[-K\mathcal{B}_K(\cdot)v] \\
&= \mathcal{B}(t)\mathcal{G}_D(t)v.
\end{aligned}$$

Also, note that $C_K(t)x_o = C(t)x_o + \mathcal{G}(t)C_D(t)x_o$. It follows that

$$\begin{aligned}
\mathcal{G}_N(t+\tau)(u \underset{\tau}{\diamond} v) &= \mathcal{G}(g \underset{\tau}{\diamond} h) \\
&= \mathcal{G}(\tau)\mathcal{G}_D(\tau)u \underset{\tau}{\diamond} C(t)B(\tau)[\mathcal{G}_D(\tau)u] + \mathcal{G}(t)[C_D(t)\mathcal{B}_K(\tau)u] + \mathcal{G}(t)[\mathcal{G}_D(t)v] \\
&= \mathcal{G}_N(\tau)u \underset{\tau}{\diamond} C(t)\mathcal{B}_K(\tau)u + \mathcal{G}(t)[C_D(t)\mathcal{B}_K(\tau)u] + \mathcal{G}_N(t)v \\
&= \mathcal{G}_N(\tau)u \underset{\tau}{\diamond} C_K(t)\mathcal{B}_K(\tau)u + \mathcal{G}_N(t)v.
\end{aligned}$$

Thus, the controlled system (11-13) is an abstract linear system with maps $(S_K, \mathcal{B}_K, \mathcal{C}_K, \mathcal{G}_N)$ as defined above. □

Theorem 4.2 *The notation of the previous theorem is used. The controlled system (11-13) has a realization as a well-posed system $(A - BK, B, C_K)$ where*

$$C_K x_o = \lim_{\tau \to 0} \left[\frac{1}{\tau} \int_0^\tau (Cx_o)(r) dr - \frac{1}{\tau} \int_0^\tau (\mathcal{G}[Kx_o])(r) dr \right] \quad (23)$$

with

$$\text{dom}(C_K) = \{x_o \in \mathcal{X} \mid \text{the above limit exists}\}.$$

If the original system (1) is regular with feedthrough operator E, and output operator with Lebesgue extension C_L, we can choose $C_K = C_L - EK$ with $\text{dom}(C_K) = \text{dom}(C_L)$.

Proof: Since $K \in \mathcal{L}(\mathcal{X}, U)$, it follows ([8], Thm. 1.3.9iii) that the infinitesimal generator of S_K is $A_K = A - BK$ with

$$\text{dom}(A_K) = \{x_o \in \mathcal{X} \mid (A - BK)x_o \in \mathcal{X}\}.$$

The fact that B is the input operator associated with $(S_K, \mathcal{B}_K)$ was established in the previous theorem. We have using (9),(20), that

$$C_K x_o = \lim_{\tau \to 0} \frac{1}{\tau} \int_0^\tau (Cx_o)(r) - \mathcal{G}[KS_K(\cdot)x_o](r) dr \quad (24)$$

with $\text{dom}(C_K) = \{x_o \in \mathcal{X} \mid \text{the above limit exists}\}$. Now,

$$\lim_{\tau \to 0} \frac{1}{\tau} \int_0^\tau \mathcal{G}[KS_K(\cdot)x_o](r) dr = \lim_{\tau \to 0} \frac{1}{\tau} \int_0^\tau \mathcal{G}[KS_K(\cdot)x_o - Kx_o](r) dr + \lim_{\tau \to 0} \frac{1}{\tau} \int_0^\tau \mathcal{G}[Kx_o](r) dr.$$

Since $S_K(\cdot)x_o$ is continuous, for $x_o \in \mathcal{X}$, and $\lim_{\tau \to 0} S_K(\tau)x_o - x_o = 0$,

$$\lim_{\tau \to 0} \frac{1}{\tau} \int_0^\tau \|KS_K(r)x_o - Kx_o\|^2 dr = 0.$$

It follows that

$$\begin{aligned}
\lim_{\tau \to 0} \frac{1}{\tau} \| \int_0^\tau \mathcal{G}(t)[KS_K(\cdot)x_o - Kx_o] dr \| &\leq \lim_{\tau \to 0} \frac{1}{\tau} \int_0^\tau \|\mathcal{G}(t)[KS_K(\cdot)x_o - Kx_o]\| dr \\
&\leq \lim_{\tau \to 0} \frac{\tau^{\frac{1}{2}}}{\tau} \left(\int_0^\tau \|\mathcal{G}(t)[KS_K(\cdot)x_o - Kx_o]\|^2 dr \right)^{\frac{1}{2}} \\
&\leq \lim_{\tau \to 0} \|\mathcal{G}(1)\| \left[\frac{1}{\tau} \int_0^\tau \|KS_K(r)x_o - Kx_o\|^2 dr \right]^{\frac{1}{2}} \\
&= 0.
\end{aligned}$$

Therefore,

$$C_K x_o = \lim_{\tau \to 0} \frac{1}{\tau} \left(\int_0^\tau (Cx_o)(r) dr - \int_0^\tau \mathcal{G}[Kx_o](r) dr \right)$$

with $\text{dom}(C_K) = \{x_o \in \mathcal{X} \mid \text{the above limit exists}\}$.

If the original system is regular, the second term has a well-defined limit for all $x_o \in \mathcal{X}$ and $C_K x_o = (C_L - EK)x_o$ with $\text{dom}(C_K) = \text{dom}(C_L)$ where C_L indicates the Lebesgue extension of C. □

Theorem 4.3 *The notation of Theorem (4.1) is used. The controlled system (11-13) is regular if and only if the system (1,4) is regular. If (1,4) is regular, with feedthrough operator E, then the controlled system (11-13) has feedthrough operator E.*

Proof: First, for $v \in U, s \geq 0$, define
$$u(s) := KB_K(s)v.$$

Since $(S_K, \mathcal{B}_K)$ satisfy (a0)-(a1), and K is a bounded operator, there exists a constant β such that for all $0 \leq s \leq 1$
$$\| u(s) \| \leq \beta \|K\| \sqrt{s} \| v \|.$$

Thus,
$$\lim_{\tau \to 0} \frac{1}{\tau} \int_0^\tau \|u(s)\|^2 ds = 0. \tag{25}$$

Now, for $v \in U$,
$$\lim_{\tau \to 0} \frac{1}{\tau} \int_0^\tau \mathcal{G}_N(r) v \, dr = \lim_{\tau \to 0} \frac{1}{\tau} \int_0^\tau \mathcal{G}(r)[\mathcal{G}_D(\cdot)v] dr$$
$$= \lim_{\tau \to 0} \frac{1}{\tau} \int_0^\tau \mathcal{G}(r)[v] dr - \lim_{\tau \to 0} \frac{1}{\tau} \int_0^\tau \mathcal{G}(r)[u(\cdot)] dr$$

Using (25) and the technique used in the previous theorem we obtain
$$\lim_{\tau \to 0} \frac{1}{\tau} \int_0^\tau \mathcal{G}_N(r)[v] dr = \lim_{\tau \to 0} \frac{1}{\tau} \int_0^\tau \mathcal{G}(r)[v] dr.$$

Thus, the controlled system is regular if and only if the original system is regular. If the original system is regular, with feedthrough operator E, then the controlled system has feedthrough operator E. $\square$

For non-regular systems, the second term in (23) may not have a limit for all $x_o \in \mathcal{X}$ and $\text{dom}(C_K)$ does not necessarily contain $\text{dom}(C)$. The triple $(A - BK, B, C)$ may not be well-posed.

For regular systems, which as discussed in section 2, include most systems of practical interest, and for bounded feedback K, the well-posedness of $(A - BK, B, C)$ is implied by the well-posedness of $(A - BK, B, C - EK)$. However, this triple is not a realization of the controlled system (11-13) unless $E = 0$.

References

[1] H.T. Banks and K.A. Morris. Input-Output Stability for Accelerometer Control Systems. Fields Institute Technical Report #FI-CT16, Waterloo, Ontario, 1992. submitted, *Control- Theory and Advanced Technology*.

[2] J. Bontsema and R.F. Curtain. Perturbation Properties of a Class of Infinite-Dimensional Systems with Unbounded Control and Observation. *IMA Journal of Mathematical Control and Information*, Vol. 5, pg. 333-352, 1988.

[3] Ruth F. Curtain. Well-posedness of infinite-dimensional linear systems in time and frequency domain. Technical Report TW-287, University of Groningen, 1988.

[4] Ruth F. Curtain, Equivalance of Input-Output Stability and Exponential Stability for Infinite-Dimensional Systems. *Mathematical Systems Theory*, Vol. 21, pg. 19-48, 1988.

[5] R.F. Curtain and A. Pritchard. *Infinite-Dimensional Linear Systems Theory*, Springer-Verlag, 1978.

[6] A. Pazy. *Semigroups of Linear Operators and Applications to Partial Differential Equations.* Springer-Verlag Inc., 1983.

[7] A.J. Pritchard and D. Salamon. The linear quadratic control problem for infinite dimensional systems with unbounded input and output operators. *SIAM Jour. on Control and Optimization*, Vol. 25, 121-144, 1987.

[8] D. Salamon. *Control and Observation of Neutral Systems.* Research Notes in Math., Vol. 91, Pitman, 1984.

[9] D. Salamon. Infinite-dimensional linear systems with unbounded control and observation: A functional analytic approach, *Trans. Amer. Math. Soc..* Vol. 300, 383-431, 1987.

[10] George Weiss. Admissibility of Unbounded Control Operators *SIAM Jour. Control and Optimization*, Vol. 27, 527:545, 1989.

[11] George Weiss. Admissible Observation Operators for Linear Semigroups. *Israel Jour. of Mathematics*, Vol. 64, 17-43, 1989.

[12] G. Weiss. Linear systems on Hilbert spaces. *Control and Estimation of Distributed Parameter Systems*, ed. F. Kappel, K. Kunisch and W. Schappacher, Birkhauser, 1988.

[13] Jan C. Willems. The Generation of Lyapunov Functions for Input-Output Stable Systems. *SIAM Jour. Control*, Vol. 9, pg. 105-134, 1971.

Chapter 11
Point Observation in Linear-Quadratic Elliptic Distributed Control Systems

Link Ji†   Goong Chen‡

Abstract

Point sensors occur naturally in the study of distributed control systems. In this paper, we study a linear-quadratic regulator problem governed by a potential equation where in the quadratic cost, point observations at sensor locations are required to be close to certain target values. A primal approach using boundary integral equations is adopted here. We prove that in space dimension N greater than or equal to three, the regulator problem generally does not have any nontrivial optimal control if some sensors are located on the *boundary* of the domain. While if the domain is two dimensional, then a unique optimal control exists with sharp Sobolev space regularity $H^{\frac{1}{2}-\epsilon}$ on the boundary. Such an optimal control will also have logarithmic singularities at some of the boundary sensor locations. One of the major advantages of our approach is that it is directly amenable to boundary element numerical calculations. Collocation-panel method error estimates are established and numerical results are included to further illustrate the theory.

1 Introduction

In this paper, we study a distributed parameter linear-quadratic regulator problem as follows:

(1)
(2)
$$\text{(LQR)} \begin{cases} \inf J(u) = \sum_{i=1}^{m} |w(P_i) - z_i|^2 + \gamma \int_{\partial\Omega} u^2(x) d\sigma \\ \text{subject to} \\ \quad \Delta w = f \in H^r(\Omega), r \geq -1, \\ \quad \frac{\partial w}{\partial n} = u \in L^2(\partial\Omega), \end{cases}$$

where

Ω is a bounded simply connected open domain in $\mathbf{R}^N$ with C^∞ boundary $\partial\Omega$,
f is a given (loading) function on Ω,
$\frac{\partial}{\partial n}$ is the normal derivative, with n pointing outward,
u is a Neumann type boundary control,
$\gamma > 0$ is a given weighting factor,
$z_i \in \mathbf{R}$, $1 \leq i \leq m$, are prescribed "target" values,
$P_i \in \overline{\Omega}, 1 \leq i \leq m$, are prescribed "sensor locations",
$H^r(M)$ is the Sobolev space of order r on a Euclidean manifold M.

The study of the above system is motivated by contemporary problems in distributed parameter control, where, e.g., *piezoelectric sensors* are installed at points P_i, $1 \leq i \leq n$,

---

† Now at Department of Oceanography, Texas A&M University, College Station, TX 77843.

‡ Department of Mathematics, Texas A&M University, College Station, TX 77843. Supported in part by AFOSR Grants 87-0334 and 91-0097.

to measure the deformation at these points. (For the case when the governing system is the linear elastostatics equations, see Ji [6].) We call the values $w(P_i)$, $1 \leq i \leq n$, *point observations*, as reflected in the title of the paper. The governing system (2) is static; nevertheless, w in (2) may be regarded as the *asymptotic state* of either the heat equation

$$\frac{\partial}{\partial t}w(x,t) - \Delta w(x,t) = -f(x), \tag{3}$$

or of the wave equation

$$\frac{\partial^2}{\partial t^2}w(x,t) - \Delta w(x,t) = -f(x), \tag{4}$$

where the transient response has died out (therefore the dependence of $w(x,t)$ on t is negligible and $\partial w/\partial t$ in (3) and $\partial^2 w/\partial t^2$ in (4) can be omitted). The boundary condition in (2) reflects the fact that the boundary control u enters either as heat flux (corresponding to (3)) or as boundary tension force (corresponding to (4)), which are natural types of controllers for engineering systems. (The basic methodology in this paper will also work if the boundary control is Dirichlet, i.e., $w = u$ on $\partial\Omega$. But the cost functional J would have to be formulated differently for the problem to be of sufficient physical interest.) At sensor locations P_i, we wish the observation values $w(P_i)$ to be close to the target values z_i, with the control cost (represented by the integral term in (1)) also minimized.

Even though the setting of Problem (LQR) falls into the general category of elliptic optimal control problems studied in Chapter 2 of Lions [7], the methods therein are not directly applicable to (LQR). In particular, the fundamental questions of existence, uniqueness and regularity of the optimal controls here are widely open. In this paper, we will give complete answers to these questions.

The basic methodology adopted by us here is the modern theory of *boundary integral equations* (BIE) [3]. This theory evolves from the classical theory of integral equations in *potential theory* and numerically leads to the *boundary element methods* (BEM) [3]. This approach has manifested certain important advantages over the traditional Galerkin variational approach in that it provides rather explicit information about the state and control, and it is amenable to direct numerical computation (through the boundary element methods). See also another recent paper by Chen and Zhou [4] displaying the same advantage of this approach.

The organization of this paper is as follows:
(i) In §2, we address the basic issues of Sobolev space regularities of state and control.
(ii) In §3, we prove the nonexistence result of optimal controls if the space dimension N is greater than or equal to 3.
(iii) In §4, we establish the existence, uniqueness and sharp regularity results for the space dimension $N = 2$.
(iv) In §5, numerical results and error estimates are discussed and illustrated for an example in a two dimensional domain.
(v) Relevant open questions are described in §6.

2 The basic question of Sobolev space regularity of state and control

In problem (LQR), first assume that $f \in C^\infty(\overline{\Omega})$. Then it is well known from the elliptic theory of PDE that no matter how u lacks smoothness on $\partial\Omega$, any solution w of (2) will satisfy

$$w \in C^\infty(\Omega). \tag{5}$$

(Here, of course, in order for (2) to have a solution, the compatibility condition

$$\int_{\partial \Omega} u d\sigma = \int_{\Omega} f dx \tag{6}$$

must be satisfied. If this condition is satisfied, then (2) has infinitely many solutions, any two of which differ by a constant. In our formulation later on, condition (6) will be taken care of automatically; see (19).) Therefore if $P_i \in \Omega$ for $1 \le i \le m$, then the observation values $w(P_i)$ are all well defined pointwise. Since the cost functional J in (1) is strictly convex, it is an almost immediate conclusion that there exists a unique optimal control $\hat{u}$ for (LQR) s.t.

$$\hat{u} \in C^{\infty}(\partial \Omega). \tag{7}$$

Troubles happen when some of the sensor locations $P_i \in \partial \Omega$. Again, assume $f \in C^{\infty}(\overline{\Omega})$. Since the control u has only base regularity $L^2(\partial \Omega)$, from the theory of elliptic boundary value problems (cf. Lions-Magenes [8], e.g.), we have

$$w \in H^{\frac{3}{2}}(\Omega). \tag{8}$$

By the trace theorem, $w|_{\partial \Omega} \in H^1(\partial \Omega)$. The Sobolev imbedding theorem ([1], [3, Theorem 2.1.3, p. 30], e.g.) states

$$H^s(M) \hookrightarrow C^{k,\alpha}(M), M \text{ is an } \widetilde{m}\text{-dimensional Euclidean manifold,} \tag{9}$$

if

$$s - \frac{\widetilde{m}}{2} = k + \alpha, \quad k \in \mathbf{Z}^+ \equiv \{0, 1, 2, \ldots\}, 0 < \alpha < 1. \tag{10}$$

(Here $C^{k,\alpha}(M)$ is the space of k-times continuously differentiable Hölder-Lipschitz continuous functions with exponent α on M.) Thus

$$\Omega \subset \mathbf{R}^2 \Rightarrow M = \partial \Omega, \widetilde{m} = 1$$
$$1 - \frac{\widetilde{m}}{2} = 1 - \frac{1}{2} = k + \alpha; \quad k = 0, \alpha = \frac{1}{2},$$

and

$$w|_{\partial \Omega} \in C^{0,\frac{1}{2}}(\partial \Omega), \quad (N = 2) \tag{11}$$

so $w(P_i)$ is pointwise well defined. However, (9) and (10) do not apply if

$$\Omega \subset \mathbf{R}^N, \text{ with } N \ge 3.$$

In this case, what we can apply is the imbedding property ([3, Theorem 2.1.7, p. 32], e.g.)

$$H^s(M) \hookrightarrow L^p(M), \tag{12}$$

where

$$p\colon 1 \le p \le p_0, \quad \text{if} \quad \frac{1}{p_0} = \frac{1}{2} - \frac{s}{\widetilde{m}} > 0,$$

and

$$p: 1 \leq p < \infty, \quad \text{if } \frac{1}{2} - \frac{s}{\tilde{m}} = 0.$$

With $M = \partial\Omega$ and $\tilde{m} = N - 1$, we obtain

(13) $$w|_{\partial\Omega} \in H^1(\partial\Omega) \hookrightarrow \begin{cases} L^p(\partial\Omega), & \forall\ p: 1 \leq p < \infty, N = 3; \\ L^p(\partial\Omega), & \forall\ p: 1 \leq p \leq p_0, \frac{1}{p_0} = \frac{1}{2} - \frac{1}{N-1}, N > 3. \end{cases}$$

Thus we see that for $N \geq 3, w$ may not be pointwise defined at $P_i \in \partial\Omega$ because in general $w|_{\partial\Omega} \notin C(\partial\Omega)$, the space of continuous functions on $\partial\Omega$.

Under such a circumstance, a person may wishfully think that perhaps the outcome may not be all that bad, at least there may exist certain *special cases* for which (LQR) may have some meaningful solutions, even though $N \geq 3$. In § 3, we will prove that the only possible such a special case is trivial, namely, the optimal control $\hat{u} \equiv 0$ on $\partial\Omega$.

3 The nonexistence result of optimal control for (LQR): $N \geq 3$

Let $N \geq 3$. According to [3, Chapter 6], any solution w of (2) can be represented as a sum of a volume potential and a simple-layer potential:

(14) $$w(x) = \left[-\int_\Omega E(x,\xi)f(\xi)d\xi\right] + \int_{\partial\Omega} E(x,\xi)\eta(\xi)d\sigma_\xi, \qquad x \in \Omega,$$

where $E(x,\xi)$ is the fundamental solution of the Laplacian ([3, p. 214]):

(15) $$\Delta_\xi E(x,\xi) = -\delta(x-\xi), \quad x,\xi \in \mathbf{R}^N,$$

(16) $$E(x,\xi) = \frac{\Gamma(N/2)}{2\pi^{N/2}(N-2)}|x-\xi|^{-(N-2)}, \qquad N \geq 3,$$

and η is called a layer density, to be determined from the BIE

(17) $$\frac{1}{2}\eta(x) + \int_{\partial\Omega} \frac{\partial E(x,\xi)}{\partial n_x}\eta(\xi)d\sigma_\xi = u(x) + \frac{\partial}{\partial n_x}\int_\Omega E(x,\xi)f(\xi)d\xi, \qquad x \in \partial\Omega,$$

implied by (14) and the boundary condition in (14). Before we begin the treatment of (LQR), it is useful to introduce the following notation and to recall a list of properties [3, Chapter 6]:

(III.A) $\mathcal{L}_1: \eta \mapsto \mathcal{L}_1\eta,\ (\mathcal{L}_1\eta)(x) = \int_{\partial\Omega} E(x,\xi)\eta(\xi)d\sigma_\xi,\ \eta \in H^s(\partial\Omega),\ x \in \partial\Omega,$
  $\mathcal{L}_1: H^s(\partial\Omega) \to H^{s+1}(\partial\Omega)$ is an isomorphism, $\forall\ s \in \mathbf{R}$;

(III.B) $\mathcal{L}_2: \eta \mapsto \mathcal{L}_2\eta,\ (\mathcal{L}_2\eta)(x) = \frac{1}{2}\eta(x) + \int_{\partial\Omega} \frac{\partial E(x,\xi)}{\partial n_x}\eta(\xi)d\sigma_\xi,\ \eta \in H^s(\partial\Omega),\ x \in \partial\Omega,$
  $\mathcal{L}_2: H^s(\partial\Omega) \to H^s(\partial\Omega)$ is continuous and Fredholm of index zero, with
  $\mathcal{N}(\mathcal{L}_2)$ (the null space of the operator $\mathcal{L}_2$) = span$\{\mathcal{L}_1^{-1}(1) \in H^s(\partial\Omega) \mid \forall\ s \in \mathbf{R}\}$,
  Coker $(\mathcal{L}_2)$ (the cokernel of the operator $\mathcal{L}_2$) = span$\{1\}$.
  $\mathcal{L}_2$ has adjoint $\mathcal{L}_2^*$ defined by

$$(\mathcal{L}_2^*\eta)(x) = \frac{1}{2}\eta(x) + \int_{\partial\Omega} \frac{\partial E(x,\xi)}{\partial n_\xi}\eta(\xi)d\sigma_\xi$$

with $\mathcal{N}(\mathcal{L}_2^*) = \text{Coker}(\mathcal{L}_2)$, $\text{Coker}(\mathcal{L}_2^*) = \mathcal{N}(\mathcal{L}_2)$. Further, the integral operator parts of $\mathcal{L}_2$ and $\mathcal{L}_2^*$, defined by

$$\left.\begin{aligned}(\mathcal{L}_2^I\eta)(x) &= \int_{\partial\Omega} \frac{\partial E(x,\xi)}{\partial n_x}\eta(\xi)d\sigma_\xi \\ (\mathcal{L}_2^{I^*}\eta)(x) &+ \int_{\partial\Omega} \frac{\partial E(x,\xi)}{\partial n_\xi}\eta(\xi)d\sigma_\xi\end{aligned}\right\} \qquad x \in \partial\Omega,$$

satisfies
$$\mathcal{L}_2^I, \mathcal{L}_2^{I^*}: H^s(\partial\Omega) \to H^{s+1}(\partial\Omega) \text{ continuous}, \forall s \in \mathbf{R}.$$

(III.C) $V: f \mapsto Vf: (Vf)(x) = \int_\Omega E(x,\xi)f(\xi)d\xi$, $f \in H^r(\Omega)$, $r \geq -1$, $x \in \Omega$,
$V: H^r(\Omega) \to H^{r+2}(\Omega)$ is continuous, $\forall r \geq -1$.
Further, $\frac{\partial}{\partial n}V(f) \in H^{r+\frac{1}{2}}(\partial\Omega)$ for $r \geq -1$; see [4]. □

Note that we have

$$(18) \quad \int_{\partial\Omega} \frac{\partial E(x,\xi)}{\partial n_x} d\sigma_x = \begin{cases} -\frac{1}{2}, & \xi \in \partial\Omega, \\ -1, & \xi \in \Omega. \end{cases}$$

Integrating (17) with respect to x on $\partial\Omega$ and applying (18), we obtain

$$(19) \quad 0 = \int_{\partial\Omega} u(x)d\sigma_x - \int_\Omega f(\xi)d\xi,$$

so the compatibility condition (6) is satisfied.

We now assume that

$$(20) \quad P_i \in \partial\Omega, \quad \forall i: 1 \leq i \leq m.$$

Equations (14) and (17) hold in the sense of distributions. However, if $w(P_i)$ is well defined at P_i, then from (14) necessarily

$$(21) \quad w(P_i) = \int_{\partial\Omega} E(P_i,\xi)\eta(\xi)d\sigma_\xi - \int_\Omega E(P_i,\xi)f(\xi)d\xi,$$

where the two distributions defined by the two integrals on the RHS of (14) have pointwise defined values at P_i as indicated in (21). We now substitute (21) and u from (17) into (1), yielding

$$(22) \quad \begin{aligned} J(u) &= \sum_{i=1}^m |[\mathcal{L}_1(\eta)(P_i) - V(f)(P_i)] - z_i|^2 + \gamma \int_{\partial\Omega} \left|\mathcal{L}_2\eta - \frac{\partial}{\partial n}V(f)\right|^2 d\sigma \\ &\equiv \tilde{J}(\eta), \end{aligned}$$

where $\tilde{J}(\eta)$ is seen as a quadratic functional depending only on unknown density η.

LEMMA 3.1. *Assume (20) and $N \geq 3$. Suppose that (LQR) has a unique optimal control $\hat{u} \in H^{s_0}(\partial\Omega)$ for some $s_0 \geq 0$, then there exists some (nonunique) $\tilde{\eta} \in H^{s_0}(\partial\Omega)$ s.t.*

$$(23) \quad \frac{1}{2}\tilde{\eta}(x) + \int_{\partial\Omega} \frac{\partial E(x,\xi)}{\partial n_x}\tilde{\eta}(\xi)d\sigma_\xi = \hat{u}(x) + \frac{\partial}{\partial n_x}\int_\Omega E(x,\xi)f(\xi)d\xi, \quad x \in \partial\Omega,$$

provided that $f \in H^r(\Omega)$ and $r \geq s_0 - \frac{1}{2}$.

Proof. Since $r \geq s_0 - \frac{1}{2}, s_0 \geq 0$, from (III.C), we get

$$\frac{\partial}{\partial n_x}\int_\Omega E(x,\xi)f(\xi)d\xi \in H^{r+\frac{1}{2}}(\partial\Omega) \subseteq H^{s_0}(\partial\Omega), \quad x \in \partial\Omega.$$

The compatibility condition (19) implies that

$$(24) \quad \text{RHS of (23)} \perp \text{Coker}(\mathcal{L}_2).$$

Therefore (23) has a solution $\tilde{\eta} \in H^{s_0}(\partial\Omega)$. (In fact, all solutions of (23) are of the form $\tilde{\eta} + c\mathcal{L}_1^{-1}(1), \forall c \in \mathbf{R}$.) □

Lemma 3.1 has established the basic fact that the simple-layer density η has the same regularity as the optimal control $\hat{u}$, if it exists, provided that f is sufficiently regular. Therefore the question of η-regularity in taking the variational derivative of $\tilde{J}(\eta)$ is no longer detrimental. It is easy to see that $\tilde{J}(\eta)$ is convex in η, and corresponding to any minimizer $\hat{u}$ of $J(u)$, there is an $\hat{\eta}$ satisfying (23) which minimizes $\tilde{J}(\eta)$, and vice versa. Let $\hat{\eta}$ be a minimizer of $\tilde{J}(\eta)$. Then, by calculus of variations,

$$\begin{aligned}
0 &= \frac{1}{2}\tilde{J}'(\hat{\eta}) \cdot \delta\eta \\
&= \sum_{i=1}^m \int_{\partial\Omega} E(P_i, y) \left\{ \int_{\partial\Omega} E(P_i, \xi)\hat{\eta}(\xi)d\sigma_\xi - [V(f)(P_i) + z_i] \right\} (\delta\eta)(y) d\sigma_y \\
&\quad + \gamma \left\langle \mathcal{L}_2^* \left[\mathcal{L}_2\hat{\eta} - \frac{\partial}{\partial n}V(f) \right], \delta\eta \right\rangle_{L^2(\partial\Omega)}, \quad \forall\, \delta\eta \in C^\infty(\partial\Omega).
\end{aligned}$$

Therefore $\hat{\eta}$ necessarily satisfies the boundary integral equation

(25)
$$\begin{aligned}
\gamma\mathcal{L}_2^*(\mathcal{L}_2\hat{\eta})(y) + \sum_{i=1}^m E(P_i, y) \int_{\partial\Omega} E(P_i, \xi)\hat{\eta}(\xi)d\sigma_\xi \\
= \sum_{i=1}^m E(P_i, y)[V(f)(P_i) + z_i] + \gamma\mathcal{L}_2^*\left(\frac{\partial}{\partial n}V(f)\right)(y), \quad y \in \partial\Omega.
\end{aligned}$$

We are now in a position to prove the first main theorem of our paper.

THEOREM 3.2. *Let $N \geq 3$, $r \geq -\frac{1}{2}$, and $P_i \in \partial\Omega$, $1 \leq i \leq m$. Then problem (LQR) does not have any nontrivial optimal control $\hat{u} \in L^2(\partial\Omega)$.*

Proof. Let $\hat{\eta}$ minimize $\tilde{J}(\eta)$. Then necessarily $\hat{\eta}$ satisfies the BIE (25). We rewrite (25) as

(26)
$$\begin{aligned}
4\left(\frac{1}{2}I + \mathcal{L}_2^{I*}\right)\left(\frac{1}{2}I + \mathcal{L}_2^I\right)\hat{\eta} &= -\frac{4}{\gamma}\sum_{i=1}^m E(P_i, \cdot)\left\{\left[\int_{\partial\Omega} E(P_i, \xi)\hat{\eta}(\xi)d\sigma_\xi \right.\right. \\
&\quad \left.\left. - V(f)(P_i)\right] - z_i\right\} + 4\mathcal{L}_2^*\left(\frac{\partial}{\partial n}V(f)\right) \\
&= -\frac{4}{\gamma}\sum_{i=1}^m E(P_i, \cdot)[w(P_i) - z_i] + 4\mathcal{L}_2^*\left(\frac{\partial}{\partial n}V(f)\right),
\end{aligned}$$

where we have used the fact that (21) is pointwise defined at P_i. Expanding the LHS of (26), we have

(27)
$$\begin{aligned}
\hat{\eta} &= -\frac{4}{\gamma}\sum_{i=1}^m E(P_i, \cdot)[w(P_i) - z_i] - 2(\mathcal{L}_2^I\hat{\eta} + \mathcal{L}_2^{I*}\hat{\eta}) - 4\mathcal{L}_2^{I*}(\mathcal{L}_2^I\hat{\eta}) \\
&\quad + 4\mathcal{L}_2^*\left(\frac{\partial}{\partial n}V(f)\right).
\end{aligned}$$

Let us inspect the regularity of the terms on the RHS of (27). If there exists a minimizer $\hat{u} \in L^2(\partial\Omega)$ for $J(u)$, then by Lemma 3.1 there corresponds an $\hat{\eta} \in L^2(\partial\Omega)$ minimizing $\tilde{J}(\eta)$. Thus

(28)
$$\left.\begin{aligned} \mathcal{L}_2^I\hat{\eta}, \mathcal{L}_2^{I*}\hat{\eta} &\in H^1(\partial\Omega), \\ \mathcal{L}_2^{I*}(\mathcal{L}_2^I\hat{\eta}) &\in H^2(\partial\Omega), \end{aligned}\right\} \quad \text{by (III.B)},$$

(29) $$\mathcal{L}_2^*\left(\frac{\partial}{\partial n}V(f)\right) \in L^2(\partial\Omega), \text{ by (III.B), (III.C) and } r \geq -\frac{1}{2}.$$

Thus $\hat{\eta} \in L^2(\partial\Omega)$, iff

(30) $$-\frac{4}{\gamma}\sum_{i=1}^{m} E(P_i, \cdot)[w(P_i) - z_i] \in L^2(\partial\Omega).$$

Each summand above is of the form

(31) $$c_i E(P_i, y) = c_i \cdot C_N \cdot \frac{1}{|P_i - y|^{N-2}}, \qquad y \in \partial\Omega,$$

$$C_N \equiv \frac{\Gamma(N/2)}{2\pi^{N-2}(N-2)}, \quad \text{see (16)}.$$

It is immediately seen that

(32) $$c_i E(P_i, \cdot) \in L^\alpha(\partial\Omega) \quad \text{for any} \quad \alpha: 1 \leq \alpha < \frac{N-1}{N-2}, \quad N \geq 3,$$

and

$$c_i E(P_i, \cdot) \notin L^2(\partial\Omega), \quad \text{if} \quad c_i \neq 0.$$

Therefore (30) holds iff

(33) $$w(P_i) - z_i = 0, \qquad 1 \leq i \leq m.$$

Therefore $\hat{\eta} \in L^2(\partial\Omega)$ iff (33) holds. When (33) holds, from (26) we have

$$4\mathcal{L}_2^*(\mathcal{L}_2\hat{\eta}) = 4\mathcal{L}_2^*\left(\frac{\partial}{\partial n}V(f)\right).$$

Hence
$$\mathcal{L}_2\hat{\eta} - \frac{\partial}{\partial n}V(f) \in \mathcal{N}(\mathcal{L}_2^*).$$

By (23) and (III.B),

$$\hat{u} = \mathcal{L}_2\hat{\eta} - \frac{\partial}{\partial n}V(f) = c, \quad c \in \mathbf{R} \text{ is arbitrary, on } \partial\Omega.$$

Since $\hat{u}$ is assumed to be optimal, the constant $c = \hat{u}$ minimizing $J(u)$ when (33) holds is $c = 0$. Therefore

$$\hat{u} \equiv 0 \quad \text{on} \quad \partial\Omega. \qquad \square$$

From Theorem 3.2, we easily conclude the following.

COROLLARY 3.3. *Let $N \geq 3$ and $r \geq -\frac{1}{2}$. Assume that m_1 sensor locations P_i are located on $\partial\Omega$, and the remaining are located on Ω, i.e.,*

(34) $$\left.\begin{aligned} P_i \in \partial\Omega, & \quad 1 \leq i \leq m_1, \\ P_i \in \Omega, & \quad m_1 + 1 \leq i \leq m_2, \quad m_2 > m_1. \end{aligned}\right\}$$

If problem (LQR) has an optimal control $\hat{u} \in L^2(\partial\Omega)$, then for this $\hat{u}$ the corresponding state w must satisfy

(35) $$w(P_i) = z_i, \qquad 1 \leq i \leq m_1.$$

This $\hat{u}$ is unique, with optimal regularity

$$\hat{u} \in H^{r+\frac{1}{2}}(\partial\Omega).$$

There exists no optimal control $\hat{u} \in L^2(\partial\Omega)$ whose corresponding state w violates (35). □

4 The existence, uniqueness and regularity of optimal control for (LQR): $N = 2$

In $\mathbf{R}^2$, the fundamental solution of the Laplacian is given by

(36) $$E(x,\xi) = -\frac{1}{2\pi}\ln|x-\xi|, x,\xi \in \mathbf{R}^2, \text{ cf. (15), (16).}$$

It is due to this logarithmic growth at infinity that some key properties stated in (III.A)-(III.C) may no longer hold, the major one of which being that $\mathcal{L}_1$ in general need not be an invertible mapping from $H^s(\partial\Omega)$ onto $H^{s+1}(\partial\Omega)$ as given in (III.A), and must be "augmented" according to [3, Theorem 6.12.1, pp. 287-288] in order to constitute an isomorphism $\mathcal{L}'_1\colon \mathbf{R} \oplus H^s(\partial\Omega) \to \mathbf{R} \oplus H^{s+1}(\partial\Omega)$. This causes some slight inconvenience, but can be easily overcome. Nevertheless, in order to keep our arguments short and uncomplicated, we make a simplifying assumption that

(37) $$\text{diameter of } \Omega \equiv \sup\{|x_1 - x_2|\ \big|\ x_1,x_2 \in \Omega\} < 1.$$

Under (37), all properties (III.A)-(III.C) remain valid [5] (after all $E(x,\xi)$ therein has been substituted by (36)). Better still, we have the infinitely smoothing property

(38) $$\mathcal{L}_2^I, \mathcal{L}_2^{I^*}\colon H^s(\partial\Omega) \to C^\infty(\partial\Omega), \quad \forall\ s \in \mathbf{R},$$

of the integral operators $\mathcal{L}_2^I$ and $\mathcal{L}_2^{I^*}$ ([3, pp. 249-250]).

We note that by scaling $x \mapsto x/k$, $x \in \mathbf{R}^2$, $k > 0$, using sufficiently large k, we can always transform Ω into a domain whose diameter is less than 1, i.e., satisfying (37). Thus (37) should not be construed as a severe restriction.

It is easy to see that Lemma 3.1 remains valid for $N = 2$. We now prove the second main theorem of the paper.

THEOREM 4.1. *Let $N = 2$, $r \geq 0$, $P_i \in \partial\Omega$ for $1 \leq i \leq m$, and (37) hold. Then problem (LQR) has a unique optimal control $\hat{u}$ with the following sharp regularity dichotomy:*
(39) $\quad$ (i) $\quad \hat{u} \in C^\infty(\partial\Omega)$ iff $\hat{u} \equiv 0$ on $\partial\Omega$,
(40) $\quad$ (ii) $\hat{u} \in H^{\frac{1}{2}-\varepsilon}(\partial\Omega)$, for any ε: $0 < \varepsilon < \frac{1}{2}$, if $\hat{u} \not\equiv 0$. Further, $\hat{u} \notin H^{\frac{1}{2}}(\partial\Omega)$.

Proof. We make the same line of arguments as in the proof of Theorem 3.2, using (36) everywhere in lieu of (16), until we have arrived at (30). Instead of (31), we now have

(41) $$c_i E(P_i, y) = -\frac{c_i}{2\pi}\ln|P_i - y|$$
$$\equiv c'_i \ln|P_i - y|, \quad y \in \partial\Omega,$$

Using Lemma 4.3 later, we have

(42) $$c_i E(P_i, \cdot) \in H^{\frac{1}{2}-\varepsilon}(\partial\Omega), \text{ for any } \varepsilon\colon 0 < \varepsilon < \frac{1}{2},$$

and the above regularity is the sharpest possible (for $c_i \neq 0$).

Return to (27), we now have

$$\hat{\eta} = \underbrace{-\frac{4}{\gamma}\sum_{i=1}^{m} E(P_i,\cdot)[w(P_i) - z_i]}_{\in H^{\frac{1}{2}-\varepsilon}(\partial\Omega),\text{ by (42)}} \underbrace{-2(\mathcal{L}_2^I\hat{\eta} + \mathcal{L}_2^{I^*}\hat{\eta}) - 4\mathcal{L}_2^{I^*}(\mathcal{L}_2^I\hat{\eta})}_{\in C^{\infty}(\partial\Omega),\text{ by (38)}}$$

(43)
$$+ \underbrace{4\mathcal{L}_2^*\left(\frac{\partial}{\partial n}V(f)\right)}_{\in H^{\frac{1}{2}}(\partial\Omega),\text{ due to }r\geq 0}.$$

There are two possibilities:
(a) $w(P_i) - z_i \neq 0$ for some i: $1 \leq i \leq m$.
   In this case, $\hat{\eta}$ will have optimal regularity $H^{\frac{1}{2}-\varepsilon}(\partial\Omega)$. By Lemma 3.1 (with $N = 2$), $\hat{u}$ will also have optimal regularity $H^{\frac{1}{2}-\varepsilon}(\partial\Omega)$.
(b) $w(P_i) - z_i = 0$ for all i: $1 \leq i \leq m$.
   In this case, we can argue as in the last part of the proof of Theorem 3.2 to show that $\hat{u} \equiv 0$ on $\partial\Omega$.

We now show that $\hat{\eta}$ satisfying the BIE (25) is unique. This follows easily from the fact that
$$\mathcal{L}_2^*\mathcal{L}_2: \mathcal{N}(\mathcal{L}_2)^\perp \longrightarrow \mathcal{N}(\mathcal{L}_2)^\perp$$
is positive definite and that the boundary integral operator
$$T: \eta \mapsto \sum_{i=1}^{m} E(P_i,\cdot)\int_{\partial\Omega} E(P_i,\xi)\eta(\xi)d\sigma_\xi$$
satisfies
$$\langle T(\mathcal{L}_1^{-1}(1)), \mathcal{L}_1^{-1}(1)\rangle_{L^2(\partial\Omega)} = \langle 1,1\rangle_{L^2(\partial\Omega)} > 0,$$
for $\mathcal{L}_1^{-1}(1) \in \mathcal{N}(\mathcal{L}_2)$.

Hence (39) and (40) have been proved. □

COROLLARY 4.2. *In Theorem 4.1, assume further that $f \in C^\infty(\overline{\Omega})$. Then $\hat{u}$ is infinitely differentiable at every point $x \in \partial\Omega$, $x \neq P_i$, $1 \leq i \leq m$. At $P_i \in \partial\Omega$, $\hat{u}$ has at worst a logarithmic singularity of magnitude $\mathcal{O}(\ln|x - P_i|)$, $x \in \partial\Omega$ in a neighborhood of P_i, for $1 \leq i \leq m$.*

Proof. Obvious from (23), (38) and (43). □

We still need to prove (42).

LEMMA 4.3. *Let $\Omega \subset \mathbf{R}^2$ and $P \in \partial\Omega$. Then*

(44)
$$\ln|x - P| \in H^{\frac{1}{2}-\varepsilon}(\partial\Omega), \text{ for any } \varepsilon > 0, \quad (x \in \partial\Omega).$$

Proof. Since $\partial\Omega$ is C^∞ with finite measure, it is equivalent to proving that

(45)
$$\ln|x| \in H^{\frac{1}{2}-\varepsilon}((-1,1)), \quad x \in (-1,1),$$

where $(-1,1)$ is the open interval with left end -1 and right end 1, and $\ln|x|$ is the logarithmic function on $\mathbf{R}$ manifesting the singularity $\ln|x - P|$ for $x \in \partial\Omega$ and $P \in \partial\Omega$ in (44). To show (45), let

(46)
$$\ln|x| = \sum_{n=-\infty}^{\infty} a_n e^{2\pi i n x}, \quad x \in (-1,1),$$

be the Fourier series of $\ln|x|$ on $(-1,1)$, where

$$(47) \qquad a_n = \frac{1}{2}\int_{-1}^{1} \ln|x| e^{-2\pi i n x} dx, \quad n \in \mathbf{Z} = \{\ldots, -2, -1, 0, 1, 2, \ldots\}.$$

Since $\ln|x| \in H^s((-1,1))$ for any $s \geq 0$ iff

$$(48) \qquad \sum_{n=-\infty}^{\infty} (1+|n|^{2s})|a_n|^2 < \infty,$$

we wish to determine the largest admissible s making (48) convergent. This requires the estimation of $|a_n|$; from (47),

$$
\begin{aligned}
a_n &= \frac{1}{2}\int_{-1}^{1} \ln|x| e^{-2\pi i n x} dx \\
&= \frac{1}{2n}\int_{-n}^{n} \ln\left|\frac{y}{n}\right| e^{-2\pi i y} dy \qquad (y = nx) \\
&= \frac{1}{2n}\left[\int_{-n}^{n} \ln|y| e^{-2\pi i y} dy - \ln|n| \int_{-n}^{n} e^{-2\pi i y} dy\right] \\
&= \frac{1}{2n}\int_{-n}^{n} \ln|y| e^{-2\pi i y} dy.
\end{aligned}
$$
(49)

We now note that

$$\frac{\ln|x|}{|x|} \leq C, \ \forall \ \text{sufficiently large } |x|, x \in \mathbf{R},$$

therefore $\ln|x|$ is a tempered distribution on $\mathbf{R}$ and admits a Fourier transform ([3, p. 66])

$$(50) \qquad \int_{-\infty}^{\infty} \ln|x| e^{-2\pi i x \xi} dx = -[\ln(2\pi) - \Gamma'(1)]\delta(\xi) - \frac{1}{2}|\xi|^{-1}, \quad \xi \in \mathbf{R},$$

where Γ' is the derivative of the gamma function, and δ is the Dirac delta distribution. (That is, the RHS of (50) is also a tempered distribution.) Letting $\xi = 1$ in (50), we obtain

$$(51) \qquad \int_{-\infty}^{\infty} \ln|x| e^{-2\pi i x} dx = -\frac{1}{2}\frac{1}{|\xi|}\Big|_{\xi=1} = -\frac{1}{2}.$$

Using (51) in (49), we get

$$(52) \qquad a_n = \frac{1}{2n}\left[-\frac{1}{2} + o(1)\right], \text{ when } |n| \text{ is large}, n \neq 0.$$

By (52), we see that (48) holds iff

$$(53) \qquad \frac{1}{16}\sum_{\substack{n=-\infty \\ n\neq 0}}^{\infty} (1+|n|^{2s})|n|^{-2} < \infty.$$

(Note that $a_0 = \int_{-1}^{1} \ln|x| dx$ is always a finite number and thus can be excluded in (53).) It is now nearly trivial to see that (53) holds iff $s < \frac{1}{2}$. Thus (45) has been established, and hence (44) follows. □

The counterpart of Corollary 3.3 for $N = 2$ can be easily given in the following.

COROLLARY 4.4. *Let $N = 2$ and $r \geq 0$. Assume that (35) holds. Then problem (LQR) has a unique optimal control $\hat{u} \in H^s(\partial\Omega)$ for any $s < \frac{1}{2}$. Further, for this $\hat{u}$ if the corresponding state w satisfies*

$$w(P_i) = z_i, \quad 1 \leq i \leq m_1,$$

then the regularity of $\hat{u}$ can be improved to

$$\hat{u} \in H^{r+1}(\partial\Omega), \quad \text{for} \quad f \in H^r(\Omega). \qquad \Box$$

5 Numerical results computed by boundary elements

We now perform boundary element numerical calculations on a computer. Here we only consider the case $P_i \in \partial\Omega$ for all i: $1 \leq i \leq m$. By Theorem 3.2, such a problem (LQR) does not have any nontrivial solution for $N \geq 3$, therefore here we treat only the *two dimensional case*. From (25), define a bilinear form a and a linear form θ:

$$(54) \quad a(\eta_1, \eta_2) = \gamma \langle \mathcal{L}_2 \eta_1, \mathcal{L}_2 \eta_2 \rangle_{L^2(\partial\Omega)} + \sum_{i=1}^{m} \int_{\partial\Omega} E(P_i, \xi) \eta_1(\xi) d\sigma_\xi$$

$$\cdot \int_{\partial\Omega} E(P_i, \xi) \eta_2(\xi) d\sigma_\xi,$$

$$(55) \quad \theta(\eta) = \left\langle \sum_{i=1}^{m} E(P_i, \cdot)[V(f)(P_i) + z_i] + \gamma \mathcal{L}_2^* \left(\frac{\partial}{\partial n} V(f) \right), \eta \right\rangle_{L^2(\partial\Omega)},$$

for η_1, η_2 and η in $L^2(\partial\Omega)$. It is a simple exercise to show that a is coercive (symmetric and continuous) on $L^2(\partial\Omega)$:

$$(56) \quad a(\eta, \eta) \geq \alpha \|\eta\|_{L^2(\partial\Omega)}^2, \quad \text{for some} \quad \alpha > 0, \ \forall \ \eta \in L^2(\partial\Omega),$$

and θ is continuous on $L^2(\partial\Omega)$ for $r \geq 0$. Therefore, by the Lax-Milgram theorem, the variational equation

$$(57) \quad a(\hat{\eta}, \eta) = \theta(\eta), \quad \forall \ \eta \in L^2(\partial\Omega),$$

has a unique solution $\hat{\eta} \in L^2(\partial\Omega)$. This unique density $\hat{\eta}$ corresponds to the optimal control $\hat{u}$ through (23). Further, from the proof of Theorem 4.1, we have the improved regularity

$$(58) \quad \hat{\eta} \in H^{\frac{1}{2}-\varepsilon}(\partial\Omega), \quad \forall \ \varepsilon : 0 < \varepsilon < \frac{1}{2}.$$

Now, let $S_h^{t,\ell}(\partial\Omega)$ be a (t,ℓ)-system of approximation subspaces on $\partial\Omega$ satisfying

$$S_h^{t,\ell}(\partial\Omega) \subset H^\ell(\partial\Omega)$$

and for any $r \geq 0$, each integer s, $0 \leq s \leq \min(r,\ell)$, and each $\eta \in H^r(\partial\Omega)$, there exists $v_h \in S_h^{t,\ell}(\partial\Omega)$, satisfying

$$(59) \quad \|\eta - v_h\|_{H^s(\partial\Omega)} \leq C h^\mu \|\eta\|_{H^r(\partial\Omega)}, \quad 0 < h < 1, \mu = \min(t-s, r-s)$$

for some $C > 0$ independent of η and h; cf. [3, §5.4]. We may now state the following error estimates.

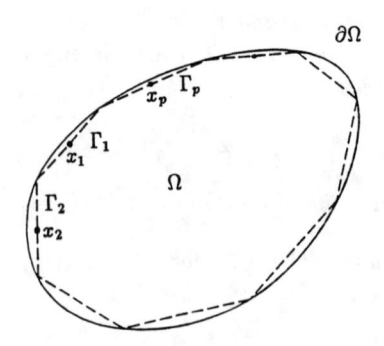

FIG. 1. *Polygonal approximation of $\partial\Omega$ by panels*

THEOREM 5.1. *Let $S_h^{t,\ell}(\partial\Omega)$ be a 1-parameter family of approximation spaces as given above with $t \geq 1$, $\ell \geq 0$. Let $\hat{\eta}_h$ be the unique solution of*

$$\begin{cases} \hat{\eta}_h \in S_h^{t,\ell}(\partial\Omega), \\ a(\hat{\eta}_h, v_h) = \theta(v_h), & \forall\, v_h \in S_h^{t,\ell}(\partial\Omega). \end{cases}$$

Let $\hat{\eta}$ be the unique solution of (57). Then

(60) $$\|\hat{\eta} - \hat{\eta}_h\|_{L^2(\partial\Omega)} \leq C h^{\frac{1}{2}-\varepsilon} \|\hat{\eta}\|_{H^{\frac{1}{2}-\varepsilon}(\partial\Omega)}, \qquad 0 < \varepsilon < \frac{1}{2},$$

for some $C > 0$ independent of h and ε.

Proof. This follows from (58), (59) and a lemma due to Céa; see [3, Chapter 5], for example. □

In practice, however, the above *Galerkin*-boundary element type approximations are replaced by *collocation*-boundary element calculations, the simplest of which is the *panel method* discretizing $\partial\Omega$ by polygonal panels as shown in Fig 5.1. On each panel Γ_j, let $\bigcup_{j=1}^{p} \Gamma_j$. χ_{Γ_j} be the characteristic function on Γ_j:

$$\chi_{\Gamma_j}(x) = \begin{cases} 1, & \text{if } x \in \Gamma_j, \\ 0, & \text{if } x \in \bigcup_{i=1}^{p} \Gamma_i \setminus \Gamma_j. \end{cases}$$

Write

(61) $$\hat{\eta}_h = \sum_{j=1}^{p} c_j \chi_{\Gamma_j},$$

and determine c_j, $1 \leq i \leq j \leq p$, from (25) by

(62) $$\gamma \sum_{j=1}^{p} c_j \mathcal{L}_2^*(\mathcal{L}_2 \chi_{\Gamma_j})(x_k) + \sum_{i=1}^{m} E(P_i, x_k) \sum_{j=1}^{p} c_j \int_{\Gamma_j} E(P_i, \xi) \chi_{\Gamma_j}(\xi) d\sigma_\xi$$
$$= \sum_{i=1}^{m} E(P_i, x_k)[V(f)(P_i) + z_i] + \gamma \mathcal{L}_2^*\left(\frac{\partial}{\partial n} V(f)\right)(x_k), \qquad 1 \leq k \leq p,$$

where x_k is the midpoint of Γ_k as shown in Fig. 1. This collocated panel method essentially corresponds to a collocation-boundary element method (when the polygonal discretization error of $\partial\Omega$ by $\bigcup_{j=1}^{p} \Gamma_j$ is neglected) with $S_h^{t,\ell}(\partial\Omega) = S_h^{1,0}(\partial\Omega)$, for which a (slightly adapted form of a) theorem due to Arnold and Wendland [2] (see also [3, Theorem 10.3.1]) is applicable, implying the unique solvability of c_j, $1 \leq j \leq p$, in (62) for all $h > 0$ sufficiently small and giving optimal asymptotic estimate

$$\|\hat{\eta} - \hat{\eta}_h\|_{L^2(\partial\Omega)} \leq Ch^{\frac{1}{2}-\varepsilon}\|\hat{\eta}\|_{H^{\frac{1}{2}-\varepsilon}(\partial\Omega)}, \qquad 0 < \varepsilon < \frac{1}{2}, \tag{63}$$

the same as (60).

Example 5.1. Let Ω be the open disk in $\mathbf{R}^2$ with radius $\frac{1}{2}$. Consider

$$\inf \sum_{i=1}^{3} |w(P_i) - z_i|^2 + \gamma \int_{\partial\Omega} u^2 d\sigma,$$

subject to

$$\begin{cases} \Delta w = 0 & \text{on } \Omega, \\ \frac{\partial w}{\partial n} = u & \text{on } \partial\Omega, \end{cases}$$

where

$$P_1 = \frac{1}{2}e^{i\frac{\pi}{2}}, \quad P_2 = \frac{1}{2}e^{i\pi}, \quad P_3 = \frac{1}{2}e^{i\frac{3}{2}\pi}, \tag{64}$$
$$\gamma = 100,$$
$$z_1 = 1, \quad z_2 = 0, \quad z_3 = 1.$$

This domain Ω just misses the condition (37); nevertheless it is easy to verify that all properties in (III.A)-(III.C) remain valid and thus the theory in §4 applies (§4 would have to be modified only when Ω is the *unit disk*.) Using the collocation-panel method as described in (62), we obtain $\hat{\eta}_h$. Using $\tilde{\eta} = \hat{\eta}_h$ in (17) (with $f \equiv 0$ on Ω therein), we obtain $\hat{u}_h$. It is easy to see that

$$\|\hat{u} - \hat{u}_h\|_{L^2(\partial\Omega)} \leq C'h^{\frac{1}{2}-\varepsilon}\|\hat{u}\|_{H^{\frac{1}{2}-\varepsilon}(\partial\Omega)}, \qquad \forall \varepsilon: 0 < \varepsilon < \frac{1}{2}, \tag{65}$$

holds for some $C' > 0$ independent of $\hat{u}$, h and ε.

In our numerical calculations, we use

$$p = 8, 16, 32, 64, 128, 256, \left(h = (2\pi)\bigg/\left(\frac{1}{2} \cdot p\right), \text{ cf. } (61)\right), \tag{66}$$

panels with uniform mesh on $\partial\Omega$. The computed $\hat{u}_h$ is illustrated in Fig. 2, where $\hat{u}_h$ is plotted against the angular variable θ (of the polar coordinates on Ω). The readers can see very discernible logarithmic singularities emerging (as $h \downarrow 0$) at $\theta = \frac{\pi}{2}$, π and $\frac{3\pi}{2}$, where the sensors P_1, P_2 and P_3 are located, see (5.11).

The rate of convergence of $\|\hat{u} - \hat{u}_h\|_{L^2(\partial\Omega)}$, measured as the slope of $\ln(\|\hat{u} - \hat{u}_h\|_{L^2(\partial\Omega)})$, is 0.5, as show in Fig. 3, confirming (65)

The computed optimal state w, obtained from (14) with $\eta = \hat{\eta}_h$ therein and $f \equiv 0$, is displayed in Fig. 4, using 256 panels.

The cost J is evaluated to be

$$0.293410 \times 10^2, \quad p = 8,$$
$$0.209673 \times 10^2, \quad p = 16,$$
$$0.178173 \times 10^2, \quad p = 32,$$
$$0.163961 \times 10^2, \quad p = 64,$$
$$0.157111 \times 10^2, \quad p = 128,$$
$$0.153727 \times 10^2, \quad p = 256,$$

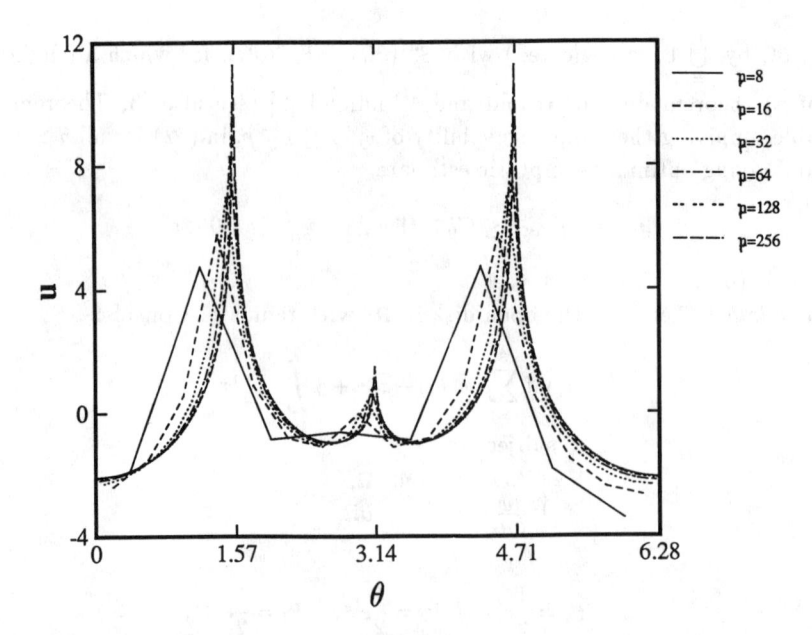

FIG 2. *The computed optimal control $\hat{u}_h$, using $8, 16, \ldots, 128, 256$ panels. Logarithmic singularities can be seen to emerge at $\theta = \frac{\pi}{2}$, π and $\frac{3\pi}{2}$, where the boundary point sensors are located.*

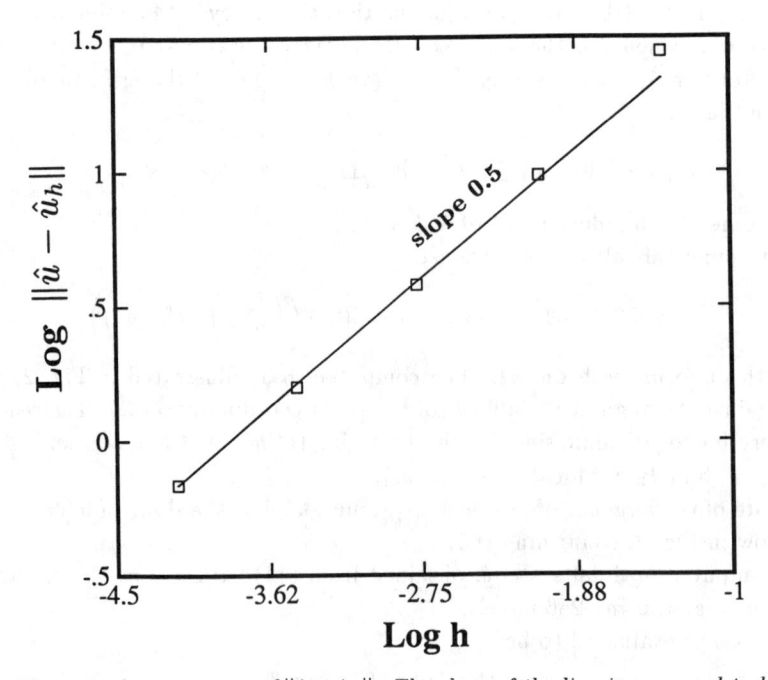

FIG 3. *The rate of convergence of $\|\hat{u} - \hat{u}_h\|$. The slope of the line is measured to be 0.5.*

6 Discussions

(1) So far, we have not been able to study the problem using the *adjoint system* corresponding to (LQR) as done by Lions in [7, Chapter 2] of his book. The approach taken here

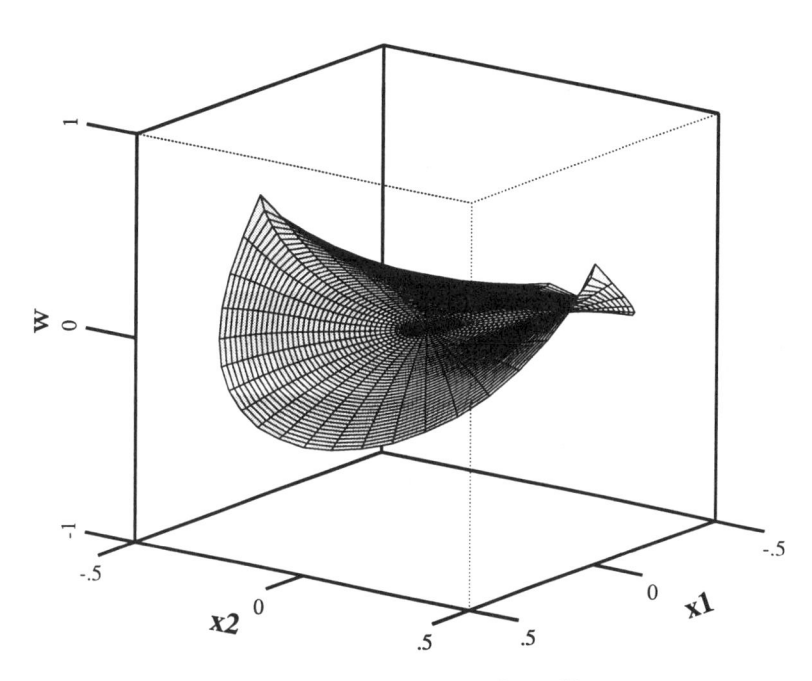

FIG. 4. *The computed optimal state w, using 256 panels on $\partial\Omega$.*

is completely *primal*. It would be interesting to see if the adjoint system can yield any extra results and information about (LQR).

(2) Theorem 3.2 *should not* be construed as saying that in space dimension $N \geq 3$, point sensors on $\partial\Omega$ cannot be used. The reason is that in practice, constraints like

(67) $$|u(x)| \leq M, \qquad x \in \partial\Omega, \text{ a.e.,}$$

are normally imposed on the boundary control u. Thus $u \in L^2(\partial\Omega) \cap L^\infty(\partial\Omega)$, and if $f \in W^{-\frac{1}{2},\infty}(\Omega)$, in this case the state w in (2) satisfies

(68) $$w \in W^{\frac{3}{2},p}(\Omega) \qquad \forall \, p \colon 1 \leq p < \infty,$$

therefore in lieu of (9), we use the Sobolev imbedding theorem [1]

$$W^{\ell,q}(M) \subset C^{k,\alpha}(M), \quad 1 > \alpha = \ell - k - \frac{\widetilde{m}}{q} > 0; \; \ell > 0; k \geq 0 \text{ is an integer.}$$

Using q very large as promised by (6.2), we see that

$$W^{\frac{3}{2},p}(\Omega) \subset C^{1,\alpha}(\Omega)$$

for all α: $0 < \alpha < 1/2$, therefore $w(P_i)$, $1 \leq i \leq m$, are all pointwise well defined. Therefore the difficulties associated with $N \geq 3$ as mentioned in §2 no longer exist.

We are now examining these relevant questions and hope to be able to report new findings in the future.

References

[1] R. Adams, *Sobolev Spaces*, Academic Press, New York, 1975.
[2] D. N. Arnold and W. L. Wendland, *The convergence of spline collocation for strongly elliptic equations on curves*, Numer. Math. 47 (1985), pp. 317-341.
[3] G. Chen and J. Zhou, *Boundary Element Methods*, Academic Press, London-San Diego, 1992.
[4] G. Chen, J. Zhou and R. McLean, *Boundary element method for shape (domain) optimization of linear-quadratic elliptic boundary control problems*, preprint. (Presented at the IFIP Workshop on "Boundary Control and Boundary Variations", Sophia-Antipolis, France, June, 1992.) Workshop Proceedings, J.-P. Zolesio ed., Springer Lecture Notes on Control and Information Science, to appear.
[5] G. Hsiao and R. C. MacCamy, *Solutions of boundary value problems by integral equations of the first kind*, SIAM Rev. 15 (1973), pp. 687-705.
[6] L. Ji, *Boundary element methods and supercomputer computations for some distributed parameter control problems*, Ph.D. Thesis, Department of Mathematics, Texas A&M University, College Station, Texas, 1991.
[7] J. L. Lions, *Contrôle Optimal de Systèmes Gouvernés par des Equations aux Derivées Partielles*, Dunod, Gauthier-Villars, Paris, 1968.
[8] J. L. Lions and E. Magenes, *Nonhomogeneous Boundary Value Problems and Applications*, Vol. 1, Springer-Verlag, New York, 1970.

Chapter 12
Boundary Control for a Viscous Burgers' Equation*

Christopher I. Byrnes [†]   David S. Gilliam[‡]   Victor I. Shubov [‡]

Abstract

In this paper we consider a boundary control problem for Burgers' equation on a finite interval. The controls enter as gain parameters in the boundary conditions as in [6, 7]. It is shown that the uncontrolled problem, obtained by equating the control parameters to zero, is not asymptotically stable. For positive gains we show that, at least locally in $L_2(0,1)$ we establish the global existence in time of solutions of the forced equation and show that the solution remains in a fixed ball in L_2 for all positive times. For zero forcing term we establish exponential stability for small initial data in L_2. We then observe that the corresponding nonlinear semigroup for the closed loop Burgers' system is compact for all $t > 0$ and are able to make several statements regarding the existence and properties of local attractors.

1 Introduction

There has been considerable attention given in the literature to the study of mathematical properties of solutions of nonlinear distributed parameter systems such as Navier-Stokes equations (see, for example [1, 4, 5, 6, 7, 11, 12], [18]- [21],[23]-[25]). Of considerable interest is the question of boundary control via point actuators and sensors. In the present paper we consider a simple example of such a boundary control problem for a viscous Burgers' equation on a finite interval. The uncontrolled problem corresponding to homogeneous Neumann boundary conditions is shown not to be asymptotically stable. Boundary control is effected by introducing radiation boundary conditions at each end of the rod with parameters in the boundary conditions considered as "gains" or control parameters. For small enough initial data in L_2 and for external source terms which are small enough in the Sobolev space H^{-1} and independent of time, we obtain global existence in time of solutions. In the case of zero forcing term we show that the system is exponential stable. Further it is shown that the corresponding nonlinear semigroups define compact operators for all positive times and based on this it is possible to obtain the existence of a local attractor.

It is well-known [16] that the one-dimensional Burgers' equation with an external forcing term can be reduced by the Hopf-Cole substitution to the one-dimensional heat equation with a "potential" term. However, in the case of Neumann or radiation boundary conditions, which we consider, this substitution does not help. The reason is that these conditions, in contrast with the Dirichlet case, are transformed into quadratically nonlinear conditions at the ends of the interval. This circumstance does not allow us to treat the above mentioned linear heat equation by standard methods. Further we comment that for

---

*This work was supported in part by grants from AFOSR and Texas Advanced Research Program.
[†]Department of Systems Science and Mathematics,Washington University, St. Louis, MO 63130.
[‡]Department of Mathematics, Texas Tech University, Lubbock, TX, 79409.

boundary conditions which are not either Dirichlet or periodic, the nonlinear term does not go away in the energy balance relation of the problem (cf. (25)). This is really the main difficultly encountered in the analysis that follows. The problem is that due to this extra term it is somewhat more difficult to obtain the necessary apriori estimates which guarantee the existence of an absorbing ball and are needed to push through the Galerkin approximation method described in [19, 25].

2 Formulation of Problem

We consider the controlled viscous Burgers' system

(1)
$$w_t - \epsilon w_{xx} + w w_x = f$$
$$-w_x(0,t) + k_0 w(0,t) = 0$$
$$w_x(1,t) + k_1 w(1,t) = 0$$
$$w(x,0) = \phi(x)$$

where k_0, $k_1 \geq 0$ are considered as boundary control parameters.

For $k_0 = k_1 = 0$ and $f = 0$ we have the uncontrolled system

$$w_t(x,t) - \epsilon w_{xx}(x,t) + w(x,t) w_x(x,t) = 0$$
$$w_x(0,t) = 0$$
$$w_x(1,t) = 0$$
$$w(x,0) = \phi(x)$$

This Burgers' system is not asymptotically stable. For example, for initial data $\phi(x) = 1$ the solution is $w(x,t) = 1$, for all $t \geq 0$. However, at least for initial data in $H^1(0,1)$ small enough, the solutions do approach constant equilibria as $t \to \infty$. This can be established by the application of an infinite dimensional version of the Center Manifold Theorem (cf. [10], [15]) together with the local existence theory developed in [22], [15].

Namely, defining $F(w) = -w w_x$ and the unbounded selfadjoint operator $B = (d/dx)^2$,

$$\mathcal{D}(B) = \{\varphi \in H^2(0,1) \mid \varphi_x(0) = \varphi_x(1) = 0\}$$

we see that
$$F: H^1(0,1) \to L_2(0,1),$$

is C^1, $F(0) = 0$, $F'(0) = 0$ and
$$0 \in \sigma_p(B)$$

so $Z \equiv L_2(0,1)$ has the natural decomposition

$$Z = Z_1 \oplus Z_2$$

where $Z_1 = \ker(B) = \mathbb{C}$, $Z_2 = Z_1^\perp$. Z_1 is invariant for both B and F.

Thus there exists (cf [15], [22]) β, $\rho > 0$ so that for initial data $\phi \in H^1(0,1)$ with $\|\phi\|_{H^1} \leq \rho$ the solution w has the decomposition

$$w = z_1 + z_2, \quad z_1 \in Z_1, \quad z_2 \in Z_2$$

and there is a constant $z_\infty \in \mathbb{C}$ so that

$$|z_1(t) - z_\infty| + \|z_2(t)\|_{H^1} = O\left(e^{-\beta t}\right).$$

As indicated in the introduction, in this paper we are interested in questions of existence, stability, etc., for L_2 initial data. To this end and in a standard notation we define the notion of a weak solution as follows (see, for example [19, 21]).

DEFINITION 2.1. *Let $\phi \in L_2(0,1)$, $f \in L_2(0,1)$, $Q_T = [0,1] \times [0,T]$, where $T > 0$. A function*

$$w \in L_\infty([0,T], L_2(0,1)) \cap L_2\left([0,T], H^1(0,1)\right) \equiv \mathcal{L}$$

is a weak solution of (1) if it satisfies the following identity

(2)
$$\begin{aligned}
\int_{Q_T} (-w\eta_t + \epsilon w_x \eta_x + w w_x \eta) \, dx \, dt + \\
\epsilon k_0 \int_0^T w(0,t)\eta(0,t) \, dt + \epsilon k_1 \int_0^T w(1,t)\eta(1,t) \, dt \\
= \int_0^1 \phi(x)\eta(x,0) \, dx + \int_{Q_T} f\eta \, dx \, dt
\end{aligned}$$

for all $\eta \in H^1(Q_T)$ such that $\eta(x,T) = 0$. Here the norm in $\mathcal{L}$ is

$$|w|_{Q_T}^2 = \text{ess sup}_{t \in [0,T]} \|w(\cdot,t)\|_{L_2(0,1)}^2 + \int_0^T \|w(\cdot,t)\|_{H^1(0,1)}^2 \, dt.$$

We define the operator

$$A_K = -\frac{d^2}{dx^2}$$

with dense domain in $L_2(0,1)$

$$\mathcal{D}(A_K) = \{\varphi \in H^2(0,1) : \varphi'(0) - k_0\varphi(0) = 0, \; \varphi'(1) + k_1\varphi(1) = 0\}.$$

Then A_K is a strictly positive selfadjoint operator for $k_0 \geq 0$, $k_1 \geq 0$ and $k_0 + k_1 > 0$. Defining $\lambda = \mu^2$ it is a simple computation to obtain the characteristic equation providing the eigenvalues (spectrum of A_K)

(3)
$$(k_0 k_1 - \mu^2)\frac{\sin(\mu)}{\mu} + (k_0 + k_1)\cos(\mu) = 0$$

This equation has infinitly many zeros $\{\mu_j\}_{j=1}^\infty$ satisfying

$$(j-1)\pi < \mu_j < j\pi, \; j = 1, 2, \cdots$$

Corresponding to the eigenvalues $\lambda_j = \mu_j^2$ for A_K we have the complete orthonormal system

(4)
$$\psi_j(x) = \kappa_j \left(k_0 \frac{\sin(\mu_j x)}{\mu_j} + \cos(\mu_j x)\right), \; j = 1, 2, \cdots$$

in $L_2(0,1)$ where

(5)
$$\kappa_j = \sqrt{\frac{2\mu_j^2(k_1^2 + \mu_j^2)}{(k_0^2 + \mu_j^2)(k_1^2 + \mu_j^2) + (k_0 + k_1)(\mu_j^2 + k_0 k_1)}}.$$

is the normalization constant. The eigenvalues $\lambda_j = \mu_j^2$, are simple and for all $\varphi \in L_2(0,1)$ we have

$$\varphi = \sum_{j=1}^{\infty} \varphi_j \psi_j, \quad \varphi_j = <\varphi, \psi_j>,$$

$$D(A_K) = \{\varphi \in L_2(0,1) : \sum_{j=1}^{\infty} \lambda_j^2 |\varphi_j|^2 < \infty\}$$

and for $\varphi \in D(A_K)$

$$A_K \varphi = \sum_{j=1}^{\infty} \lambda_j \varphi_j \psi_j.$$

For $\varphi \in L_2(0,1)$ we have the representation for the resolvent

$$(\lambda I - A_K)^{-1} \varphi = \sum_{j=1}^{\infty} \frac{1}{(\lambda - \lambda_j)} \varphi_j \psi_j.$$

The selfadjoint operator $(-A_K)$ is the infinitesimal generator of an analytic semigroup $T_K(t)$ given by

$$T_K(t)\varphi = \sum_{j=1}^{\infty} e^{-\lambda_j t} \varphi_j \psi_j$$

and for every $t > 0$, $T_K(t)$ is a compact operator satisfying

$$\|T_K(t)\| \leq e^{-\lambda_1 t}.$$

We can further define the square root of A_k by first defining the domain

$$D(A_K^{1/2}) = \{\varphi \in L_2(0,1) : \sum_{j=1}^{\infty} \lambda_j |\varphi_j|^2 < \infty\}$$

and for $\varphi \in D(A_K^{1/2})$

(6) $$(A_K)^{1/2}\varphi = \sum_{j=1}^{\infty} \mu_j \varphi_j \psi_j$$

Put in slightly different words and in the usual way, the operator A_K defines an infinite scale of Hilbert spaces $\mathcal{H}^s$ ($s \in \mathbb{R}$) with norms

(7) $$\|w\|_s = \|A_K^{s/2} w\|$$

where $\|\cdot\|$ denotes the norm in $L_2(0,1)$. It is easy to see that the norm in $\mathcal{H}^1$ can be represented by

(8) $$\|\varphi\|_1^2 = \int_0^1 |\varphi_x|^2 \, dx + k_0 |\varphi(0)|^2 + k_1 |\varphi(1)|^2$$

LEMMA 2.1. *For each $\varphi \in \mathcal{H}^1$ we have*

(9) $$\|\varphi\|^2 \leq \lambda_1^{-1} \|\varphi\|_1^2,$$

where $0 < \lambda_1 < \pi^2$ is the first eigenvalue of A_K. Furthermore, the norm (8) is equivalent to the $H^1(0,1)$ Sobolev norm and, therefore $\mathcal{H}^1 = H^1(0,1)$. In fact we have

(10) $$\left(1 + \lambda_1^{-1}\right)^{-1} \|\varphi\|_{H^1}^2 \leq \|\varphi\|_1^2 \leq [\max(2(k_0 + k_1), (1 + k_0 + k_1))] \|\varphi\|_{H^1}^2$$

which from now on we write as

(11)
$$a\|\varphi\|_{H^1} \leq \|\varphi\|_1 \leq b\|\varphi\|_{H^1}$$

Proof: The inequality (9) is a simple consequence of (6) and (8) and the estimate is sharp. Now using the fact that $\|\varphi'\|^2 \leq \|\varphi\|_1^2$ and (9), we obtain

$$\|\varphi\|_{H^1}^2 = \|\varphi\|^2 + \|\varphi'\|^2 \leq \left(1 + \lambda_1^{-1}\right)\|\varphi\|_1^2$$

and this estimate is not sharp. For the inverse estimate write

$$\varphi(0) = \varphi(x) - \int_0^x \varphi'(t)\, dt$$

so that with the Cauchy-Schwartz inequality we have

$$\begin{aligned}|\varphi(0)|^2 &\leq 2|\varphi(x)|^2 + 2\left(\int_0^x \varphi'(t)\, dt\right)^2 \\ &\leq 2|\varphi(x)|^2 + 2x\int_0^x |\varphi'(t)|^2\, dt\end{aligned}$$

Integrating over $(0,1)$ both sides of this inequality gives

$$\begin{aligned}|\varphi(0)|^2 &\leq 2\int_0^1 |\varphi(x)|^2\, dx + 2\int_0^1 x \int_0^x |\varphi'(t)|^2\, dt\, dx \\ &= 2\int_0^1 |\varphi(x)|^2\, dx + 2\int_0^1 |\varphi'(t)|^2 \left(\int_t^1 x\, dx\right) dt \\ &= 2\int_0^1 |\varphi(x)|^2\, dx + \int_0^1 |\varphi'(t)|^2(1-t^2)\, dt \\ &\leq 2\int_0^1 |\varphi(x)|^2\, dx + \int_0^1 |\varphi'(t)|^2\, dt \\ &= 2\|\varphi\|^2 + \|\varphi'\|^2\end{aligned}$$

In exactly the same way we obtain

$$|\varphi(1)|^2 \leq 2\|\varphi\|^2 + \|\varphi'\|^2.$$

So,

$$\begin{aligned}\|\varphi\|_1^2 &= \|\varphi'\|^2 + k_0|\varphi(0)|^2 + k_1|\varphi(1)|^2 \\ &\leq (1 + k_0 + k_1)\|\varphi'\|^2 + 2(k_0 + k_1)\|\varphi\|^2 \\ &\leq \max\{2(k_0 + k_1), 1 + k_0 + k_1\}\|\varphi\|_{H^1}^2\end{aligned}$$

$\square$

In the sequel we also need an extension of a classical inequality (cf. [21]) for $\varphi \in W_m^1[a,b]$, $m \geq 1$ with $\varphi(a) = 0$

(12)
$$\|\varphi\|_{L_q} \leq \beta\left(\|\varphi_x\|_{L_m}\right)^\alpha \left(\|\varphi\|_{L_r}\right)^{1-\alpha}$$

where $r \leq q \leq \infty$ ($q = \infty$ is valid in 1-dimension),

$$\alpha = \left(\frac{1}{r} - \frac{1}{q}\right)\left(1 - \frac{1}{m} + \frac{1}{r}\right)^{-1}$$

and

$$\beta = \left(1 + \frac{m-1}{m}r\right)^{\alpha}.$$

LEMMA 2.2. *For any $\varphi \in H^1(0,1)$ we have*

$$\|\varphi\|_{L_q} \leq \gamma \|\varphi\|_1^{\alpha} \|\varphi\|^{1-\alpha} \tag{13}$$

where $2 \leq q \leq \infty$, $\alpha = 1/2 - 1/q$, $\gamma = 2^{(1+\alpha)/2}\left(3 + 2\lambda_1^{-1}\right)^{\alpha/2}$, λ_1 is the first eigenvalue of A_K and $\|\varphi\|_1$ is the norm given in (8).

Proof: Applying the inequality (12) with $m = r = 2$ on the interval $[-1,1]$, we have

$$\|\tilde{\varphi}\|_{L_q[-1,1]} \leq 2^{\alpha}\left(\|\tilde{\varphi}_x\|_{L_2[-1,1]}\right)^{\alpha}\left(\|\tilde{\varphi}\|_{L_r[-1,1]}\right)^{1-\alpha} \tag{14}$$

where $\alpha = 1/2 - 1/q$ and $2 \leq q \leq \infty$ and $\tilde{\varphi} \in H^1[-1,1]$ with $\tilde{\varphi}(-1) = 0$.

Now consider an arbitrary function $\varphi \in H^1(0,1)$ which we extend to $[-1,1]$ by

$$\tilde{\varphi} = \begin{cases} \varphi(x), & x \in [0,1] \\ (x+1)\varphi(-x), & x \in [-1,0] \end{cases} \tag{15}$$

It is clear that $\tilde{\varphi} \in H^1[-1,1]$ and $\tilde{\varphi}(-1) = 0$. Note that we have the following estimates relating the norms in $L_2[-1,1]$ and $L_2[0,1]$ for $\tilde{\varphi}$.

$$\|\tilde{\varphi}\|_{L_2[-1,1]}^2 = \|\varphi\|^2 + \int_{-1}^{0}(x+1)^2|\varphi(-x)|^2\,dx \leq 2\|\varphi\|^2 \tag{16}$$

$$\begin{aligned}
\|\tilde{\varphi}_x\|_{L_2[-1,1]}^2 &= \|\varphi_x\|^2 + \int_{-1}^{0}|\varphi(-x) - (x+1)\varphi_x(-x)|^2\,dx \\
&\leq \|\varphi_x\|^2 + 2\int_{-1}^{0}\left(|\varphi(-x)|^2 + (x+1)^2|\varphi_x(-x)|^2\right)dx \\
&\leq \|\varphi_x\|^2 + 2\left(\|\varphi\|^2 + \|\varphi_x\|^2\right) \\
&= 3\|\varphi_x\|^2 + 2\|\varphi\|^2 \\
&\leq \left(3 + 2\lambda_1^{-1}\right)\|\varphi\|_1^2
\end{aligned} \tag{17}$$

where in the last step we used (9) and the fact that $\|\varphi_x\|^2 \leq \|\varphi\|_1^2$. Now substituting (16) and (17) into (14) and taking into account that

$$\|\varphi\|_{L_q[0,1]} \leq \|\tilde{\varphi}\|_{L_q[-1,1]}$$

we immediately arrive at (13). □

THEOREM 2.3. *Let*

$$\rho = \left(\frac{\epsilon^4 \lambda_1}{8\tilde{\gamma}}\right)^{1/2} \tag{18}$$

where $\tilde{\gamma} = 2^8(3 + 2\lambda_1^{-1})$. Assume that $\phi \in L_2(0,1)$ and $f \in \mathcal{H}^{-1}$ (note that $\mathcal{H}^{-1} = H^{-1}(0,1)$) satisfy the conditions

(19) $$\|\phi\|^2 \le \rho$$
(20) $$\|f\|_{-1}^2 \le \epsilon^2 \lambda_1 \rho$$

Notice that $L_2(0,1) \subset \mathcal{H}^{-1}$ and
(21) $$\|f\|_{-1}^2 \le \lambda_1 \|f\|^2.$$

So, if $f \in L_2(0,1)$ we replace (20) by the condition
(22) $$\|f\|^2 \le \epsilon^2 \rho.$$

1. For any $T > 0$ there exists a unique weak solution of (1) in the sense of definition 2.1. This solution is a continuous function of t in the following sense
$$\|w(\cdot, t + \Delta t) - w(\cdot, t)\| \to 0, \ \Delta t \to 0$$
i.e., $w \in C([0,T], L_2(0,1)) \cap L_2([0,T], H^1(0,1))$.

2. For all $t > 0$, the above solution has the estimate
(23) $$\|w(t)\|^2 \le \max\left\{\rho, \frac{2\|f\|_{-1}^2}{\epsilon^2 \lambda_1}\right\} \equiv \rho_0$$

For $f \in L_2(0,1)$ this estimate is replaced by
(24) $$\|w(t)\|^2 \le \max\left\{\rho, \frac{2\|f\|^2}{\epsilon^2}\right\} \equiv \rho_0$$

3. The solution $w(x,t)$ depends continuously on the initial condition ϕ in the norm of $L_2(0,1)$, moreover
$$\|w_1(\cdot, t) - w_2(\cdot, t)\| \le C(t)\|\phi_1 - \phi_2\|,$$
where $C(t) \ge 0$ is a continuous function of t.

REMARK 2.1. The result expressed in part 2. of the Theorem estimate shows that for sufficiently small initial data and forcing term (as described by (19) and (20)) the dynamical semigroup generated by (1) is bounded dissipative, i.e., it has an absorbing ball. Each trajectory with initial point satisfying (19) does not leave the ball of radius ρ_0 in (23). The proofs of statements 1. and 3. are a bit more lengthy and technical and are postponed to a full account of the results described here and contained in [9]. In [9], we show that the results can be obtained by a combination of the estimates we derive below with standard techniques that are well known, for example, in the case of the Navier-Stokes equation (see [19]). Thus the essential new results are the estimates derived in the proof that follows.

Proof of Theorem 2.3: Take the L_2-inner product of (1) with w to obtain
$$\int_0^1 \frac{\partial w}{\partial t} w \, dx = \epsilon \int_0^1 w_{xx} w \, dx - \int_0^1 w^2 w_x \, dx + \int_0^1 fw \, dx$$

Integrating by parts yields
$$\frac{1}{2}\frac{d}{dt}\|w\|^2 + \epsilon \left(\|w_x\|^2 + k_0|w(0,t)|^2 + k_1|w(1,t)|^2\right)$$
$$+ \int_0^1 w^2 w_x \, dx = \int_0^1 fw \, dx$$

which is the same as

$$\text{(25)} \quad \frac{1}{2}\frac{d}{dt}\|w\|^2 + \epsilon\|w\|_1^2 = -\int_0^1 w^2 w_x \, dx + \int_0^1 fw \, dx$$

We estimate the first term on the right in (25) using the simple inequality $ab \leq a^2/(4\delta) + \delta b^2$

$$\text{(26)} \quad \left|\int_0^1 w^2 w_x \, dx\right| \leq \frac{1}{4\delta}\int_0^1 w^4 \, dx + \delta \int_0^1 w_x^2 \, dx$$

With $q = 4$, so that $\alpha = 1/4$ in (13)

$$\text{(27)} \quad \|w\|_{L_4} \leq \gamma (\|w\|_1)^{1/4} (\|w\|)^{3/4}$$

Raising both sides to the fourth power, defining

$$\text{(28)} \quad \beta = \gamma^4 = 2^{5/2}(3 + 2\lambda_1^{-1})^{1/2}$$

where γ is the constant from (13) corresponding to $q = 1/4$ and again applying the inequality $ab \leq \delta' a^2 + b^2/(4\delta')$ we have

$$\text{(29)} \quad \begin{aligned} \|w\|_{L_4}^4 &\leq \beta \|w\|_1 (\|w\|)^3 \\ &\leq \delta'\|w\|_1^2 + \frac{\beta^2}{4\delta'}\|w\|^6 \end{aligned}$$

Combining (26) and (29) we obtain

$$\text{(30)} \quad \left|\int_0^1 w^2 w_x \, dx\right| \leq \delta\|w\|_1^2 + \frac{1}{4\delta}\left(\frac{\beta^2}{4\delta'}\|w\|^6 + \delta'\|w\|_1^2\right)$$

For the second term on the right in (25) we have

$$\text{(31)} \quad \left|\int_0^1 fw \, dx\right| \leq \|f\|_{-1}\|w\|_1 \leq \delta''\|w\|_1^2 + \frac{1}{4\delta''}\|f\|_{-1}^2$$

If $f \in L_2(0,1)$ then we can replace this with

$$\text{(32)} \quad \begin{aligned} \left|\int_0^1 fw \, dx\right| &\leq \|f\|\|w\| \leq \lambda_1^{-1/2}\|f\|\|w\|_1 \\ &\leq \delta''\|w\|_1^2 + \frac{\lambda_1^{-1}}{4\delta''}\|f\|^2 \end{aligned}$$

From (11) and (25) we have

$$\text{(33)} \quad \begin{aligned} \frac{1}{2}\frac{d}{dt}\|w\|^2 + \epsilon\|w\|_1^2 &\leq \delta''\|w\|_1^2 + \frac{1}{4\delta''}\|f\|_{-1}^2 + \\ &+ \delta\|w\|_1^2 + \frac{1}{4\delta}\left(\delta'\|w\|_1^2 + \frac{\beta^2}{4\delta'}\|w\|^6\right) \\ &= \left(\delta'' + \delta + \frac{\delta'}{4\delta}\right)\|w\|_1^2 + \frac{\beta^2}{16\delta'\delta}\|w\|^6 + \frac{1}{4\delta''}\|f\|_{-1}^2 \end{aligned}$$

With the choices
$$\delta'' = \frac{\epsilon}{8}, \quad \delta = \frac{\epsilon}{8}, \quad \frac{\delta'}{4\delta} = \frac{\epsilon}{4}$$
we obtain
$$\delta' = \epsilon\delta = \frac{\epsilon^2}{8}, \quad 16\delta'\delta = 16\frac{\epsilon^2}{8}\frac{\epsilon}{8} = \frac{\epsilon^3}{4}$$
and
$$(\delta'' + \frac{\delta'}{4\delta} + \delta) = \frac{\epsilon}{2}$$
and (33) can be written as
$$\frac{1}{2}\frac{d}{dt}\|w\|^2 + \frac{\epsilon}{2}\|w\|_1^2 \leq \frac{4\beta^2}{\epsilon^3}\|w\|^6 + \frac{1}{\epsilon}\|f\|_{-1}^2$$
which on multiplying by 2 gives

(34) $$\frac{d}{dt}\|w\|^2 + \epsilon\|w\|_1^2 \leq \frac{8\beta^2}{\epsilon^3}\|w\|^6 + \frac{1}{\epsilon}\|f\|_{-1}^2$$

Now define $\tilde{\gamma} = 8\beta^2$ (note that $8\beta^2 = 2^8(3 + 2\lambda_1^{-1})$ coincides with $\tilde{\gamma}$ from (18)) and let
$$y(t) \equiv \|w(t)\|^2.$$

It follows from (34) that

(35) $$\frac{dy}{dt} \leq \frac{\tilde{\gamma}}{\epsilon^3}y^3 + \frac{1}{\epsilon}\|f\|_{-1}^2$$

which implies

(36) $$\frac{dy}{dt} \leq \frac{\tilde{\gamma}}{\epsilon^3}\left(y^3 + \frac{\epsilon^2}{\tilde{\gamma}}\|f\|_{-1}^2\right)$$

and also

(37) $$\frac{dy}{dt} \leq \frac{\tilde{\gamma}}{\epsilon^3}(y+a)^3, \quad a = \left(\frac{\epsilon^2}{\tilde{\gamma}}\|f\|_{-1}^2\right)^{1/3}$$

Integrating both sides of (37) from 0 to t we have
$$\frac{-1}{2(y(t)+a)^2} + \frac{1}{2(y(0)+a)^2} \leq \frac{\tilde{\gamma}}{\epsilon^3}t.$$

which implies

(38) $$y(t) \leq \frac{1}{((y(0)+a)^{-1} - 2\tilde{\gamma}\epsilon^{-3}t)^{1/2}} - a$$

The apriori estimate (38) allows us to conclude that the L_2 norm of solutions is bounded at least on the time interval
$$0 \leq t \leq \frac{\epsilon^3}{2\tilde{\gamma}(y(0)+a)^2}.$$

Now let $\rho > 0$ be given and assume that

(39) $$y(0) \leq \rho$$

Define

(40) $$t_1(\rho) = \sup\{t : y(\tau) \leq 2\rho, \forall \tau \in [0,t]\}$$

It follows from (38) that $t_1(\rho) > 0$.

For $t \in [0, t_1(\rho)]$, we have $y^3(t) \leq 4\rho^2 y(t)$ and since $\|w\|_1^2 \geq \lambda_1 \|w\|^2$ by (9), we see that (34) implies

(41) $$\frac{dy}{dt} + \epsilon\lambda_1 y \leq \frac{4\rho^2 \widetilde{\gamma}}{\epsilon^3} y + \frac{1}{\epsilon}\|f\|_{-1}^2$$

Assume now that $\frac{4\rho^2 \widetilde{\gamma}}{\epsilon^3} \leq \frac{\epsilon\lambda_1}{2}$ or

(42) $$\rho^2 \leq \frac{\epsilon^4 \lambda_1}{8\widetilde{\gamma}}$$

Then (41) implies
(43) $$\frac{dy}{dt} + \frac{\epsilon\lambda_1}{2} y \leq \frac{1}{\epsilon}\|f\|_{-1}^2$$

and from Gronwall's Lemma we have

(44) $$y(t) \leq \left(y(0) - \frac{2}{\epsilon^2 \lambda_1}\|f\|_{-1}^2\right) e^{-\epsilon\lambda_1 t/2} + \frac{2}{\epsilon^2 \lambda_1}\|f\|_{-1}^2$$

So, if $t \in [0, t_1(\rho)]$ and ρ satisfies (42) then

(45) $$y(t) \leq \left(\rho - \frac{2}{\epsilon^2 \lambda_1}\|f\|_{-1}^2\right) e^{-\epsilon\lambda_1 t/2} + \frac{2}{\epsilon^2 \lambda_1}\|f\|_{-1}^2$$

If now we have that $(\rho - 2/(\epsilon^2 \lambda_1)\|f\|_{-1}^2) \geq 0$ which means

(46) $$\|f\|_{-1}^2 \leq \frac{\epsilon^2 \lambda_1 \rho}{2}$$

then (45) implies that
(47) $$y(t) \leq \rho, \quad \forall t \in [0, t_1(\rho)]$$

and therefore
(48) $$y(t) \leq \rho, \quad \forall t \in [0, \infty)$$

If we have
$$\frac{2}{\epsilon^2 \lambda_1}\|f\|_{-1}^2 \leq 2\rho,$$

which means
(49) $$\|f\|_{-1}^2 \leq \epsilon^2 \lambda_1 \rho$$

then (45) implies
$$y(t) \leq \max\left\{\rho, \frac{2\|f\|_{-1}^2}{\epsilon^2 \lambda_1}\right\}.$$

The last constant does not exceed 2ρ and again all the above estimates hold for all $0 < t < \infty$, i.e., $t_1(\rho) = \infty$. □

COROLLARY 2.4. *Assume that $f = 0$ and $\phi \in L_2(0,1)$ satisfies (19) with ρ given by (18). Then the weak solution w of (1) satisfies*

(50) $$\|w(t)\|^2 \leq \rho e^{-(\epsilon\lambda_1 t)/2}$$

Proof: The proof follows directly from the proof of Theorem 2.3 with $f = 0$. □

We now derive an apriori estimate that allows us to conclude that the evolution operators generated by the problem (1) are compact for all $t > 0$. In this proof we follow the idea suggested in [20] in which the existence of the global attractor for 2-dimensional Navier-Stokes was first established. Since we deal with apriori estimates we do not justify the existence of all derivatives we use in this proof. The existence of these derivatives in the present case are exhibited in the paper [9] as indicated earlier for the existence of solutions.

THEOREM 2.5. *Assume that $f \in L_2(0,1)$ satisfies (22) with ρ given in (18) and $\phi \in L_2(0,1)$ satisfies (19). Then there exists a continuous function $M(\xi, t)$, in $\xi \geq 0$, $t \geq 0$ such that*

(51) $$\|w(t)\|_1 \leq t^{-1/2} M(\|\phi\|, t)$$

Proof:

Take the L_2-inner product of (1) with $-w_{xx}$ to obtain

$$-\int_0^1 \frac{\partial w}{\partial t} w_{xx}\, dx = -\epsilon \int_0^1 |w_{xx}|^2\, dx + \int_0^1 w w_x w_{xx}\, dx - \int_0^1 f w_{xx}\, dx$$

Integrating by parts yields

$$\frac{1}{2}\frac{d}{dt}\left(\|w_x\|^2 + k_0|w(0,t)|^2 + k_1|w(1,t)|^2\right) + \epsilon\|w_{xx}\|^2$$
$$= \int_0^1 w w_x w_{xx}\, dx - \int_0^1 f w_{xx}\, dx$$

which is the same as

(52) $$\frac{1}{2}\frac{d}{dt}\|w_x\|_1^2 + \epsilon\|w_{xx}\|^2 = \int_0^1 w w_x w_{xx}\, dx - \int_0^1 f w_{xx}\, dx$$

Recall that by lemma 2.1 there exist $a, b > 0$ such that

$$a\|w\|_{H^1}^2 \leq \|w\|_1^2 \leq b\|w\|_{H^1}^2.$$

We now proceed to estimate the first term on the right in (52)

(53) $$\left|\int_0^1 w w_x w_{xx}\, dx\right| \leq (\|w\|_{L_\infty})(|<|w_x|,|w_{xx}|>|) \leq \|w\|_{L_\infty}\|w_x\|\|w_{xx}\|$$

Now using the extended inequality obtain in Lemma 2.2

(54) $$\|w\|_{L_q} \leq \gamma (\|w\|_1)^\alpha (\|w\|)^{1-\alpha}$$

where $2 \leq q \leq \infty$ ($q = \infty$ is valid in 1-dimension) and with $q = \infty$ so that $\alpha = 1/2$ we have

(55) $$\|w\|_{L_\infty} \leq \gamma (\|w\|_1)^{1/2} (\|w\|)^{1/2}$$

where

$$\gamma = 2^{3/4}(3 + 2\lambda_1^{-1})^{1/4}$$

is the constant in (13) corresponding to $q = \infty$. The estimates (53) and (55) provide

(56) $$\left|\int_0^1 w w_x w_{xx}\, dx\right| \leq \gamma (\|w\|_1)^{1/2} (\|w\|)^{1/2} \|w_x\|\|w_{xx}\|$$
$$\leq \gamma (\|w\|_1)^{3/2} (\|w\|)^{1/2} \|w_{xx}\|$$

So,

(57) $$\left|\int_0^1 w w_x w_{xx} \, dx\right| \leq \delta \|w_{xx}\|^2 + \frac{\gamma^2}{4\delta} \|w\| (\|w\|_1)^3$$

By the proof of theorem 2.1, we have

(58) $$\|w(t)\| \leq \rho_0^{1/2}$$

(ρ_0 from (24)) and we can write

(59) $$\left|\int_0^1 w w_x w_{xx} \, dx\right| \leq \delta \|w_{xx}\|^2 + \frac{\gamma^2 \rho_0^{1/2}}{4\delta} (\|w\|_1)^3$$

As for the second term on the right in (52), we have

(60) $$|<f, w_{xx}>| \leq \delta' \|w_{xx}\|^2 + \frac{1}{4\delta'} \|f\|^2.$$

(52), (59), and (60) imply

(61) $$\frac{d}{dt}\|w\|_1^2 + 2\epsilon \|w_{xx}\|^2 \leq 2(\delta + \delta') \|w_{xx}\|^2 +$$
$$+ \frac{\gamma^2 \rho_0^{1/2}}{2\delta} (\|w\|_1)^3 + \frac{1}{2\delta'} \|f\|^2$$

Let $2\delta = \epsilon$, $2\delta' = \epsilon$ so that $\dfrac{\gamma^2 \rho_0^{1/2}}{2\delta} = \dfrac{\gamma^2 \rho_0^{1/2}}{\epsilon}$ and we have

(62) $$\frac{d}{dt}\|w\|_1^2 \leq \frac{\gamma^2 \rho_0^{1/2}}{\epsilon} (\|w\|_1)^3 + \epsilon^{-1} \|f\|^2$$

Multiplying (62) by t we have

(63) $$\frac{d}{dt}\left(t\|w\|_1^2\right) \leq \frac{\gamma^2 \rho_0^{1/2}}{\epsilon} t\|w\|_1^3 + \|w\|_1^2 + \frac{t}{\epsilon}\|f\|^2$$

Let $y(t) = t\|w(t)\|_1^2$, $a(t) = \dfrac{\gamma^2 \rho_0^{1/2}}{\epsilon} \|w\|_1$, and $b(t) = \|w\|_1^2 + \epsilon^{-1} t \|f\|^2$, then (63) can be written as

(64) $$\frac{d}{dt} y(t) \leq a(t) y(t) + b(t)$$

and by Gronwall's inequality

(65) $$y(t) \leq \exp\left(\int_0^t a(\tau) \, d\tau\right) \int_0^t b(\tau) \, d\tau$$

since $y(0) = 0$. So,

(66) $$t\|w(t)\|_1^2 \leq \exp\left(\int_0^t \frac{\gamma^2 \rho_0^{1/2}}{\epsilon} \|w(\tau)\|_1 \, d\tau\right) \int_0^t \left(\|w(\tau)\|_1^2 + \epsilon^{-1} \tau \|f\|^2\right) d\tau$$

By Cauchy-Schwartz inequality we have for each $t > 0$

(67) $$\int_0^t \|w(t)\|_1 \, d\tau \leq \sqrt{t} \left(\int_0^t \|w(\tau)\|_1^2 \, d\tau\right)^{1/2}$$

But according to the proof of theorem 2.1 there exists a continuous function $C(\cdot, t)$ such that

(68) $$\int_0^t \|w(\tau)\|_1^2 \, d\tau \leq C(\|\phi\|, t)$$

Indeed, we had

(69) $$\frac{d}{dt}\|w\|^2 + \epsilon\|w\|_1^2 \leq \frac{8\beta^2}{\epsilon^3}\|w\|^6 + \frac{1}{\epsilon}\|f\|^2$$

and by (58)
$$\|w(t)\|^2 \leq \rho_0.$$

So,

(70) $$\frac{d}{dt}\|w\|^2 + \epsilon\|w\|_1^2 \leq \frac{8\beta^2 \rho_0^3}{\epsilon^3} + \frac{1}{\epsilon}\|f\|^2$$

and hence integrating from 0 to t

(71) $$\|w(t)\|^2 + \epsilon \int_0^t \|w(\tau)\|_1^2 \, d\tau \leq \frac{8\beta^2 \rho_0^3}{\epsilon^3} t + \frac{3}{\epsilon}\|f\|^2 t + \|\phi\|^2$$

and finally

(72) $$\int_0^t \|w(\tau)\|_1^2 \, d\tau \leq \frac{8\beta^2 \rho_0^3}{\epsilon^4} t + \frac{3}{\epsilon^2}\|f\|^2 t + \frac{1}{\epsilon}\|\phi\|^2 \equiv C_3(\|\phi\|, t)$$

Thus, combining (66) and (72) we see that there exists a continuous function $M(\cdot, t)$ so that
$$t\|w(t)\|_1^2 \leq M^2(\|\phi\|, t)$$
and the result of the theorem is obtained. $\square$

As a result of theorem 2.1, for $\|f\|^2 \leq (\epsilon^2 \rho)$ and defining $B_0 = \{\phi \in L_2(0,1) : \|\phi\|^2 \leq \rho\}$, we can define a 1-parameter family of mappings
$$S_t : B_0 \to L_2(0,1) : \phi \mapsto S_t(\phi) = w(t), \ t \geq 0.$$

These mappings are continuous. Now consider a set
$$\gamma_0(B_0) = \cup_{t \geq 0} S_t(B_0).$$

S_t can be naturally extended to $\gamma_0(B_0)$ and the family $\{S_t, \ t \geq 0\}$ becomes a semigroup of transformations on $\gamma_0(B_0)$ which can be extended to $\overline{\gamma_0(B_0)}$ by continuity. Thus in this way we obtain a semigroup $\{S_t, \ t \geq 0\}$ with phase space $\overline{\gamma_0(B_0)}$. From theorem 2.2 we see that all the mappings S_t with $t > 0$ are compact, i.e., if $B \subset \overline{\gamma_0(B_0)}$ then $S_t(B)$ is relatively compact.

THEOREM 2.6. *Define*

(73) $$\mathcal{A} = \cap_{\tau \geq 0} \overline{\cup_{t \geq \tau} S_t(B_0)}$$

Then $\mathcal{A}$ is a local attractor:
  1. $\mathcal{A}$ is nonempty

  2. $\mathcal{A}$ is connected

  3. $\mathcal{A}$ is compact

4. $\mathcal{A}$ attracts all bounded subsets of $\overline{\gamma_0(B_0)}$:

$$\lim_{t\to\infty} d(S_t(B), \mathcal{A}) = 0, \ \forall B \subset \overline{\gamma_0(B_0)}$$

Here $d(B, A) = \sup_{b \in B} \inf_{a \in A} \|b - a\|$.

5. $\mathcal{A}$ is invariant: $S_t(\mathcal{A}) = \mathcal{A}$ for all $t \geq 0$.

6. $S_t|_\mathcal{A}$ can be extended to a group.

Proof: The proof of statements 1.-5. are simple consequences of Theorem 2.1, 2.2 and the general results on attractors found in, for example, [24] or [23]. Statement 6. follows from the fact that all mappings $S_t|_\mathcal{A}$ are invertible. This can be shown, for example, by the method that was used in [20] for 2-dimensional Navier-Stokes equations. □

Notice, that Statement 3 of Theorem 2.3 can be reformulated in the following way: The semigroup operators S_t, $t \geq 0$ are Lipschitz as mappings from $L^2(0,1)$ to $L^2(0,1)$. Our next theorem shows that this result can be strengthed. Namely, the semigroup operators S_t for $t > 0$ are Lipschitz as mappings from $L^2(0,1)$ to $\mathcal{H}^1 = H^1(0,1)$.

THEOREM 2.7. *For any initial data ϕ_1, $\phi_2 \in B_0$ and corresponding solutions w_1, w_2 given in Theorem 2.1 we have*

(74) $$\|w_1(t) - w_2(t)\|_1 \leq C(t)\|\phi_1 - \phi_2\|$$

for $t > 0$ where $C(t) > 0$ is a continuous function such that $\lim_{t \downarrow 0} C(t) = \infty$.

COROLLARY 2.8. *The fractal and hence Hausdorf dimensions of $\mathcal{A}$ are finite.*

This corollary can be derived from (74) in a standard way which is explained in detail, for example, in [23]. It is possible to derive explicit estimates for the above dimensions. However, we do not do it here, because the structure of the attractor will be studied in detail in our forthcoming paper of this topic.

References

[1] Babin, A.V., Vishik, I.M. "Attractors of partial differential equations in an unbounded domain," *Proceedings of the Royal Society of Edinburgh*, 116A (1990) 221-243.

[2] Babin, A.V., Vishik, M.I. "Attractors of nonlinear evolution equations and estimates of their dimensions," Russian Mathematical Surveys. 38 (1983) 151-213.

[3] J. Burgers, "Application of a model system to illustrate some points of the statistical theory of free turbulence," *Nederl. Akad. Wefensh. Proc.*, 43 (1940), 2-12.

[4] J.A. Burns and S. Kang, "A control problem for Burgers' equation with bounded input/output," *Nonlinear Dynamics*, (1991), Vol. 2, 235-262.

[5] J.A. Burns and S. Kang, "A Stabilization problem for Burgers' equation with unbounded control and observation," *International Series of Numerical Mathematics*, (1991), Vol. 100, 51-72.

[6] C.I. Byrnes, D.S. Gilliam, "Boundary feedback design for nonlinear distributed parameter systems," *Proc. of IEEE Conf. on Dec. and Control*, Britton, England 1991.

[7] C.I. Byrnes and D.S. Gilliam, "Stability of certain distributed parameter systems by low dimensional controllers: a root locus approach," *Proceedings 29th IEEE International Conference on Decision and Control*.

[8] C.I. Byrnes and D.S. Gilliam, J. He, "Root Locus and Boundary Feedback design for a Class of Distributed Parameter Systems," submitted to *SIAM J. Control and Opt.*

[9] C.I. Byrnes, D.S. Gilliam and V.I. Shubov, "Boundary control, stabilization and attractors for a viscous Burgers' equation," preprint.

[10] J. Carr, *Applications of Centre Manifold Theory*, Springer-Verlag, 1980.

[11] Constantin, P. Foias, C. Temam, R. *Attractors representing turbulent flows*, Memoirs of A.M.S. 53 (1985) No. 314.
[12] Foias, C., Temam, R. "Some analytic and geometric properties of the solutions of the Navier-Stokes equations," *J. Math. Pures Appl.* 58 (1979) 339-368.
[13] A. Friedman *Partial Differential Equation*, Holt, Rinehart and Winston, Inc., 1969.
[14] Hale, J. *Asymptotic Behavior of Dissipative Systems. Mathematical Surveys and Monographs*, 25. AMS (1988).
[15] D. Henry *Geometric Theory of Semilinear Parabolic Equations*, Springer-Verlag, Lec. Notes in Math., Vol 840, 1981.
[16] E. Hopf, "The partial differential equation $u_t + uu_x = \mu u_{xx}$," *Comm. Pure Appl. Math.*, 3 (1950), 201-230.
[17] T. Kato, *Perturbation theory for linear operators*, Springer-Verlag, New York, (1966).
[18] Ladyzhenskaya, O.A. "On the determination of minimal global attractors for the Navier-Stokes and other partial differential equations," Russian Mathematical Surveys. 42 (1987) 27-73.
[19] Ladyzhenskaya, O.A. *The Mathematical Theory of Viscous Incompressible flow*, 2nd ed. Gordon and Breach, New York, 1969.
[20] Ladyzhenskaya, O.A. "The dynamical system that is generated by the Navier-Stokes equations," *Journal of Soviet Mathematics*, 3 (1975) 458-479.
[21] Ladyzhenskaya, O.A., Solonnikov, V.A., Ural'ceva, N.N. *Linear and Quasilinear Equations of Parabolic Type*, Translations of AMS, Vol. 23, 1968.
[22] A. Pazy, *Semigroups of Linear Operators and Applications to Partial Differential Equations*, Springer-Verlag, 1983.
[23] Shubov, V. "Long time behavior of infinite dimensional dissipative dynamical systems," *Journal of Math. Systems, Estimation and Control*, 2 (1992), 381-427.
[24] Temam, R. *Infinite Dimensional Dynamical Systems in Mechanics and Physics*, Applied Mathematical Sciences, 68. Springer-Verlag (1988).
[25] Temam, R. *Navier-Stokes Equations, Theory and Numerical Analysis*, 3rd rev. ed., North-Holland, Amsterdam, 1984.

Chapter 13
Feedback Control of Singular Integro-Differential Systems: An Input/Output Approach*

Hitay Özbay[†]   Janos Turi[‡]

Abstract

In this chapter we consider a class of systems described by singular integro-differential equations. It can be shown that such equations form a mathematical model for certain aeroelastic systems. For example Theodorsen's classical model of a thin airfoil fits in this framework. We study these systems in frequency domain from the input/output point of view and show the existence of finite dimensional stabilizing feedback controllers. An algorithmic procedure is outlined for the construction of such controllers.

1 Introduction

Active control of aeroelastic systems in state space setting has been the subject of several research articles in recent years (see e.g. [2], [1], [4] and [7]). The mathematical models considered in these studies can be described as systems of singular integro-differential equations of Volterra ([1]) or neutral type ([4]). Note that both of these models provide "input/output" characterization of the underlying systems and appear to be feasible for control design purposes. Also, Theodorsen's classical formulation of input/output dynamical behavior of a thin airfoil can be cast into the framework of such mathematical models, [23] [17]. The extension of state space based (semigroup framework) control design methods, needed for the class of infinite dimensional systems represented by these singular integro-differential equations, received considerable attention. Along this line of research new results concerning well-posedness ([5], [6],[12], [14], [16]) and approximation ([15], [11]) were obtained for systems governed by a class of singular integro-differential equations.

In this chapter we consider a class of singular integro-differential equations possessing the essential features of the models describing aeroelastic systems, as in [1], [4] and [17]. Our main goal is to demonstrate the applicability of recently developed frequency domain based theory, for the robust control of distributed parameter systems, to the class of systems governed by such singular integro-differential equations. In particular we are interested in the construction of finite dimensional stabilizing controllers for these systems that correspond to flutter suppression in the aeroelastic problem. The main result of our study is the *existence* of finite dimensional stabilizing controllers. This is accomplished by recognizing that the original infinite dimensional linear input/output map can be represented as a coprime factor perturbations of a finite dimensional system, and by using

---

*The authors were supported in part by the National Science Foundation under grants no. DMS-8907019 and MSS-9203418.
[†]Department of Electrical Engineering, The Ohio State University, Columbus, OH 43210
[‡]Programs in Mathematical Sciences, University of Texas at Dallas, Richardson, TX 75083

the theory of robustness optimization in the gap metric (see e.g. [8]). An algorithm for the *construction* of such controllers is also outlined.

The rest of this chapter is organized as follows. In §2 we describe the system to be studied in the form of a certain singular integro-differential equation, and formulate our control problem. Also in §2 we show the relationship between a mathematical model of a thin airfoil, and the class of systems we consider. Main results of this chapter are derived in §3 where we show the existence of finite dimensional controllers stabilizing the feedback system. In §4 we outline an algorithmic procedure for constructing such controllers. In §5 we discuss the gust alleviation problem for a thin airfoil, which corresponds to disturbance attenuation problem for our mathematical model. In the same section we show that for Theodorsen's formulation the gust alleviation problem corresponds to a mixed H^2/H^∞ problem. We make our concluding remarks in §6.

2 Problem statement and preliminary remarks

In this chapter we are interested in stabilization of systems described by the following integro-differential equations:

$$(2.1) \qquad \frac{d}{dt}x_1(t) = Ax_1(t) + b_1 x_2(t) + b_2 \int_{-\infty}^0 \kappa_2(\tau)x_2(\tau + t)d\tau + bu(t)$$

$$(2.2) \qquad \frac{d}{dt}\int_{-\infty}^0 \kappa_1(\tau)x_2(\tau + t)d\tau = c_0 x_1(t)$$

where $x_1(t) \in \mathbf{R}^n$, $n \geq 1$, $x_2(t) \in \mathbf{R}$, $t \geq 0$; and A $(n \times n)$, b, b_1, b_2 $(n \times 1)$, and c_0 $(1 \times n)$ are known constant matrices/vectors. We assume that (A, b), (A, b_1), (A, b_2) are stabilizable and (c_0, A) is detectable, in finite dimensional system theoretic sense. The kernel κ_1 has the following special structure (see e.g. [4])

$$(2.3) \qquad \kappa_1(\tau) = \sqrt{1 - \frac{\alpha}{\tau}} \qquad -\infty < \tau < 0,$$

where $0 < \alpha$ is a constant. The kernel κ_2 is such that the convolution operator, defined by the integral

$$v(t) = \int_{-\infty}^0 \kappa_2(\tau)x_2(\tau + t)d\tau,$$

is bounded (as an operator acting on signals $x_2 \in L^2[0, \infty)$ generating signals $v \in L^2[0, \infty)$). Note that this is slightly more general than assuming $\kappa_2 \in L^1(-\infty, 0)$.

We want to find a command signal $u(t) \in \mathbf{R}$, $t \geq 0$, such that the system is stable, in the sense defined below. Given all the matrices A, b, b_1, b_2, c_0, and the kernels $\kappa_1(\tau)$, $\kappa_2(\tau)$, $\tau < 0$ we can find the solution of (2.1-2.2) from the initial conditions $x_1(0)$, $x_2(\tau)$, $\tau < 0$, and the input $u(t)$, $t \geq 0$. The above system is a linear time invariant system whose "state space" is infinite dimensional. Here we will use frequency domain methods to analyze the system and synthesize a controller generating u. We will not discuss the state space realizations of the plant.

2.1 On the kernel functions κ_1 and κ_2:

Note that if we define $k_i(t) = \kappa_i(-t)$, $t > 0$, $i = 1, 2$, then we have

$$\int_{-\infty}^0 \kappa_i(\tau)x_2(\tau + t)d\tau = \int_0^\infty k_i(t)x_2(t - \tau)dt = (k_i * x_2)(t)$$

where $*$ denotes the usual convolution operator. In order to apply frequency domain design techniques we will transform the system equations (2.1-2.2) into frequency domain by taking Laplace transforms. Let s be the Laplace transform variable and $\hat{}$ denote the Laplace transform of a time signal; in particular we have

$$\hat{k}_i(s) := \int_0^\infty e^{-st} k_i(t) dt, \quad i = 1, 2.$$

By the assumption that the convolution operator with kernel κ_2 is bounded we have that $\hat{k}_2 \in H^\infty$, i.e. $\hat{k}_2$ is bounded and analytic in the right half plane Re $s > 0$. We will need one more assumption on $\hat{k}_2$: that the boundary value function $\hat{k}_2(j\omega)$ is continuous on the extended imaginary axis, i.e. on $\omega \in R \cup \{\infty\}$.

THEOREM 2.1. *(The existence of $\hat{k}_1(s)$):* Let $k_1(t) = 1 + \alpha(t)$, where $\alpha(t) := -1 + \sqrt{1 + a/t}$. Then,

$$\hat{\alpha}(s) = \int_0^\infty \alpha(t) e^{-st} dt$$

exists for all $s \in RHP \backslash \{0\}$, where RHP denotes the closed right half plane, $Re(s) \geq 0$.

Proof. It is sufficient to show that the integrals

$$\mathrm{Re}(\hat{\alpha}(j\omega)) = \int_0^\infty \alpha(t) \cos(\omega t) dt,$$

and

$$\mathrm{Im}(\hat{\alpha}(j\omega)) = \int_0^\infty \alpha(t) \sin(\omega t) dt,$$

exist for all $\omega \in R \backslash \{0\}$. But note that $\sin(\omega t), \cos(\omega t)$ are periodic with zero mean, $\alpha(\cdot)$ is a monotone decreasing function with $\alpha(t) \to 0$ as $t \to \infty$. These imply that $\mathrm{Re}(\hat{\alpha}(j\omega))$ and $\mathrm{Im}(\hat{\alpha}(j\omega))$ exist for all $\omega \neq 0$.

Remark that

$$\frac{1}{s\hat{k}_1(s)} = \frac{1}{1 + s\hat{\alpha}(s)}$$

which is undefined at $s = 0$. This function has the following properties.

THEOREM 2.2. *The function*

$$\hat{h}(s) = \begin{cases} \frac{1}{1 + s\hat{\alpha}(s)} & \text{if } s \neq 0 \\ 1 & \text{if } s = 0 \end{cases}$$

is in H^∞ and $\hat{h}(j\omega)$ is uniformly continuous on $\omega \in R \cup \{\infty\}$.

Proof. First we show that $1 + s\hat{\alpha}(s) \neq 0$ for $s \in C \backslash R^-$, $(R^- = \{\sigma : \sigma \in R, \sigma \leq 0\})$ and conclude that $\hat{h}(s)$ is analytic in $C \backslash R^-$, which contains RHP. For $s = \sigma + j\omega$ we have

$$1 + s\hat{\alpha}(s) = \left[1 + \sigma \int_0^\infty \alpha(t) e^{-\sigma t} \cos(\omega t) dt + \omega \int_0^\infty \alpha(t) e^{-\sigma t} \sin(\omega t) dt \right]$$
$$+ j \left[\omega \int_0^\infty \alpha(t) e^{-\sigma t} \cos(\omega t) dt - \sigma \int_0^\infty \alpha(t) e^{-\sigma t} \sin(\omega t) dt \right]$$

The fact that $\hat{h} \in H^\infty$ follows from $\mathrm{Re}(1 + s\hat{\alpha}(s)) \geq 1$ for all $s \in C \backslash R^-$, which comes from

(2.4) $$\omega \int_0^\infty \alpha(t) e^{-\sigma t} \sin(\omega t) dt \geq 0,$$

and
(2.5) $$\sigma \int_0^\infty \alpha(t) e^{-\sigma t} \cos(\omega t) dt \geq 0.$$

We now prove (2.4) and (2.5). For notational convenience we first define $\alpha_\sigma(t) = \alpha(t) e^{-\sigma t}$, $\sigma \geq 0$. Note that $\alpha_\sigma(t)$ is a monotone decreasing function and $\alpha_\sigma(t) \to 0$ as $t \to \infty$. We can re-write (2.4) as follows

$$\omega \int_0^\infty \alpha_\sigma(t) \sin(\omega t) dt = \left[\omega \int_0^{\pi/\omega} \alpha_\sigma(t) \sin(\omega t) dt + \omega \int_{\pi/\omega}^{2\pi/\omega} \alpha_\sigma(t) \sin(\omega t) dt \right]$$
$$+ \left[\omega \int_{2\pi/\omega}^{3\pi/\omega} \alpha_\sigma(t) \sin(\omega t) dt + \omega \int_{3\pi/\omega}^{4\pi/\omega} \alpha_\sigma(t) \sin(\omega t) dt \right] + \ldots$$

Each bracketed term is positive, so (2.4) holds.

Now for (2.5) we will show that

$$\omega \int_0^\infty \alpha_\sigma(t) \cos(\omega t) dt \geq 0,$$

for all $\omega > 0$. Since $\sigma \geq 0$ this will give (2.5). We first show that

(2.6) $$\int_{\frac{\pi}{2\omega}}^\infty \alpha_\sigma(t) \cos(\omega t) dt \geq -\frac{1}{\omega} \alpha_\sigma(\frac{\pi}{2\omega}) \quad \omega \neq 0.$$

This can be verified as follows. First note that

$$\int_{\frac{\pi}{2\omega}}^{\frac{5\pi}{2\omega}} \alpha_\sigma(t) \cos(\omega t) dt = \int_{\frac{\pi}{2\omega}}^{\frac{3\pi}{2\omega}} (\alpha_\sigma(t) - \alpha_\sigma(t + \frac{\pi}{\omega})) \cos(\omega t) dt$$

(2.7) $$\geq \frac{1}{2} \int_{\frac{\pi}{2\omega}}^{\frac{3\pi}{2\omega}} \left[(\alpha_\sigma(\frac{\pi}{2\omega}) - \alpha_\sigma(\frac{\pi}{2\omega} + \frac{\pi}{\omega})) \right.$$
$$\left. + (\alpha_\sigma(\frac{\pi}{2\omega} + \frac{\pi}{\omega}) - \alpha_\sigma(\frac{\pi}{2\omega} + \frac{2\pi}{\omega})) \right] \cos(\omega t) dt$$

(2.8) $$= -\frac{2}{\omega} \frac{1}{2} \left[\alpha_\sigma(\frac{\pi}{2\omega}) - \alpha_\sigma(\frac{\pi}{2\omega} + \frac{2\pi}{\omega}) \right]$$

where we have used the convexity of $\alpha_\sigma(t) - \alpha_\sigma(t + \frac{\pi}{\omega})$.

Similarly, for $k = 1, 2, \ldots$, we have

(2.9) $$\int_{\frac{\pi}{2\omega} + \frac{2k\pi}{\omega}}^{\frac{5\pi}{2\omega} + \frac{2k\pi}{\omega}} \alpha_\sigma(t) \cos(\omega t) dt \geq -\frac{1}{\omega} \left[\alpha_\sigma(\frac{\pi}{2\pi} + \frac{2k\pi}{\omega}) - \alpha_\sigma(\frac{\pi}{2\pi} + \frac{2(k+1)\pi}{\omega}) \right].$$

Finally, by (2.6) and (2.9)

$$\int_{\frac{\pi}{2\omega}}^\infty \alpha_\sigma(t) \cos(\omega t) dt \geq -\frac{1}{\omega} \sum_{k=0}^\infty \left[\alpha_\sigma(\frac{\pi}{2\omega} + \frac{2k\pi}{\omega}) - \alpha_\sigma(\frac{\pi}{2\omega} + \frac{2(k+1)\pi}{\omega}) \right]$$
$$= -\frac{1}{\omega} \alpha_\sigma(\frac{\pi}{2\omega})$$

as claimed.

To finish the proof of (2.5) we note that

$$\omega \int_0^\infty \alpha_\sigma(t)\cos(\omega t)dt \geq \omega \int_0^{\frac{\pi}{4\omega}} \alpha_\sigma(\frac{\pi}{4\omega})\cos(\omega t)dt + \omega \int_{\frac{\pi}{4\omega}}^{\frac{\pi}{2\omega}} \alpha_\sigma(\frac{\pi}{2\omega})\cos(\omega t)dt - \alpha_\sigma(\frac{\pi}{2\omega})$$

$$= (1 - \frac{\sqrt{2}}{2})(\alpha_\sigma(\frac{\pi}{4\omega}) - \alpha_\sigma(\frac{\pi}{2\omega}))$$

(2.10)
$$> 0.$$

To recapitulate, we have established (2.4) and (2.5) which imply that $\text{Re}(1+s\hat{a}(s)) \geq 1$, and hence $1 + s\hat{a}(s) \neq 0$, for all $s \in \mathbb{C}\backslash\mathbb{R}^-$. This also shows that

$$\operatorname*{ess\,sup}_\omega |\hat{h}(j\omega)| = 1, \quad \text{i.e.} \quad \|\hat{h}\|_\infty = 1.$$

We now show the continuity of $\hat{h}(j\omega)$. Recall that

$$\hat{h}(j\omega) = \begin{cases} (1 + \omega\int_0^\infty \alpha(t)\sin(\omega t)dt + j\omega\int_0^\infty \alpha(t)\cos(\omega t)dt)^{-1} & \text{for } \omega \neq 0 \\ 1 & \text{for } \omega = 0 \end{cases}$$

It is obvious that $\hat{h}(j\omega) \to 0$, as $\omega \to \infty$, and at any $\omega \neq 0$ the function $\hat{h}(j\omega)$ is continuous. Now we need to show the continuity at $\omega = 0$. We have the following estimates

$$\left|\omega\int_0^\infty \alpha(t)\sin(\omega t)dt\right| \leq \left|\omega\int_0^{\frac{\pi}{\omega}} \alpha(t)\sin(\omega t)dt\right|$$

$$\leq \left|\omega\int_0^{\frac{\pi}{\omega}} \sqrt{\frac{\alpha}{t}}dt\right|$$

$$= 2\sqrt{\alpha\omega\pi}$$

which tends to 0 as $\omega \to 0$. Similarly,

$$\left|\omega\int_0^\infty \alpha(t)\cos(\omega t)dt\right| \leq \left|\omega\int_0^{\frac{\pi}{2\omega}} \alpha(t)\cos(\omega t)dt\right|$$

$$\leq \left|\omega\int_0^{\frac{\pi}{2\omega}} \sqrt{\frac{\alpha}{t}}dt\right|$$

$$= 2\sqrt{\alpha\omega\pi}$$

which tends to 0 as $\omega \to 0$. Thus, $\hat{h}(j\omega) \to 1$ as $\omega \to 0$.

The above theorem gives the properties of the complex function $\hat{h}(s)$, in the frequency domain. From the system theoretic point of view one has to make sure that this function corresponds to the transfer function of a linear time invariant system. This can be checked by verifying that there exists an impulse response $h(t)$ corresponding to $\hat{h}(s)$. This is accomplished by the following

THEOREM 2.3. *The complex function $\hat{h}(s)$ is the Laplace transform of a real function $h(t)$, $t \in (0, \infty)$, which is locally integrable on $(0, \infty)$.*

Proof. The claim follows by Theorem 2.6 on p. 144 of [9], noting that $\hat{h}(\cdot)$ satisfies

(i) $\hat{h}$ has an analytic extention to $\mathbb{C}\backslash\mathbb{R}^-$,

(ii) $\hat{h}(\sigma)$ is real valued for $\sigma \in (0, \infty)$.

(iii) $\lim_{\sigma\to\infty} \hat{h}(\sigma) = 0$.

(iv) $\text{Im}\{s\hat{h}(s)\} \geq 0$ for $\text{Im } s > 0$.

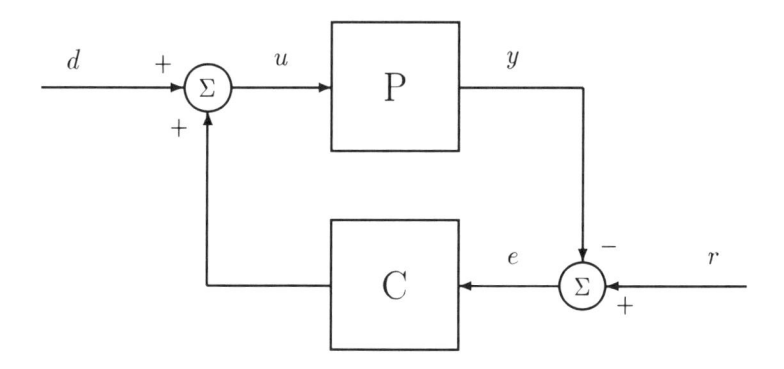

FIG. 1. *Standard feedback configuration*

2.2 Feedback control system:

The above section developed the necessary tools to transform the time domain description (1-2) of the plant into the frequency domain. Assuming zero initial conditions ($x_2(t) = 0, t \leq 0$) and taking the Laplace transform of both sides of (2.2) we obtain

$$\hat{x}_2(s) = \frac{1}{s\hat{k}_1(s)} c_0 \hat{x}_1(s). \tag{2.11}$$

Similarly, taking the Laplace transform of (2.1) we get (assuming $x_1(0) = 0$)

$$s\,\hat{x}_1(s) = A\hat{x}_1(s) + b_1 \hat{x}_2(s) + b_2 \hat{k}_2(s)\hat{x}_2(s) + b\hat{u}(s). \tag{2.12}$$

Note that (2.11) and (2.12) are well defined because the terms $\hat{k}_2(s)$ and $(s\hat{k}_1(s))^{-1}$ are well defined, as shown in the above section. Now, substituting (2.11) into (2.12) we obtain an expression relating the input, $\hat{u}(s)$, to the "states" $\hat{x}_1(s)$, and $\hat{x}_2(s)$:

$$\hat{x}_1(s) = \left(sI - A - \frac{b_1 c_0 + b_2 c_0 \hat{k}_2(s)}{s\hat{k}_1(s)} \right)^{-1} b\hat{u}(s) \tag{2.13}$$

$$\hat{x}_2(s) = \frac{1}{s\hat{k}_1(s)} c_0 \hat{x}_1(s). \tag{2.14}$$

Equations (2.13-2.14) are the frequency domain equivalents of (2.1-2.2), and these will be used to find a "stabilizing" controller which generates an appropriate command signal $\hat{u}(s)$.

In our control design we consider a feedback scheme which is shown in Figure 1. In this configuration y represents the output of the plant P, whose dynamics are described by (2.13-2.14). The output y is a measured physical quantity, usually a combination of the states, x_1 and x_2. The signal u is the command input to the plant, d is the disturbance, r is the reference input, (r may also represent the measurement noise) and e is the measured error between the reference and the output. The controller to be designed is represented by the block C.

According to our zero initial conditions assumption the plant P responds to u only, and the closed loop system has two exogenous inputs, namely r and d. We will consider *finite energy* reference and disturbance signals, i.e. $r, d \in L^2[0, \infty)$.

DEFINITION 2.1. *The closed loop system shown in Figure 1 is stable if all external inputs $r, d \in L^2[0, \infty)$ give rise to signals $e, u, y \in L^2[0, \infty)$, (i.e. finite energy inputs generate finite energy outputs), and the maximum energy amplification in the system is finite, i.e.*

$$\max\{ \sup_{0 \neq r \in L^2} \frac{\|e\|_2 + \|u\|_2}{\|r\|_2}|_{d=0}; \sup_{0 \neq d \in L^2} \frac{\|e\|_2 + \|u\|_2}{\|d\|_2}|_{r=0} \} < \infty.$$

It is well known that we have closed loop stability if and only if the entries of the transfer function from $\begin{bmatrix} r \\ d \end{bmatrix}$ to $\begin{bmatrix} e \\ u \end{bmatrix}$:

$$T_{(P,C)} := \begin{bmatrix} (I + PC)^{-1} & -P(I + CP)^{-1} \\ C(I + PC)^{-1} & (I + CP)^{-1} \end{bmatrix}$$

are in H^∞, i.e. analytic in the right half plane Re $s > 0$ and bounded on the extended imaginary axis: $\{j\omega : \omega \in \mathbb{R}\} \cup \{\infty\}$. All these transfer functions relate Laplace transforms of the inputs to the Laplace transforms of the outputs, so they are functions of the complex variable s.

Note that the plant transfer function $P(s)$ (which gives the output $\hat{y}(s)$ as $P(s)\hat{u}(s)$) is not a rational function. This means that the plant cannot be described by a finite dimensional state space realization. Therefore, we cannot use standard finite dimensional control system theory to design an appropriate controller for this system.

It was shown that (cf. [22]) the plant is *stabilizable* (i.e. there exists a controller C, whose transfer function is a ratio of two H^∞ functions, satisfying the above definition of the stability) if and only if there exists a strong coprime factorization for P, in H^∞, i.e. $P = N/M$, for some $N, M \in H^\infty$ and $\begin{bmatrix} M \\ N \end{bmatrix}$ has a left inverse whose entries are in H^∞. Moreover, if the plant is stabilizable then there exist (cf. [22]) functions $U, V \in H^\infty$ such that
(2.15) $$MV + NU = 1,$$
and the set of all stabilizing controllers is given by

$$\{ C = \frac{U + MQ}{V - NQ} : Q \in H^\infty \}$$

For the system described by (2.13-2.14) we will assume that the output of the plant P consists of a combination of the finite dimensional part of the "states" x_1, i.e. $y = c_1 x_1$ with $c_1 : 1 \times n$ non-zero constant vector. For simplicity we assume $c_1 = c_0$, i.e. the right hand side of (2.2) is the measured output y. Then, the plant transfer function is of the form

$$P(s) := c_1 \left(sI - A - \frac{b_1 c_1 + b_2 c_1 \hat{k}_2(s)}{s\hat{k}_1(s)} \right)^{-1} b.$$

Simple algebraic manipulations yield the equivalent form,

(2.16) $$P(s) = \frac{c_1(sI - A)^{-1} b}{1 - (1/s\hat{k}_1(s))(\hat{k}_2(s) c_1(sI - A)^{-1} b_2 + c_1(sI - A)^{-1} b_1)}.$$

Note that the transfer function $P(s)$, from u to y contains strictly proper rational terms $c_1(sI - A)^{-1} b$, $c_1(sI - A)^{-1} b_1$ and $c_1(sI - A)^{-1} b_2$, along with the terms $\frac{1}{s\hat{k}_1(s)}$ and $\hat{k}_2(s)$.

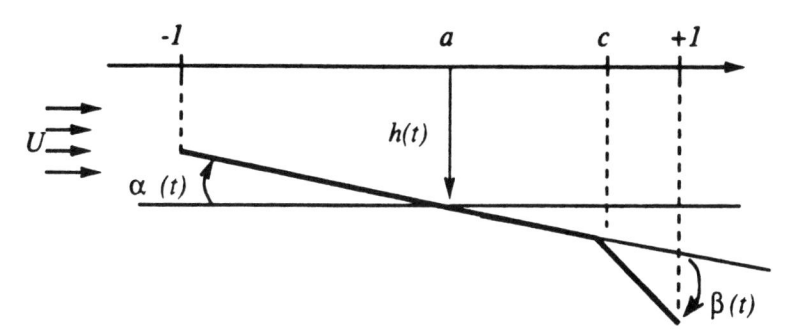

FIG. 2. *Thin Airfoil*

We conclude from this discussion that the plant transfer function $P(s)$ is a non-rational function of s. This illustrates the infinite dimensionality of the plant. Our objective is to find a finite dimensional controller, (i.e. $C(s)$ rational), stabilizing this infinite dimensional plant.

We now would like to make a few remarks to put our work here in perspective for the reader. The system described by (2.1) and (2.2) has been proposed by several authors (see e.g. [3], [2], [4], [5]) to model the elastic motions of three degrees of freedom thin airfoils in two dimensional unsteady flows, see e.g. [19] and §2.3 below. Our zero initial conditions assumption corresponds to the so-called *indicial problem* (cf. [3] p. 291). For the sake of completeness we want to mention some relevant work on the finite delay version of the system (2.1)-(2.2) (the integrals are taken from a finite time, say $-r$ to 0, instead of $-\infty$ to 0). For example well-posedness of these types of systems in different state spaces are studied in [5], [11]. Approximation and control issues are discussed in detail in [15], using semigroup framework and non-zero initial conditions. Also, recently a well-posedness result for scalar equations of the type (2.2), in weighted L^1 spaces, is obtained in [14].

2.3 Special case: Theodorsen's model of a thin airfoil:

In this subsection we summarize Theodorsen's classical model for a thin airfoil shown in Figure 2. We will see that this model is a special case of the class of plants whose transfer functions are of the form (2.16).

This system can be modeled by the following equation

$$(2.17) \qquad M_s \ddot{z}(t) + B_s \dot{z}(t) + K_s z(t) = \frac{1}{m_s} F(t) + G u(t),$$

where $z(t) = [h(t), \alpha(t), \beta(t)]^T$, $u(t)$ represents the control input (torque applied at the flap), and constant matrices M_s, B_s, K_s are of size 3×3, and G is 3×1. Aeroelastic loads are represented by $F(t) = [P(t), M_\alpha(t), M_\beta(t)]^T$, which can be expressed as

$$(2.18) \qquad F(t) = M_a \ddot{z}(t) + B_a \dot{z}(t) + K_a z(t) + F_c(t)$$

where $F_c(t)$ is the "circulatory" part of $F(t)$. The matrices M_a, B_a and K_a can be computed in terms of the problem data see e.g. [21] [23].

According to Theodorsen's formulation, $F_c(t)$ can be expressed in the frequency domain as (see e.g. [21] pp. 395–396)

$$(2.19) \qquad \widehat{F}_c(s) = C_t(s)(B_{c1} + s B_{c2}) \widehat{z}(s)$$

where $C_t(j\omega)$ is the Theodorsen's function, and B_{c1}, B_{c2} are constant matrices given by

$$B_{c1} = \hat{b}_1 \hat{c}_1 , \quad \text{and} \quad B_{c2} = \hat{b}_1 \hat{c}_2$$

for some constant vectors $\hat{c}_1, \hat{c}_2$ 1×3 and $\hat{b}_1$ 3×1.

If we define the measured output as

$$y(t) := \hat{c}_1 z(t) + \hat{c}_2 \dot{z}(t),$$

it can be shown that (see [17] and [18]) the plant transfer function is of the form

$$(2.20) \qquad \frac{\hat{y}(s)}{\hat{u}(s)} = P_t(s) = \frac{\hat{C}_o(sI - \hat{A})^{-1}\hat{B}_o}{1 - \hat{C}_o(sI - \hat{A})^{-1}\hat{B}_1 \, C_t(s)}$$

where $C_t(s)$ is the Theodorsen's function, and

$$\hat{A} = \begin{bmatrix} 0_{3\times 3} & I_{3\times 3} \\ (M_s - \frac{M_a}{m_s})^{-1}(\frac{K_a}{m_s} - K_s) & (M_s - \frac{M_a}{m_s})^{-1}(\frac{B_a}{m_s} - B_s) \end{bmatrix}$$

$$\hat{C}_o = [\hat{c}_1 \; \hat{c}_2] , \quad \hat{B}_1 = \begin{bmatrix} 0_{3\times 1} \\ (M_s - \frac{M_a}{m_s})^{-1}\hat{b}_1 \end{bmatrix} , \quad \hat{B}_o = \begin{bmatrix} 0_{3\times 1} \\ (M_s - \frac{M_a}{m_s})^{-1}G \end{bmatrix}.$$

Now, let us compare the plant transfer function $P(s)$ described by (2.16) and the one obtained using Theodorsen's formulation, $P_t(s)$. In our derivation if we define

$$c_1 := \hat{C}_o \qquad b := \hat{B}_o \qquad b_2 := \hat{B}_1 \text{ (or } b_2 := 0) \qquad b_1 := 0 \text{ (or } b_1 := \hat{B}_1)$$

and $\hat{k}_1, \hat{k}_2$ are such that

$$\frac{\hat{k}_2(s)}{s\hat{k}_1(s)} = C_t(s), \quad (\text{or } \frac{1}{s\hat{k}_1(s)} = C_t(s))$$

then $P(s) = P_t(s)$. In other words transfer function of a thin airfoil derived from Theodorsen's approach is a special case of the class of plants whose transfer functions are of the form $P(s)$ given by (2.16).

3 Existence of a finite dimensional stabilizing controller

In this section we consider the plant described by equations (2.1)-(2.2) with zero initial conditions. In the frequency domain the plant transfer function, $P(s)$, is given by (2.16). We will now discuss in detail the problem of stabilizing P by a finite dimensional controller C.

Let us begin by studying the stabilizability conditions for the plant P. The issue is to find coprime factorizations for $P(s)$, i.e. we need to find (see e.g. [22]) $N, M \in H^\infty$ such that $P(s) = N(s)/M(s)$ and

$$\inf_{\text{Re } s > 0} (|N(s)| + |M(s)|) > 0.$$

As shown in §2.1 the function $(s\hat{k}_1(s))^{-1}$ is in H^∞. Also note that, since the kernel $k_2(t), t \geq 0$, generates a bounded convolution operator from $L^2[0,\infty)$ to $L^2[0,\infty)$, we have $\hat{k}_2 \in H^\infty$. We will use the following notation for these H^∞ functions:

$$(s\hat{k}_1(s))^{-1} =: N_{k_1}(s) \quad \text{and} \quad \hat{k}_2(s) =: N_{k_2}(s).$$

On the other hand by stabilizability of the pairs $(A, b), (A, b_1), (A, b_2)$ and detectability of (c_1, A), we can find rational functions $N_b, N_{b_1}, N_{b_2}, M_a \in H^\infty$ such that

$$\frac{N_b(s)}{M_a(s)} = c_1(sI - A)^{-1}b \text{ and } \frac{N_{b_i}(s)}{M_a(s)} = c_1(sI - A)^{-1}b_i, \quad i = 1, 2,$$

and the pairs $(N_b, M_a), (N_{b_1}, M_a), (N_{b_2}, M_a)$ are coprime. Thus, $P(s)$ can be rewritten as

$$(3.21) \qquad P(s) = \frac{N_b(s)}{M_a(s) - N_{k_1}(s)(N_{b_1}(s) + N_{k_2}(s)N_{b_2}(s))}.$$

Defining
$$(3.22) \qquad N(s) := N_b(s)$$

and
$$(3.23) \qquad M(s) := M_a(s) - N_{k_1}(s)(N_{b_1}(s) + N_{k_2}(s)N_{b_2}(s))$$

we see that stabilizability of P is equivalent to having the pair (N, M) coprime.

Recall that $N_b(s)$ is rational, so it has zeros in the extended closed right half plane at only finitely many points, say $s_1, \ldots, s_\ell$. Then, for (N, M) to be coprime we have the following necessary condition:

$$(3.24) \qquad \lim_{s \to s_j} \left| M_a(s) - N_{k_1}(s)\left(N_{b_1}(s) + N_{k_2}(s)N_{b_2}(s)\right) \right| > 0$$

for all $j = 1, \ldots, \ell$.

It is obvious from (3.21) that if (3.24) is satisfied then the pair (N, M) is coprime hence P is stabilizable. Thus, we have proven the following

LEMMA 3.1. *The plant P is stabilizable if and only if the condition (3.24) is satisfied. Moreover, if (3.24) is satisfied then $P(s) = N(s)/M(s)$ is a coprime factorization, where N and M are as defined in (3.22) and (3.23) respectively and*

$$(3.25) \qquad \lambda := \inf_{\operatorname{Re} s > 0} \sqrt{|N(s)|^2 + |M(s)|^2} > 0.$$

At this point we would like to remark that the condition (3.24) is rather easy to check. One simply has to evaluate certain H^∞ functions at finitely many points. However, stabilizability of P implies only that there exists a stabilizing controller which is in the form $C(s) = N_c(s)/M_c(s)$ where $N_c \in H^\infty$ and $M_c \in H^\infty$ are coprime. That is, the stabilizing controller may have to be infinite dimensional (i.e. $C(s)$ is possibly irrational). We are mainly interested in existence of *finite dimensional* stabilizing controllers. We now want to discuss this issue. The main idea is to "approximate" the irrational terms in the expression for $P(s)$, in an appropriate manner, by some rational functions. This way we obtain a finite dimensional approximation of the plant which is "close" to the original plant in the so-called *gap metric* sense, (see e.g. [8] for a precise definition and details). We can then find a rational controller (from the new plant) which stabilizes the approximate plant as well as the original plant. The details on how this method works are given below.

Recall that the functions $N_{k_1}(j\omega)$ and $N_{k_2}(j\omega)$ are continuous for all $\omega \in \mathbb{R} \cup \{\infty\}$. Therefore, these functions are uniformly approximable by rational functions in H^∞ (see e.g. [13]), i.e. given arbitrary small $\epsilon > 0$, there exists rational functions $N_{k_1}^f, N_{k_2}^f \in H^\infty$ such that
$$(3.26) \qquad \|N_{k_1} - N_{k_1}^f\|_\infty + \|N_{k_2} - N_{k_2}^f\|_\infty < \epsilon.$$

Hence, the denominator, M, of P is uniformly approximable in H^∞ by rational functions. Thus, for any $\epsilon > 0$ we can find a rational transfer function

(3.27) $$P_f(s) = \frac{N^f(s)}{M^f(s)}$$

such that

(3.28) $$P(s) = \frac{N^f(s)(s)}{M^f(s) + \Delta_M(s)}$$

with

(3.29) $$\|\Delta_M\|_\infty < \epsilon,$$

where $N^f = N = N_b$ (recall that N_b is already rational) and

$$M^f = M_a - N^f_{k_1}(N_{b_1} + N^f_{k_2} N_{b_2}),$$

is a rational function approximating M. We shall make use of the following.

LEMMA 3.2. *Consider the equations (3.27), (3.28) and (3.29). Then, for $\epsilon > 0$ sufficiently small, P is stabilizable implies that P_f is also stabilizable.*

Proof. Define

$$\lambda_f := \inf_{\mathrm{Re}\ s > 0} \sqrt{|N^f(s)|^2 + |M^f(s)|^2}.$$

Clearly by choosing $\epsilon > 0$ sufficiently small we can make $\lambda_f \geq \frac{\lambda}{2}$. On the other hand, by the fact that P is stabilizable we have $\lambda > 0$. Therefore, $\lambda_f > 0$, which means that P_f is stabilizable (see [22]).

Although we have a coprime factorization for P_f from Lemma 3.2, we need to have a *normalized* coprime factorizations in order to apply the theory of robustness optimization in the gap metric, [8]. So, we now find a normalized coprime factorization for P_f by constructing a rational function $G^f \in H^\infty$, with $(G^f)^{-1} \in H^\infty$, satisfying

(3.30) $$(G^f)^* G^f = |G^f(j\omega)|^2 = (|N^f(j\omega)|^2 + |M^f(j\omega)|^2)^{-1}$$

Note that since $\lambda_f > 0$ such a function exists, and it can be computed using finite dimensional spectral factorization techniques. Then, P_f can be rewritten as

(3.31) $$P_f = \frac{N_f}{M_f}$$

where

(3.32) $$N_f = N^f G^f, \quad \text{and} \quad M_f = M^f G^f.$$

Since we have

$$N_f^* N_f + M_f^* M_f = 1,$$

the vector valued function $\begin{bmatrix} N_f \\ M_f \end{bmatrix}$ is inner. Moreover, we can find $U_f, V_f \in H^\infty$ rational functions such that

$$V_f M_f + U_f N_f = 1.$$

See for example [24], pp. 82–84, on how to compute these matrices. The set of all stabilizing controllers for P_f is given by

$$\left\{ C_f = \frac{U_f + M_f Q}{V_f - N_f Q} \ : \ Q \in H^\infty \right\}.$$

It is also known, (see e.g. [8] and references therein) that the controller $C_f = (U_f + M_f Q)(V_f - N_f Q)^{-1}$ stabilizes all plants of the form

$$P_\delta = \frac{N_f + \Delta_{N_f}}{M_f + \Delta_{M_f}}$$

where $\Delta_{P_\delta} := [\Delta_{M_f}, \Delta_{N_f}]$ has arbitrary entries in H^∞, subject to $\|\Delta_{P_\delta}\|_\infty < \epsilon$, if and only if

$$\left\| \begin{bmatrix} U_f \\ V_f \end{bmatrix} - \begin{bmatrix} M_f \\ N_f \end{bmatrix} Q \right\|_\infty \leq \frac{1}{\epsilon}.$$

In view of the above fact one can define a quantity γ_f, for the plant P_f, as

$$\gamma_f^{-1} := \inf_{Q \in H^\infty} \left\| \begin{bmatrix} U_f \\ V_f \end{bmatrix} - \begin{bmatrix} M_f \\ N_f \end{bmatrix} Q \right\|_\infty.$$

The quantity γ_f characterizes the largest amount of uncertainty level ϵ tolerated by P_f, in the sense that there exists a controller stabilizing P_f and all plants of the form P_δ if and only if $\epsilon \leq \gamma_f$. Since P_f is stabilizable we have $\lambda_f \geq \gamma_f > 0$. (We refer to [8] for all the details of these facts and other related references.) In a similar fashion one can define the quantity γ for the plant P as the largest uncertainty level tolerated by P. Again, since P is stabilizable we have $\lambda \geq \gamma > 0$.

Note that the original plant can be written as

$$P = \frac{N^f}{M^f + \Delta_M} = \frac{N^f G^f}{M^f G^f + G^f \Delta_M}.$$

Accordingly we set

$$\Delta_P := G^f \Delta_M.$$

We are now ready to state our main result.

THEOREM 3.1. *Consider the plant described by (2.1-2.2), (or equivalently by its transfer function (3.21)). Suppose that all the properties assumed, in §2, for $\hat{k}_1(s)$ and $\hat{k}_2(s)$ hold. Then, there exists a controller C_f, finite dimensional, (i.e. $C_f(s)$ is rational) stabilizing the closed loop system with plant P.*

Proof. First recall that the property (3.24) is necessary and sufficient for *stabilizability* of P. Also from the discussion following Lemma 3.2 we know that if $\|\Delta_P\|_\infty < \gamma_f$ then there exists a controller stabilizing both P_f and P. Such a controller can be obtained from the finite dimensional plant P_f in an optimal way. Moreover, this controller, $C_{f,opt}$, is rational, i.e. finite dimensional. This is essentially a consequence of the well known result that H^∞ optimal controllers for finite dimensional plants are finite dimensional. Now, we know that, by uniform approximability of N_{k_1} and N_{k_2} we can make $\epsilon > 0$, in (3.29), as small as we wish. However, we want to prove that we can make $\|\Delta_P\|_\infty$ arbitrarily small. To show this it is sufficient to prove that there exists $\eta' > 0$ and $\epsilon' > 0$ such that for all $\epsilon' > \epsilon > 0$ we have $\|G^f\|_\infty < \eta'$. This is indeed true because from the proof of the Lemma 3.2 we know that $\|G^f\| \leq \frac{\lambda}{2}$ for some sufficiently small ϵ, say for $\epsilon \leq \epsilon'$. Thus, for any $0 < \epsilon \leq \epsilon'$ fixed we can make $\|\Delta_P\|_\infty < \epsilon$, by a suitable choice of the rational functions $N_{k_1}^f, N_{k_2}^f \in H^\infty$. Finally in order to prove the existence of a finite dimensional controller stabilizing P, we need to ensure that as $\epsilon \to 0$, γ_f is bounded below by a strictly positive number. Again, we know that as $\epsilon \to 0$ the distance, in the gap metric, between P and P_f approaches to zero and hence the quantity γ_f approaches to $\gamma > 0$, see [8]. This concludes the proof of the existence of a finite dimensional controller which stabilizes the closed loop system with the infinite dimensional plant P.

4 Construction of the controller $C_{f,opt}$

In this section we summarize the procedure described in the previous section as an algorithm for the construction of a stabilizing finite dimensional controller.

Algorithm: Given data A, b, b_1, b_2, c_1, $\hat{k}_2(s) = N_{k_2}(s)$, $(s\hat{k}_1(s))^{-1} = N_{k_1}(s)$.

Step 1: Find rational functions N_b, N_{b_1}, N_{b_2}, $M_a \in H^\infty$ such that

$$\frac{N_b(s)}{M_a(s)} = c_1(sI - A)^{-1}b, \quad \frac{N_{b_i}(s)}{M_a(s)} = c_1(sI - A)^{-1}b_i, \quad i = 1, 2;$$

and the pairs (N_b, M_a), (N_{b_1}, M_a) and (N_{b_2}, M_a) are coprime.

Step 2: Pick a small number $\epsilon > 0$, and find $N_{k_1}^f$, $N_{k_2}^f \in H^\infty$, rational functions such that

$$\|N_{k_1}^f - N_{k_1}\|_\infty < \epsilon$$
$$\|N_{k_2}^f - N_{k_2}\|_\infty < \epsilon$$
$$\|N_{k_1}^f N_{k_2}^f - N_{k_1} N_{k_2}\|_\infty < \epsilon.$$

Step 3: Define

$$N^f := N_b \qquad M^f := M_a - N_{k_1}^f(N_{b_1} + N_{k_2}^f N_{b_2})$$

and check that $N^f(s)$ and $M^f(s)$ do not have common zeros in the right half plane, i.e.

$$\lambda_f := \inf_{\mathrm{Re}\, s > 0} \sqrt{|N^f(s)|^2 + |M^f(s)|^2} > 0,$$

otherwise go to Step 1 and decrease ϵ until this is satisfied.

Step 4: Find the rational function $G^f \in H^\infty$ such that $(G^f)^{-1} \in H^\infty$ and

$$|G^f(j\omega)|^2 = (|N^f(j\omega)|^2 + |M^f(j\omega)|^2)^{-1}.$$

Then, define $N_f = N^f G^f$, $M_f = M^f G^f$.

Step 5: Compute $\|M_a\|_\infty$, $\|N_{b_2}\|_\infty$, $\|N_{b_1}\|_\infty$, $\|G^f\|_\infty$ (or find upper bounds for each of these norms) and check that

$$\|M^f - M\|_\infty < (\|M_a\|_\infty + \|N_{b_2}\|_\infty + \|N_{b_1}\|_\infty)\,\epsilon.$$

(Note that from Step 2 these are automatically satisfied.) Using the above bounds find a real number η such that

$$\eta \geq \|G^f\|_\infty(\|M_a\|_\infty + \|N_{b_2}\|_\infty + \|N_{b_1}\|_\infty).$$

Then we have $\|\Delta_P\|_\infty < \eta\,\epsilon$.

Step 6: Compute γ_f for the plant $P_f = N_f/M_f$ from the formula

$$\gamma_f = \sqrt{1 - \|\Gamma_f\|^2},$$

where Γ_f is the Hankel operator with symbol $[M_f^* \quad N_f^*]$.

Step 7: Check if $\gamma_f \geq \eta\epsilon > \|\Delta_P\|_\infty$

* True: go to next step

* False: go to Step 1, decrease ϵ and repeat the procedure.

Step 8: From $P_f := N_f/M_f$ compute the optimal controller $C_{f,opt}$ which robustly stabilize the gap ball around P_f of radius γ_f, see e.g. [8]. Note that $C_{f,opt}$ is rational.

End of the algorithm: $C_{f,opt}$ found in Step 8 stabilizes the original plant $P = N/M$ □.

We now want to make some remarks about the above algorithm. There are several ways to perform the computations required in Step 2; see for example [10] for different approximation schemes and further references on this subject. Though we should mention that for a particular approximation scheme some extra assumptions on N_{k_1} and N_{k_2} may be needed, besides uniform continuity of these functions on the imaginary axis, [10].

From Lemma 3.2 we have stabilizability of P_f for $\epsilon > 0$ sufficiently small, so it is guaranteed that the algorithm will pass the test in Step 3.

There are also several methods to perform a spectral factorization which gives G^f in Step 4, see for example [24] pp. 82–84. Computing the H^∞ norm of the scalar rational transfer functions N_{b_1}, N_{b_2}, M_a and G^f is not difficult. For example a simple plot of their magnitudes on the imaginary axis (Bode plots) would give these norms. Therefore, the constant in Step 5 can be found easily. Similarly computing γ_f for a given rational plant P_f is rather easy, see for example [20] for an operator theoretic method and further references on different methods. We know that for $\epsilon > 0$ sufficiently small we have $\gamma_f \geq \gamma/2 > 0$, therefore the algorithm will eventually pass the test in Step 7.

In Step 8 computing $C_{f,opt}$ requires finding the singular values and vectors of the Hankel operator of Step 6. These also can be obtained from the standard methods of H^∞ optimal control theory, see e.g. [20] for further details and references.

Finally we want to make a remark on the order of the controller. As ϵ decreases the order of P_f, hence the dimension of the controller $C_{f,opt}$ increases. Therefore, to have a reasonably low order controller one may want to find the smallest η in Step 5, and the largest $\epsilon > 0$ satisfying $\gamma_f \geq \eta\epsilon$ (note that γ_f also depends on ϵ).

5 Disturbance attenuation problem

In previous sections we were only interested in stabilization of the closed loop system using a finite dimensional controller. Although stability is the most important design specification in feedback control, there are other design goals to be achieved, such as disturbance attenuation, sensitivity minimization, etc. In this section we would like to discuss briefly the disturbance attenuation problem which corresponds to the gust alleviation problem for the aero-elastic system described in §2.3.

Let us consider the thin airfoil model of §2.3, i.e. $P(s) = P_t(s)$. For this system we can model the gust as a disturbance in the flow, and hence we can think that (see e.g. [1]) $F(t)$ is perturbed by a term $n_g(t)$ which is the output of a filtered white noise, i.e.

$$n_g(t) = \int_0^t W_g(t-\tau)w(\tau)d\tau,$$

where w is white Gaussian with unit spectral density, and $\widehat{W}_g(s)$ is a 3×1 filter shaping the spectral density of the gust. The term $n_g(t)$ modifies the equation (2.17) governing the airfoil motion in such a way that we now have

$$\hat{y}(s) = P_t(s)\hat{u}(s) + T_g(s)\hat{w}(s)$$

where

$$T_g(s) = \frac{\hat{C}_o(sI - \hat{A})^{-1}\hat{A}_1\widehat{W}_g(s)}{1 - \hat{C}_o(sI - \hat{A})^{-1}\hat{B}_1 C_t(s)}, \quad \text{with} \quad \hat{A}_1 = \frac{1}{m_s}\begin{bmatrix} 0_{3\times 3} \\ (M_s - \frac{M_a}{m_s})^{-1} \end{bmatrix}.$$

The term $T_g(s)$ can be seen as a filter generating the output disturbance in the closed loop system.

We would like to "minimize" the effect of the gust on the system output, i.e. the output energy is to be minimized when the system is excited by the gust. The feedback controller generates the command signal: $\hat{u}(s) = -C(s)\hat{y}(s)$. Therefore, we want to minimize the energy of y, i.e. $\|y\|_2$,

(5.33) $$\|y\|_2 = \|(1 + P_t C)^{-1} T_g\|_2 =: \gamma_2(C)$$

over all controllers C stabilizing the plant P_t.

Let us define the finite dimensional plant

$$P_{t,f} = \frac{\hat{C}_o(sI - \hat{A})^{-1}\hat{B}_o}{1 - \hat{C}_o(sI - \hat{A})^{-1}\hat{B}_1 \, C_{t,f}(s)}$$

where $C_{t,f}$ is a finite dimensional approximate of the Theodorsen's function $C_t(s)$. Also define an L^∞ approximation error bound ε_f for the Theodorsen's function:

$$\|C_{t,f} - C_t\|_\infty \leq \varepsilon_f.$$

Then it can be shown that a finite dimensional controller C_f stabilizing $P_{t,f}$ also stabilizes P_t if

$$\varepsilon_f \, \|W_1(1 + P_{t,f}C_f)^{-1}\|_\infty < 1,$$

where W_1 is a rational weighting function, which can be explicitly computed in terms of the problem data, see [17] and [18] for details. Furthermore, it can also be shown that (cf. [18])

(5.34) $$\|y\|_2 \leq \frac{1}{1 - \varepsilon_f \, \|W_1(1 + P_{t,f}C_f)^{-1}\|_\infty} \, \|W_2(1 + P_{t,f}C_f)^{-1}\|_2$$

where W_2 is a rational weighting function which can be expressed in terms of the problem data. The inequality (5.34) implies that the gust alleviation problem for the plant P_t amounts to finding a solution to a mixed H^2/H^∞ control problem with a finite dimensional plant $P_{t,f}$. Furter details on this issue can be found in [18].

6 Conclusions

In this paper we have considered a system described by the singular integro-differential equations (2.1)-(2.2). The aeroelastic model derived in §2.3 provides a motivation to study this type of systems. Our main focus was on the stabilization of the plant by a finite dimensional controller. By approximating the infinite dimensional parts of the coprime factors of the original plant the problem is put in the framework of the theory of robust stabilization in the gap metric. An algorithm is given to construct a stabilizing finite dimensional controller. Important steps of the algorithm are finding rational approximates of certain H^∞ functions that are continuous on the boundary, computing the norm of a certain Hankel operator whose symbol is rational, and constructing the controller from the singular values and vectors of this Hankel operator. These are straightforward computations; there are several techniques and software packages available for such operations.

In the case of a noisy measurement (or a disturbance) one may want to find a controller which not only stabilize the system but also minimizes the effect of the noise/disturbance on certain signals of interest. This problem is defined and its solution is briefly discussed in §5.

We would like to mention that for a system similar to the one considered here Balakrishnan, (see §4 of [1]), considers a state space approach with non-zero initial conditions and minimizes a quadratic cost involving the "energy" of the states and the command signal while keeping the system stable. An important open problem in relation with this method is to study robustness of approximations of the optimal infinite dimensional controller obtained in [1].

References

[1] A. V. Balakrishnan, "Active control of airfoils in unsteady aerodynamics," *Applied Math. and Optimization*, **4** (1978), pp. 171-195.
[2] A. V. Balakrishnan and J. W. Edwards, "Calculation of the transient motion of elastic airfoils forced by control surface motion and gusts," NASA JM-81-351, 1980.
[3] R. L. Bisplinghoff, H. Ashley and R. Halfman, *Aeroelasticity*, Addison-Wesley, Cambridge, 1955.
[4] J. A. Burns, E. M. Cliff and T. L. Herdman, "A state-space model for an aeroelastic system," Proc. of the 22nd IEEE Conference on Decision and Control, San Antonio TX, (1983), pp. 1074-1077.
[5] J. A. Burns, T. L. Herdman and J. Turi, "Neutral functional integro-differential equations with weakly singular kernels," *J. Math. Anal. Appl.*, **145** (1990), pp. 371-401.
[6] J. A. Burns and K. Ito, "On well-posedness of solutions to integro-differential equations of neutral type in a weighted L^2 space," Tech. Report No. 11-91, Center for Applied Mathematical Sciences, University of Southern California, CA.
[7] J. W. Edwards, "Unsteady aerodynamic modelling and active aeroelastic control," *SUDAAR*, **504** (1977), Stanford University.
[8] T. T. Georgiou and M. C. Smith, "Optimal robustness in the gap metric," *IEEE Trans. Automat. Control*, **35** (1990), pp. 673-686.
[9] G. Gripenberg, S. O. Londen and O. Staffans, *Volterra Integral and Functional Equations*, Cambridge University Press, New York, 1990.
[10] G. Gu, P. P. Khargonekar and E. B. Lee, "Approximation of infinite dimensional systems," *IEEE Trans. Automat. Control*, **34** (1989), pp. 610-618.
[11] T. L. Herdman and J. Turi, "Singular neutral equations," in *Distributed Parameter Control Systems*, G. Chen, E. B. Lee, W. Littman, L. Markus, eds., Marcel Dekker Inc., New York, 1991, pp. 501-511.
[12] T. L. Herdman and J. Turi, "An application of finite Hilbert transforms in the derivation of a state space model for an aeroelastic system," *J. Int. Eqns. Appl.*, **3** (1991), pp. 271-287.
[13] K. Hoffman, *Banach Spaces of Analytic Functions*, Dover Publications, New York, 1988.
[14] K. Ito and F. Kappel, "On integro-differential equations with weakly singular kernels," in *Differential Equations with Applications*, J. Goldstein, F. Kappel, and W. Schappacher, eds., Marcel Dekker, 1991, pp. 209-218.
[15] K. Ito and J. Turi, "Numerical methods for a class of singular integro-differential equations based on a semigroup approximations," *SIAM J. Numerical Analysis*, **28** (1991), pp. 1698-1722.
[16] F. Kappel and K. P. Zhang "On neutral functional differential equations with nonatomic

difference operator," *J. of Math. Analysis and Appl.*, **113** (1986), pp. 311–343.

[17] H. Özbay and G. R. Bachmann, "Robust control design techniques for active flutter suppression," Proceedings of NASA Workshop on Distributed Parameter Modeling and Control of Flexible Aerospace Systems, Williamsburg VA, June 1992, to appear.

[18] H. Özbay and G. R. Bachmann, "Active feedback controller design for a thin airfoil," submitted to American Control Conference, San Francisco CA, June 1993.

[19] H. Özbay and J. Turi, "Robust stabilization of an aeroelastic system," Proc. of the IEEE Internat. Conf. on Systems Eng., Dayton OH, August 1991, pp. 424–427.

[20] H. Özbay and A. Tannenbaum, " A skew Toeplitz approach to the H^∞ control of multivariable distributed systems," *SIAM J. Cont. and Opt.*, **28** (1990), pp. 653–670.

[21] R. H. Scanlan and R. Rosenbaum, *Introduction to the study of aircraft vibration and flutter*, Macmillan, New York, 1951.

[22] M. C. Smith, "On stabilization and existence of coprime factorizations," *IEEE Trans. Automat. Control*, **34** (1989), pp. 1005–1007.

[23] T. Theodorsen, "General theory of aerodynamic instability and the mechanism of flutter," *National Advisory Committee for Aeronautics Report*, **496** (1935).

[24] M. Vidyasagar, *Control System Synthesis: A factorization approach*, MIT Press, Cambridge, 1985.

Chapter 14
Stable and Unstable Zero Dynamics of Infinite Dimensional Systems [*]

Tzyh-Jong Tarn[†]   Antal K. Bejczy[‡]   Chuanfan Guo[†]

Abstract

A one-link flexible robot arm is formulated as a nonlinear infinite dimensional system. For this system we proved that when the output is chosen as the joint angle, the zero-dynamics is stable so that a nonlinear feedback control can be used, while if the output is chosen as the arc-length of the tip or the projection of the tip on the axis, the zero-dynamics is unstable. For such a system with unstable zero-dynamics, a control technique, the sampled output feedback control with periodic gain, can be applied. This control scheme differs from classical control schemes in that the output of the system need not be monitored continuously but only monitored at sampling instant.

1 Introduction

In analytical studies of distributed parameter systems, the distributed system model is often approximated by a lumped parameter model having finite degrees of freedom. This lumped parameter model is then analysed by using known techniques for lumped parameter systems. In our one link flexible robot arm case, the model most often used is based on finite element method or assumed mode method. Those modeling methods, although reasonable from a practical point of view in many cases, do not always lead to satisfactory results and sometimes may not be able to reflect the real properties of the system. Therefore it is desirable to formulate and study the problem directly in the framework of partial differential equations without resorting to further approximations.

For a one-link flexible robot arm, many studies have been done in the past decade. To address the nonlinear properties of the dynamics of a flexible robot arm, a number of control schemes have been developed using various nonlinear finite-dimensional dynamic models[1]-[8]. Chedmail and Khalil[8] used the nonlinear inversion for input-output linearization of a rigid-body robot arm with PID compensation on the joint angle errors, and then added to this control an auxiliary term to damp out the vibration. De Luca and Siciliano[1][7] used the nonlinear feedback to achieve input-output linearization, and the closed-loop system was found locally stable for a one-link flexible robot arm with structural damping. However, using a different finite-dimensional dynamic model, De Luca et al[9] reached the conclusion that the closed-loop system was unstable while input-output linearization was achieved. Wang and Wei[2] found that the extended motion of a flexible link with prismatic joint could enhance the arm vibration. A feedback control for damping was derived by considering the

---

[*]This research was supported in part by NSF Grant IRI-9106317, and by Sandia National Laboratories Contract No. AC-3752-C.
[†]Dept. of Systems Science and Math., Campus Box 1040, Washington University, St. Louis, MO 63130
[‡]Jet Propulsion Lab., Mail Stop 198-219, 4800 Oak Grove, Pasadena, CA 91109

time rate-of-change of the total vibrational energy of the flexible arm. With a distributed-parameter dynamic model, Khorrami[10] decomposed the system into a slow part and a fast part, using singular perturbation techniques. Control was designed to drive the slow subsystem, while a passive/active damping was required for the fast subsystem to be stable. Robert H. Cannon et al[11] used linear finite dimensional model and find that when the output is chosen as the arc-length of the tip(they called this kind of output as tip position which is different from our definition of the tip position) the system is non-minimum phase. De Luca[1] also verified this fact by using different model. But there is still some confusion about whether the non-minimum phase property is inherent or due to the truncation procedure of modeling. Another point need to be mentioned here is that many people[11]-[14] did not consider the gravity in their model.

In this paper, we begin with a nonlinear infinite dimensional model without constraint on the robot workspace which means that the robot is assumed working in any plane, especially in the vertical plane. First, we analyze the system with different output. For system with colocated control and sensor, the zero-dynamics is stable, while for system with noncolocated control and sensor, the zero-dynamics is unstable. Therefore the nonlinear feedback control scheme can be used for the first case, while for the second case it can not. Second, for system with unstable zero-dynamics, we prove that the sampled output feedback with periodic gain proposed in [19] can be applied. Periodic sampled output feedback is unusual among feedback control schemes in that the control action is determined solely through information garnered in the output of the system at isolated sample times. This unusual property is advantageous in cases where output information is not available at every instant of time because of practical constraints.

2 Zero Dynamics of Nonlinear Infinite Dimensional System

In this section, we mention briefly some results about nonlinear feedback control and zero-dynamics for infinite dimensional systems. For more details please refer to [17].

Consider the following distributed-parameter system:

(2.1)
$$\dot{q} = f(q) + g(q)u$$
$$y = h(q)$$

where $q = (q_1, q_2, ..., q_n)'$, and $\dot{q}_i$ represents dq_i/dt if q_i depends only on t and represents $\partial q_i/\partial t$ if q_i is a function of both t and a spacial variable, say x. $f = [f_1, ..., f_n]'$ and $g = [g_1, g_2, ... g_n]'$ are smooth operators on the state space $\mathbf{X}$, $u \in \mathbf{R}$ is the input and the product of g and u is a map from $\mathbf{U} \times \mathbf{X}$ to $\mathbf{X}$, y is the output and h is a scalar function which is also a smooth operator on $\mathbf{X}$.

DEFINITION 2.1. *A nonlinear system has a relative degree r at $q°$ if*

(2.2)
$$\mathcal{L}_g \mathcal{L}_f^k h(q) = 0$$

for all $k < r - 1$, and for all q in a neighborhood of $q°$; and

(2.3)
$$\mathcal{L}_g \mathcal{L}_f^{r-1} h(q) \neq 0$$

where

$$\mathcal{L}_f h(q) = \frac{\partial h(q)}{\partial q} f(q),$$
$$\mathcal{L}_f^k h(q) = \frac{\partial \mathcal{L}_f^{k-1} h(q)}{\partial q} f(q),$$
$$\mathcal{L}_g \mathcal{L}_f^{r-1} h(q) = \frac{\partial \mathcal{L}_f^{r-1} h(q)}{\partial q} g(q).$$

The relative degree r of the system given above is with the consideration that f, g and h are operators and the Lie derivatives should be calculated using formulas given in[17]. Choose $h, \mathcal{L}_f h, \ldots, \mathcal{L}_f^{r-1} h$ as part of the new state variables, denoted by $\xi_1, \ldots, \xi_r$. Set

(2.4)
$$\begin{aligned} \xi_1(q) &= h(q) \\ \xi_2(q) &= \mathcal{L}_f h(q) \\ &\vdots \\ \xi_r(q) &= \mathcal{L}_f^{r-1} h(q) \end{aligned}$$

Apply the identity mapping to the rest of the state variable q to get η. If the mapping ϕ so obtained from q to $\begin{bmatrix} \xi \\ \eta \end{bmatrix}$ and its inverse are both sufficiently smooth around q°, then it will be qualified as a state transformation. The system in the new coordinates $\begin{bmatrix} \xi \\ \eta \end{bmatrix}$ shall be in the following form, with the understanding that $\tilde{f}$ and $\tilde{g}$ are operators,

(2.5)
$$\begin{aligned} \dot{\xi}_1 &= \xi_2 \\ &\vdots \\ \dot{\xi}_{r-1} &= \xi_{r-2} \\ \dot{\xi}_r &= \mathcal{L}_f^r h + \mathcal{L}_g \mathcal{L}_f^{r-1} h \, u \\ \dot{\eta} &= \tilde{f}(\xi, \eta) + \tilde{g}(\xi, \eta) u \end{aligned}$$

Static state-feedback

(2.6) $$u = \alpha(q) + \beta(q) v$$

can be employed where $\alpha(q)$ and $\beta(q)$ are given by

(2.7)
$$\begin{aligned} \alpha(q) &= -(\mathcal{L}_g \mathcal{L}_f^{r-1} h)^{-1} \mathcal{L}_f^r h \\ \beta(q) &= (\mathcal{L}_g \mathcal{L}_f^{r-1} h)^{-1} \end{aligned}$$

Then, we have

$$\dot{\xi}_r = \mathcal{L}_f^r h + (\mathcal{L}_g \mathcal{L}_f^{r-1} h) u = v$$

and the closed-loop system becomes

(2.8)
$$\begin{aligned} \dot{\xi} &= A\xi + Bv \\ y &= C\xi \\ \dot{\eta} &= \hat{f}(\xi, \eta) + \hat{g}(\xi, \eta) v \end{aligned}$$

where $\hat{f}, \hat{g}$ are operators,

$$A = \begin{bmatrix} 0 & 1 & . & . & . & 0 \\ & & . & & & . \\ & & & . & & . \\ & & & & . & 1 \\ & \bigcirc & & & & 0 \end{bmatrix}_{(r \times r)}$$

$$B = \begin{bmatrix} 0 \\ \vdots \\ 0 \\ 1 \end{bmatrix}$$

and

$$C = \begin{bmatrix} 1 & 0 & \cdots & 0 \end{bmatrix}.$$

Having decomposed the system into two subsystems via state-feedback, as indicated by Eq (2.8), one can design the reference input v to replace poles for the linear subsystem which is controllable and observable and of finite dimension. The output can be regulated in this case. However, in order for the full closed-loop system to be stable, one has to consider the second subsystem which is of infinite dimension. The stability of the full system is related to the spectrum of the zero dynamics of the system.

DEFINITION 2.2. *The zero dynamics of system (2.8) corresponding to y° is described by:*

(2.9) $$\dot{\eta} = \hat{f}(\xi^\circ, \eta)$$

where ξ° is given by

$$\xi^\circ := \xi(0) = [y^\circ \ 0 \ldots 0]'$$

Notice that the zero dynamics of a system really describes the dynamic behavior of the system when feedback u and initial condition of ξ are chosen to keep the output of the system fixed at some value y°. Zero dynamics is not an autonomous system.

REMARK 2.1. *Imposing the output to remain at y° for all times means that the state will be confined within the submanifold*

$$M^\circ = \{q \in \mathbf{R}^n : h(q) = y^\circ\}.$$

Therefore the zero dynamics gives the vector field of the closed-loop system restricted to M°.

3 Zero Dynamics for One-Link Flexible Robot Arm

3.1 Dynamic Model of One-Link Flexible Robot.

Consider the gravity effect, the dynamic model of a flexible robot arm with one flexible thin beam and rigid joint is given by[20]

(3.1) $$[J_m + \int_0^L (x^2 + w^2)\rho dx]\ddot{\theta} + \int_0^L x\rho \ddot{w} dx + 2\int_0^L w\dot{w}\rho dx \dot{\theta} + g_r \rho \int_0^L (x\cos\theta - w\sin\theta)dx = \tau$$

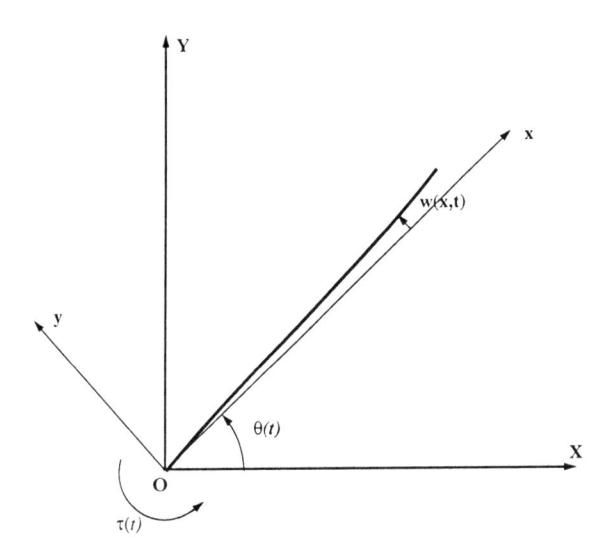

FIG. 1. *One Link Flexible Robot Arm.*

(3.2) $$\ddot{w} + x\ddot{\theta} + \frac{EI}{\rho}w_{xxxx} + g_r \cos\theta - w\dot{\theta}^2 = 0$$

with boundary conditions

(3.3) $$w(0,t) = w_x(0,t) = w_{xx}(L,t) = w_{xxx}(L,t) = 0,$$

where $w(x,t)$ is the displacement of a point on the beam in y direction(Figure 1), $\theta(t)$ is the joint angle, $(\dot{\ })$ denotes $\frac{\partial}{\partial t}(\ )$, $(\ )_x$ denotes $\frac{\partial}{\partial x}(\ )$, EI is the stiffness of the beam (E: Young's modulus; I: inertia), J_m is the motor inertia, ρ is the material density of the link, g_r is a constant which depends on the gravity constant and the robot workspace, τ is the motor torque, L is the length of the beam, m is the mass of the beam. For simplicity, we assume that EI and ρ are constants across the beam.

We see that the model of the one-link flexible robot arm is nonlinear and infinite dimensional. In equations (3.1) and (3.2), all terms containing g_r is due to gravity and g_r depends on robot workspace. If the robot workspace is in the horizontal plane, $g_r = 0$, otherwise $g_r \neq 0$. For example $g_r = g_0 \sin\gamma_0$ if the angle between the robot workspace and the horizontal plane is γ_0, where g_0 is the gravity constant. Especially $g_r = g_0$ if $\gamma_0 = \frac{\pi}{2}$, i.e., the robot workspace is in the vertical plane.

For convenience, replacing $\ddot{w}$ in equation (3.1) by using equation (3.2), the dynamic model of one-link flexible robot arm can be expressed as

$$[J_m + \int_0^L w^2 \rho dx]\ddot{\theta} - EI w_{xx}(0,t) + 2\int_0^L w\dot{w}\rho dx \dot{\theta} - g_r \rho \int_0^L w\sin\theta dx + \rho \int_0^L xw dx \dot{\theta}^2 = \tau$$
(3.4)

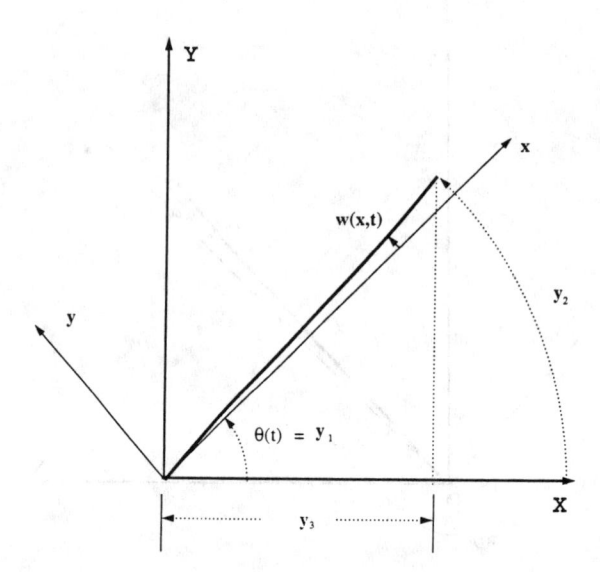

FIG. 2. *The Output Measurements.*

3.2 State Space Expression.

Let $q_1 = \theta(t)$, $q_2 = w(x,t)$, $q_3 = \dot\theta(t)$, $q_4 = \dot w(x,t)$, and $u = \tau$, the above dynamic model may be shown to take in the state space expression as follows

(3.5)
$$\begin{bmatrix} \dot q_1 \\ \dot q_2 \\ \dot q_3 \\ \dot q_4 \end{bmatrix} = \begin{bmatrix} q_3 \\ q_4 \\ f_3(q) \\ f_4(q,x) \end{bmatrix} + \begin{bmatrix} 0 \\ 0 \\ g_3(q_2) \\ g_4(q_2,x) \end{bmatrix} u := f + gu$$

where $q = [q_1, q_2, q_3, q_4]'$. f_3, f_4, g_3 and g_4 all involve operators which are specified by

(3.6)
$$f_3(q) = \frac{EI q_{2,xx}(0,t) - \int_0^L \rho q_2 (2 q_4 q_3 - S_1 + x q_3^2)\,dx}{J_m + \int_0^L q_2^2 \rho\,dx}$$

$$f_4(q,x) = -x f_3(q) + q_2 q_3^2 - C_1 - \frac{EI}{\rho} q_{2,xxxx}$$

$$g_3(q_2) = \frac{1}{J_m + \int_0^L q_2^2 \rho\,dx}$$

$$g_4(q_2,x) = -\frac{x}{J_m + \int_0^L q_2^2 \rho\,dx} = -x g_3(q_2)$$

where $S_1 = g_r \sin q_1$, $C_1 = g_r \cos q_1$.

3.3 The Output Measurements.

For a one-link flexible robot arm, one would consider the following three kinds of output measurement(Figure 2.):

(1) Joint angle :
(3.7) $$y_1 = \theta = q_1 = h(q)$$

(2) Arclength of the tip :
(3.8) $$y_2 = L\theta(t) + w(L,t) = Lq_1 + q_2(L,t) = h(q)$$

(3) The projection of the tip on the X-coordinate:
(3.9) $$y_3 = L\cos\theta(t) - w(L,t)\sin\theta(t) = L\cos q_1 - q_2(L,t)\sin q_1 = h(q)$$

The first one is called colocated (output measurement) because control and output measurement are at the same place: the joint, while the last two are called noncolocated.

We know that for a finite dimensional linear system, the zeros depend on the output. For a nonlinear infinite dimensional system, the zero-dynamics which plays the similar role obviously depends on the output too. Similar to the finite dimensional linear system, the system with stable zero-dynamics is called the minimum phase system, and the system with unstable zero-dynamics the nonminimum phase system.

3.4 Zero Dynamics for Colocated system.

Now, let us discuss zero-dynamics for the first output measurement, i.e.,

(3.10) $$y = \theta(t) = q_1 = h(q)$$

Calculation shows
(3.11) $$\begin{aligned}\mathcal{L}_g h &= 0, \\ \mathcal{L}_g \mathcal{L}_f h &= g_3 \neq 0\end{aligned}$$

therefore the relative degree is 2. Applying a control torque

(3.12) $$u = \alpha(q) + \beta(q)v$$

where
$$\begin{aligned}\alpha(q) &= -(\mathcal{L}_g \mathcal{L}_f h)^{-1} \mathcal{L}_f^2 h = -f_3/g_3 \\ \beta(q) &= (\mathcal{L}_g \mathcal{L}_f h)^{-1} = 1/g_3\end{aligned}$$

the system becomes

(3.13) $$\begin{aligned}\dot{q}_1 &= q_3 \\ \dot{q}_2 &= q_4 \\ \dot{q}_3 &= v \\ \dot{q}_4 &= f_4 + xf_3 - xv\end{aligned}$$

The zero-dynamics is given by

(3.14) $$\begin{aligned}\dot{q}_2 &= q_4 \\ \dot{q}_4 &= f_4 + xf_3|_{q_1 \equiv 0}\end{aligned}$$

or in the following form
(3.15) $$\ddot{w} + \frac{EI}{\rho} w_{xxxx} + g_r = 0$$

with boundary conditions same as (3.3).

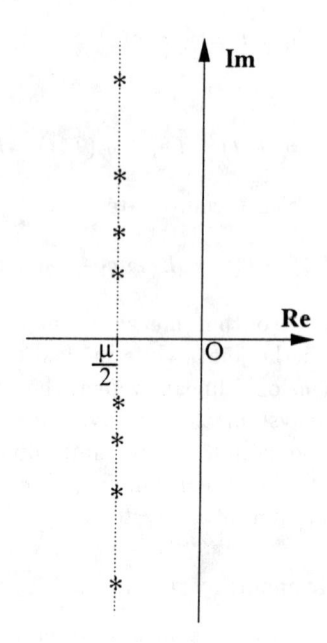

FIG. 3. *The spectrum of zero-dynamics for colocated system.*

Considering any kind of damping, say viscous damping, equation (3.15) can be represented by

$$\ddot{w} + \mu\dot{w} + \frac{EI}{\rho}w_{xxxx} + g_r = 0 \tag{3.16}$$

To solve the equation (3.16), we first let

$$w(x,t) = w_1(x,t) + \varphi(x) \tag{3.17}$$

where $\varphi(x)$ satisfies the following equations

$$\begin{aligned}&\frac{EI}{\rho}\varphi_{xxxx} + g_r = 0 \\ &\varphi(0) = \varphi_x(0) = \varphi_{xx}(L) = \varphi_{xxx}(L) = 0\end{aligned} \tag{3.18}$$

Then equation (3.16) becomes

$$\ddot{w}_1 + \mu\dot{w}_1 + \frac{EI}{\rho}w_{1,xxxx} = 0 \tag{3.19}$$

Using the separation of variables method, let $w_1(x,t) = \Phi(x)q(t)$, we can obtain two equations which are given by

$$\Phi_{xxxx}(x) = \lambda\Phi(x) \tag{3.20}$$

$$\ddot{q}(t) + \mu\dot{q}(t) + \frac{EI}{\rho}\lambda q(t) = 0 \tag{3.21}$$

with boundary conditions

$$\Phi(0) = \Phi_x(0) = \Phi_{xx}(L) = \Phi_{xxx}(L) = 0 \tag{3.22}$$

where λ is an eigenvalue corresponding to the eigenvector $\Phi(x)$. All the eigenvalues satisfying equation (3.20) and boundary conditions in (3.22) can be specified by $\lambda = \lambda_1, \lambda_2, \ldots,$ where $\lambda_i > 0, \forall i$. Therefore by equation (3.21), all the poles of the zero-dynamics should be in the left hand side of the complex plane(Figure 3). Therefore by equation (3.21) the zero-dynamics is stable so that the nonlinear feedback control can be applied[17].

3.5 Zero Dynamics for Noncolocated System.

3.5.1 The Projection of the Tip as output.

Now we choose the output of the system to be the projection of the tip on X-coordinate,

(3.23) $\qquad y = L\cos\theta(t) - w(L,t)\sin\theta(t) = L\cos q_1 - q_2(L,t)\sin q_1 = h(q)$

Using the scheme described in §2, we may obtain

(3.24) $\qquad \begin{aligned} \mathcal{L}_g h &= 0 \\ \mathcal{L}_g \mathcal{L}_f h &= -q_2(L,t)\cos q_1 g_3(q_2) \neq 0 \quad generically. \end{aligned}$

One can see that the relative degree here is 2.

Now let the state transformation $\phi: X \to \xi$ be given by

(3.25) $\qquad \begin{aligned} \xi_1 &= h(q) = L\cos q_1 - q_2(L,t)\sin q_1 \\ \xi_2 &= \mathcal{L}_f h(q) = (-L\sin q_1 - q_2(L,t)\cos q_1)q_3 - q_4(L,t)\sin q_1 \\ \xi_3 &= xq_1 + q_2(x,t) \\ \xi_4 &= \dot\xi_3 = xq_3 + q_4(x,t) \end{aligned}$

Using the new state variables ξ, the system can be expressed as

(3.26) $\qquad \begin{aligned} \dot\xi_1 &= \xi_2 \\ \dot\xi_2 &= \mathcal{L}_f^2 h + \mathcal{L}_g \mathcal{L}_f h\ u \\ \dot\xi_3 &= \xi_4 \\ \dot\xi_4 &= -\frac{EI}{\rho}\xi_{3,xxxx} + \bar f_4(\xi,x) \\ y &= \xi_1 \end{aligned}$

where $\bar f_4(\xi,x) = [-g_r\cos q_1 + q_2 q_3^2][\xi]$ which comes from inverse transformation.

Let the output be kept at zero for all time. Then the zero-dynamics of the system is given by

(3.27) $\qquad \begin{aligned} \dot\xi_3 &= \xi_4 \\ \dot\xi_4 &= -\frac{EI}{\rho}\xi_{3,xxxx} + \bar f_4(\xi,x)|_{\xi_1=\xi_2=0} \end{aligned}$

with boundary conditions

(3.28) $\qquad \xi_3(0,t) = \xi_3(L,t) = \xi_{3,xx}(L,t) = \xi_{3,xxx}(L,t) = 0.$

The boundary condition $\xi_3(L,t) = 0$ is due to the assumption that the output y is kept at zero for all time. Back to the original notations $\theta(t)$ and $w(x,t)$, the zero-dynamics shown by equation (3.27) cab be represented by

(3.29) $\ddot w(x,t) - \frac{x}{L}\ddot w(L,t) + \frac{EI}{\rho}w_{xxxx}(x,t) + g_r\cos(\arctan\frac{L}{w(L,t)}) - w(x,t)\frac{\dot w^2(L,t)}{L^2} = 0$

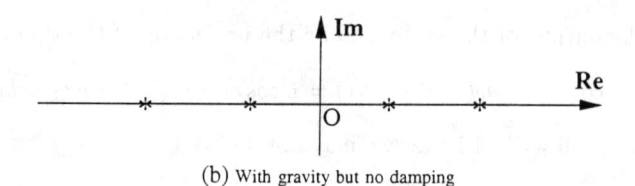

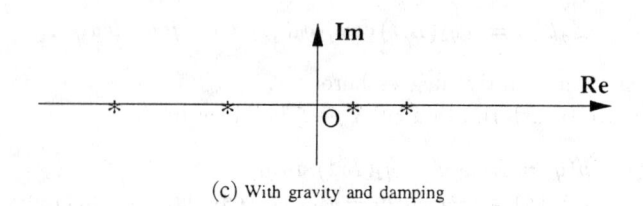

FIG. 4. *The spectrum of zero-dynamics for noncolocated system.*

with the boundary conditions same as (3.3).

Linearize equation (3.29) around an equilibrium point, the zero dynamics can be expressed by

$$\ddot{w}(x,t) - \frac{x}{L}\ddot{w}(L,t) + \frac{EI}{\rho}w_{xxxx}(x,t) + g_r\frac{w(L,t)}{L} = 0 \tag{3.30}$$

Using the separation of variables method, let $w(x,t) = \Phi(x)p(t)$ we can obtain

$$\Phi(x)\ddot{p}(t) - \frac{x}{L}\Phi(L)\ddot{p}(t) + \frac{EI}{\rho}\Phi_{xxxx}(x)p(t) + \frac{g_r}{L}\Phi(L)p(t) = 0 \tag{3.31}$$

To solve equation (3.31), we need to solve the following equations

$$\Phi_{xxxx}(x) = \lambda(\Phi(x) - \frac{x}{L}\Phi(L)) - \frac{g_r\rho}{EIL}\Phi(L) \tag{3.32}$$

and

$$\ddot{p}(t) + \frac{EI}{\rho}\lambda p(t) = 0 \tag{3.33}$$

with boundary conditions

$$\Phi(0) = \Phi_x(0) = \Phi_{xx}(L) = \Phi_{xxx}(L) = 0 \tag{3.34}$$

and some suitable initial conditions for equation (3.33).

The equation (3.33) shows that if $\lambda < 0$ then the zero-dynamics will have pole(s) in the right hand side of the complex plane. For $\lambda < 0$, the solution $\Phi(x)$ of equation (3.32) should take the following form

$$\Phi(x) = (B_1 \sin \kappa x + B_2 \cos \kappa x) \sinh \kappa x + (B_3 \sin \kappa x + B_4 \cos \kappa x) \cosh \kappa x + B_5 x + B_6 \tag{3.35}$$

where $\kappa > 0$, and $-4\kappa^4 = \lambda$. Let $V = [B_1, B_2, ..., B_6]^T$, then $V \neq 0$ means that equation (3.32) has a nontrivial solution $\Phi(x)$ corresponding to a negative eigenvalue λ.

Using equation (3.32) and the boundary conditions in (3.34), we can prove that $V \neq 0$ only if $\exists \kappa > 0$ so that the following equation is satisfied

$$\cos(\kappa L) \sinh(\kappa L) + \sin(\kappa L) \cosh(\kappa L) + \frac{Lg_r \rho}{4\kappa^3 EI}(cosh(\kappa L) - \cos(\kappa L)) = 0 \tag{3.36}$$

If $g_r = 0$(no gravity), there are infinite many κ's satisfy equation (3.36), otherwise there are finite number of κ's. For example, suppose $L = 1.0(m)$, $g_r = 9.8(m/s^2)$, $\rho = 0.4366(Kg/m)$, $EI = 229.86(Nm^2)$, we can obtain that there are at least two κ's: $\kappa_1 = 2.4, \kappa_2 = 5.5$, therefore $\lambda_1 = -133$, $\lambda_2 = -3660$.

Figure 4 shows the spectrum of the zero-dynamics. Figure 4(a) is for $g_r = 0$; Figure 4(b) shows some part of the spectrum of the zero-dynamics for $g_r \neq 0$ and without damping while Figure 4(c) is for $g_r \neq 0$ and with damping. Obviously, we can see that the zero-dynamics is unstable.

3.5.2 The Arc Length of the Tip as Output.

Now we change the output to the arc length of the tip:

$$y = L\theta(t) + w(L, t) = Lq_1 + q_2(L, t) = h(q) \tag{3.37}$$

One can check that

$$\begin{aligned} \mathcal{L}_g h &= 0, \quad \mathcal{L}_g \mathcal{L}_f h = 0, \\ \mathcal{L}_g \mathcal{L}_f^2 h &= 2q_2(L, t)q_3 g_3(q_2) \neq 0 \quad \text{generically}. \end{aligned} \tag{3.38}$$

We see that the relative degree here is 3 which is different from the case where the output is chosen as tip position. This is also different from the case when the assumed mode method is used, where for the same output, the relative degree is 2 and becomes undefined when the number of modes goes to infinity[12]. This means that using the original nonlinear infinite dimensional model is more accurate and sometimes even more convenient.

Now let the state transformation $\phi: X \to \xi$ be given by

$$\begin{aligned} \xi_1 &= h(X) = Lq_1 + q_2(L, t) \\ \xi_2 &= \mathcal{L}_f h(X) = Lq_3 + q_4(L, t) \\ \xi_3 &= \mathcal{L}_f^2 h(X) = q_2(L, t)q_3^2 - g_r \cos q_1 - \frac{EI}{\rho} q_{2,xxxx}(L, t) \\ \xi_4 &= q_2(L, t) \\ \xi_5 &= xq_1 + q_2(x, t), \quad 0 \leq x < L \\ \xi_6 &= \dot{\xi}_3 = xq_3 + q_4(x, t), \quad 0 \leq x < L \end{aligned} \tag{3.39}$$

In the new state variable ξ, the system can be expressed by

(3.40)
$$\begin{aligned}
\dot{\xi}_1 &= \xi_2 \\
\dot{\xi}_2 &= \xi_3 \\
\dot{\xi}_3 &= \mathcal{L}_f^3 h + \mathcal{L}_g \mathcal{L}_f^2 h \, u \\
\dot{\xi}_4 &= f_5(\xi) \\
\dot{\xi}_5 &= \xi_6 \\
\dot{\xi}_6 &= -\frac{EI}{\rho}\xi_{5,xxxx} + \bar{f}_4(\xi, x) \\
y &= \xi_1
\end{aligned}$$

where $f_5(\xi) = [q_2(L,t)][\xi]$ which is due to the inverse transformation.

Similar to §3.5.1, we can obtain the linearized zero-dynamics

(3.41)
$$\ddot{w}(x,t) - \frac{x}{L}\ddot{w}(L,t) + \frac{EI}{\rho}w_{xxxx}(x,t) + g_r\frac{w(L,t)}{L} = 0$$

which is the same as equation (3.23), and the boundary conditions are the same as in equation (3.22). Therefore the zero dynamics generated by equation (3.28) is also unstable.

4 Control with Periodic Feedback Gain and Sampled Output for Non-colocated Robot System

In §3, we discussed the stable and unstable zero-dynamics for the one-link flexible robot arm with different output measurements. For the system with stable zero-dynamics, the nonlinear feedback control can be applied, while for the system with unstable zero-dynamics it can not. In this section, the feedback control with periodic gain and sampled output proposed in [19] will be used. The basic idea is that we first linearize the nonlinear infinite dimensional system around any equilibrium point, then based on this linearized system design a feedback control with periodic gain and sampled output.

4.1 The Equilibrium Points of the System.

Assume $\theta = \theta_0$(constant), $w(x,t) = w_0(x)$ which is a constant with respect to time t for $0 \leq x \leq L$, then equations (3.1) and (3.2) can be reduced to

(4.1)
$$g_r\rho \int_0^L (x\cos\theta_0 - w_0\sin\theta_0)dx = \tau_0$$

(4.2)
$$\frac{EI}{\rho}w_{0,xxxx}(x) + g_r\cos\theta_0 = 0$$

with boundary conditions

(4.3)
$$w_0(0) = w_{0,x}(0) = w_{0,xx}(L) = w_{0,xxx}(L) = 0.$$

DEFINITION 4.1. *By an equilibrium point we mean a point $(\theta_0, w_0(x))'$ which is in the set $\mathcal{E}$ defined by*

$$\mathcal{E} = \{(\theta_0, w_0(x))' : \exists \tau_0 \text{ such that equations } (4.1), (4.2) \text{ and } (4.3) \text{ hold}\}.$$

We can check that for the robot working in the horizontal plane ($g_r = 0$)

$$\mathcal{E} = \mathcal{E}_0 = \{(\theta_0, 0)' : \tau_0 = 0, \ \forall \theta_0 \in \mathbf{R}\},$$

while for a robot working in any other plane

$$\mathcal{E} = \left\{ \begin{pmatrix} \theta_0 \\ -\dfrac{G_2}{24EI}(x^2 - 4Lx + 6L^2)x^2 \end{pmatrix} : \tau_0 = \left(\dfrac{L^2}{2} + \dfrac{G_1 L^5}{24EI}\right) G_2, \ \forall \theta_0 \in \mathbf{R} \right\}$$

where $G_1 = \rho g_r \sin\theta_0$, $G_2 = \rho g_r \cos\theta_0$. $\mathcal{E}_0$ is a special case of $\mathcal{E}$ for $g_r = 0$.

4.2 Linearization Around an Equilibrium Point.

For a robot works in some plane, we can find an equilibrium point $(\theta_0, w_0(x))'$ as described in equation (4.4). Linearize equations (3.4) and (3.2) around this equilibrium point, then

$$(4.4) \quad J_a \ddot{\theta}(t) - EI w_{xx}(0, t) - G_1 \int_0^L w(x,t)dx - G_2 \int_0^L w_0(x)dx \theta(t) = \tau(t)$$

$$(4.5) \quad \ddot{w}(x,t) + x\ddot{\theta}(t) + \dfrac{EI}{\rho} w_{xxxx}(x,t) - \dfrac{G_1}{\rho}\theta(t) = 0$$

with boundary conditions same as equation (3.3), where

$$J_a = J_m + \int_0^L w_0^2(x)\rho dx,$$

and $w(x,t), \theta(t)$ and τ are used here instead of $\delta w(x,t), \delta \theta(t)$ and $\delta \tau$ for simplicity.

Let $y(x,t) = w(x,t) + x\theta(t)$, equations (4.4) and (4.5) can be expressed by

$$(4.6) \quad J_a \ddot{\theta}(t) - EI y_{xx}(0,t) - G_1 \int_0^L y\,dx - \left(G_2 \int_0^L w_0(x)dx - \dfrac{1}{2}G_1\right)\theta(t) = \tau(t)$$

$$(4.7) \quad \ddot{y} + \dfrac{EI}{\rho} y_{xxxx} - \dfrac{G_1}{\rho}\theta(t) = 0$$

with boundary conditions

$$(4.8) \quad y(0,t) = y_{xx}(L,t) = y_{xxx}(L,t) = 0, \ y_x(0,t) = \theta(t).$$

Notice that the joint angle is involved in the last boundary condition.

4.3 Properties of the Linearized System.

Now let us write equations (4.6) and (4.7) in the second order form

$$(4.9) \quad \ddot{Z} + AZ = B\tau$$

where

$$(4.10) \quad Z = \begin{pmatrix} y(t) \\ \theta(t) \end{pmatrix}$$

and

(4.11)
$$A = \begin{pmatrix} \dfrac{EI}{\rho}\dfrac{\partial^4}{\partial x^4} & -\dfrac{G_1}{\rho} \\ -\dfrac{EI}{J_a}\dfrac{\partial^2}{\partial x^2}\bigg|_{x=0} - \dfrac{G_1}{J_a}\int_0^L (.)dx & -\dfrac{G_3}{J_a} \end{pmatrix},$$
$$B = \begin{pmatrix} 0 \\ \dfrac{1}{J_a} \end{pmatrix}$$

where $G_3 = G_2\int_0^L w_0(x)dx - \frac{1}{2}G_1$ and any element in A which involves derivative or integral is an operator.

The domain of A is given by

$$D(A) = \{h \in \mathcal{Z} | h_1 \in \mathbf{H}^4(0,L), h_1(0) = 0, h_{1,x}(0) = h_2,$$
$$h_{1,xx}(L) = h_{1,xxx}(L) = 0\}$$

where

$$\mathbf{H}^4(0,L) = \{f \in \mathbf{L}_2(0,L) | \frac{df}{dx}, \frac{d^2f}{dx^2}, \frac{d^3f}{dx^3}, \text{ and } \frac{d^4f}{dx^4} \in \mathbf{L}_2(0,L)\}$$

which is a Sobolev space, and

$$\mathcal{Z} = \mathbf{L}_2(0,L) \oplus \mathbf{R}$$

which is a Hilbert space with the inner product

$$\left\langle \begin{pmatrix} e_1 \\ e_2 \end{pmatrix}, \begin{pmatrix} f_1 \\ f_2 \end{pmatrix} \right\rangle_{\mathcal{Z}} = \frac{\rho}{EI}\langle e_1, f_1\rangle_{\mathbf{L}_2(0,L)} + \frac{J_a}{EI}e_2 f_2$$

for any $e, f \in \mathcal{Z}$.

Furthermore, we can check that

(4.12)
$$<Ae,f>_{\mathcal{Z}} = <e,Af>_{\mathcal{Z}}$$

therefore the operator A is symmetric. We can also prove that A is a closed, densely defined self-adjoint operator. The point spectrum of A is given by

(4.13)
$$\sigma(A) = \lambda_i, i = 1, 2, ..., n, ...$$

where $\lambda_1 = -4\beta^4 (\beta > 0)$ and β is the solution of the following equation

(4.14) $$\beta(\frac{EI\lambda_1^2 J_a}{\rho} + \frac{G_1^2}{EI} + G_3\lambda_1)(c^2 + ch^2) + EI\beta^2\lambda_1(cs - shch) + \frac{G_1^2}{EI}(cs + shch) = 0$$

where $c = \cos\beta x$, $s = \sin\beta x$, $ch = \cosh\beta x$, $sh = \sinh\beta x$, while if $\lambda_i = \gamma_i^4(\gamma_i > 0)$, $i = 2, 3, ..., n, ...$ and γ_i' are solutions of the following equation

(4.15) $$\gamma_i(-\frac{EI\lambda_i^2 J_a}{\rho} + \frac{G_1^2}{EI} - G_3\lambda_i)(cch + 1) - EI\gamma_i^2\lambda_i(sch - csh) + \frac{G_1^2}{EI}(sch + csh) = 0$$

where $c, ch, s,$ and sh are the same functions as defined above except using γ_i instead of β

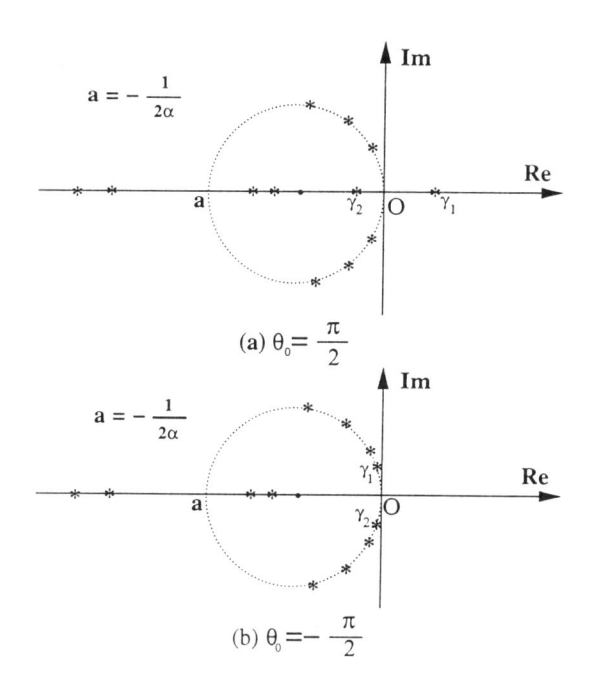

FIG. 5. *The point spectrum of $\mathcal{A}$ which involves structural damping.*

4.4 State Space Model with Damping.

Before we write down the state space representation of the linearized infinite dimensional system, we first would like to discuss the damping terms. There is no well defined representation for a damping term in a flexible structrue. The most often used damping terms are the so-called viscous damping and structural damping[13][21]-[23]. In this paper, we are going to consider a kind of structural damping and viscous damping which are similar to the ones described in [13].

Physically, we can interpret the spectrum of operator A given by equation (4.13) as following: the eigenvalue λ_1 of A is corresponding to the rigid body motion, while all the others are due to flexible property of the link.

Considering a kind of structural damping, $\alpha A \dot{Z}$, the linearized system can be rewritten as

(4.16) $$\ddot{Z} + \alpha A \dot{Z} + AZ = B\tau$$

If the arc-length of the tip is chosen as the output, the state space model is given by

(4.17) $$\begin{aligned}\dot{X} &= \mathcal{A}X + \mathcal{B}u \\ Y &= \mathcal{C}X\end{aligned}$$

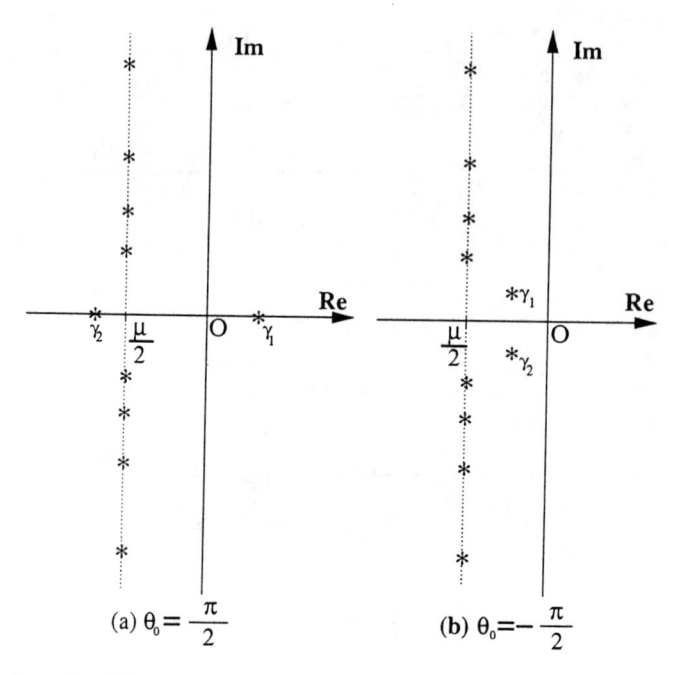

FIG. 6. *The point spectrum of $\mathcal{A}$ which involves viscous damping.*

where

$$X = \begin{pmatrix} Z \\ \dot{Z} \end{pmatrix}$$

$$\mathcal{A} = \begin{pmatrix} 0 & I \\ -A & -\alpha A \end{pmatrix}$$

$$\mathcal{B} = \begin{pmatrix} 0 \\ 0 \\ B \end{pmatrix}$$

$$\mathcal{C} = \begin{pmatrix} <\delta(L), .> & 0 & 0 & 0 \end{pmatrix}$$

We can prove that the point spectrum of $\mathcal{A}$ is

$$\sigma_P(\mathcal{A}) = \{\gamma \in \mathbf{C} | \gamma^2 + \alpha\lambda\gamma + \lambda, \lambda \in \sigma_P(A)\}$$

If we consider viscous damping, we only need to replace $\alpha A \dot{Z}$ in equation (4.17) by $\mu \dot{Z}$. Similarly, the point spectrum of $\mathcal{A}$ in this case is

$$\sigma_P(\mathcal{A}) = \{\gamma \in \mathbf{C} | \gamma^2 + \mu\gamma + \lambda, \lambda \in \sigma_P(A)\}$$

Figure 5 and Figure 6 show the point spectrum of one link flexible robot arm with structural damping and viscous damping, respectively. Figure 5(a) and Figure 6(a) are for the robot working at $(\theta_0, w_0(x))' = (\pi/2, 0)'$, while Figure 5(b) and Figure 6(b) are at $(\theta_0, w_0(x))' = (-\pi/2, 0)'$. For the case $\theta_0 \neq \pm\pi/2$, all the eigenvalues of the zero-dynamics

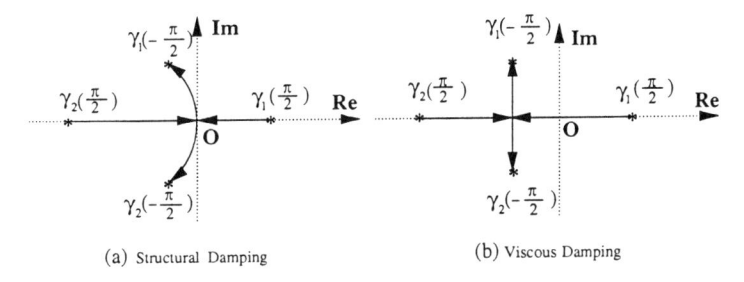

FIG. 7. *The change of the first pair of eigenvalues.*

(a) Structural Damping

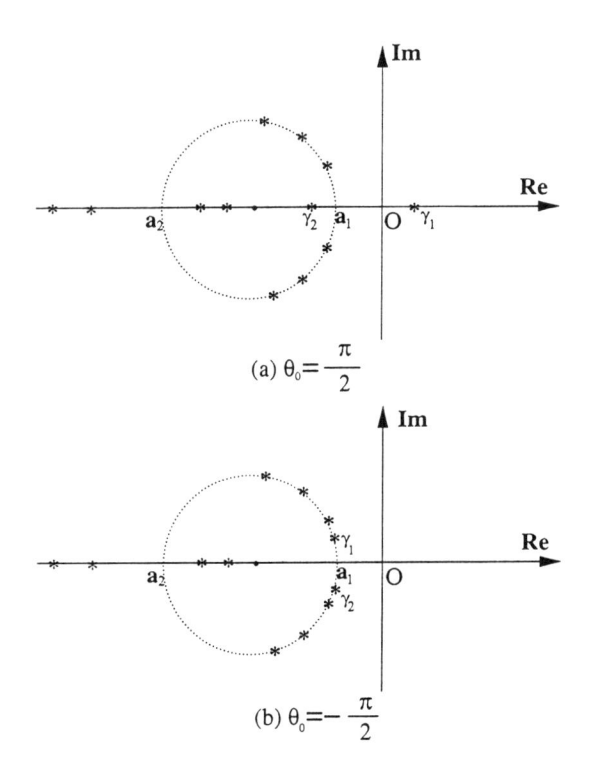

(b) Viscous Damping

(a) $\theta_0 = \dfrac{\pi}{2}$

(b) $\theta_0 = -\dfrac{\pi}{2}$

FIG. 8. *The point spectrum of $\mathcal{A}$ which involves both dampings.*

will keep in the same profile as above two cases except the first pair γ_1 and γ_2. Since γ depends on θ_0, we can let $\gamma_1 = \gamma_1(\theta_0)$, $\gamma_2 = \gamma_2(\theta_0)$. Figure 7 shows the change of γ_1 and γ_2 with θ_0. Where the arrow implies the increase direction of θ_0 from $\pi/2$ to π then to $3\pi/2$ or the decrease direction of θ_0 from $\pi/2$ to 0 then to $-\pi/2$.

Now, if we combine both dampings together, the operator $\mathcal{A}$ is given by

(4.18)
$$\mathcal{A} = \begin{pmatrix} 0 & I \\ -A & -(\alpha A + \mu) \end{pmatrix}$$

It can be proved that the point spectrum of $\mathcal{A}$ is given by

$$\sigma_P(\mathcal{A}) = \{\gamma \in \mathbf{C} | \gamma^2 + (\alpha\lambda + \mu)\gamma + \lambda, \lambda \in \sigma_P(A)\}$$

Figure 8 shows the profile of the point spectrum of $\mathcal{A}$ for $(\theta_0, w_0(x))' = (\pm\pi/2, 0)'$. Where $a_1 = -\frac{2+\alpha\mu-\sqrt{4-\alpha^2\mu^2}}{2\alpha}$ and $a_2 = -\frac{2+\alpha\mu+\sqrt{4-\alpha^2\mu^2}}{2\alpha}$.

REMARK 4.1. *Some authors[13,21,22] have studied different kind of dampings in flexible structures and $\alpha A^{\frac{1}{2}}$ is one of them. For our control problem, this kind of damping can not be used since A is not necessarily positive.*

4.5 Control Design for Noncolocated Systems.

For the system described by equation (4.17), the operator $\mathcal{B}$ is just a constant vector therefore it is bounded, while the operator $\mathcal{C}$ is unbounded which makes the control more complicated. For this kind of system, Tarn et al[19] proposed a control scheme called periodic feedback control with sampled output. The theorem is quoted here.

THEOREM 4.1. *[19] Whenever the linear system described by equation (4.17) with $\mathcal{A}$ described by (4.18) satisfies $(A_1)-(A_5)$, there exists $\Delta > 0$ and $K(t) \in \mathbf{L}^2_{loc}([0,\infty) : \mathbf{R}^{p \times m})$ such that $K(t + \Delta) = K(t)$ for all $t > 0$ and the closed loop system with the periodic output feedback control $u(t) = K(t)Y(n\Delta)$, $n\Delta \leq t \leq (n+1)\Delta$ is stable in the sense:* $\lim_{t\to\infty} \|Z(t)\| = 0$.

Where the assumptions are given by

(A_1) The infinitesimal generator A of the strongly continuous semigroup $\{T(t); t \geq 0\}$ satisfies the spectrum decomposition assumption (Kato,1966) in the sense that $\sigma_u = \{\lambda \in \sigma_P(\mathcal{A}) | Re\lambda \geq -\delta\}$ has the property that for some $\delta > 0$, the dimension of the union of the generalized eigenspaces generated from point in σ_u, $\mathcal{Z}_u$ is finite. For any such δ we define the following subspaces, semigroups, infinitesimal generators and operators:

$$\mathcal{Z} = \mathcal{Z}_u \oplus \mathcal{Z}_s$$
$$T_u(t) = P_u T(t) = T(t) P_u$$
$$T_s(t) = (I - P_u)T(t) = T(t)(I - P_u)$$
$$\mathcal{A}_u = \mathcal{A} P_u = P_u \mathcal{A}$$
$$\mathcal{A}_s = \mathcal{A}(I - P_u) = (I - P_u)\mathcal{A}$$

(A_2) $\{T_s(t); t \geq 0\}$ satisfies the spectrum determined growth condition, i.e.

$$Sup\, Re\, \sigma(\mathcal{A}_s) = \inf_{t>0} \frac{ln\|T_s(t)\|}{t} = \lim_{t\to\infty} \frac{ln\|T_s(t)\|}{t}$$

(A_3) $(\mathcal{A}_u, \mathcal{C}_u)$ is an observable pair on the space $\mathcal{Z}_u$ where $\mathcal{C}_u$ is defined by $\mathcal{C}_u = \mathcal{C}|_{\mathcal{Z}_u}$

(A_4) $\overline{R(T_0)} \supseteq \mathcal{Z}$ where $R(T_0)$ is the reachable set of $(\mathcal{A}, \mathcal{B})$ on $[0, T_0]$ for some $T_0 > 0$ (the overbar indicates closure).

(A_5) If $\mathcal{C}$ is unbounded but there exists a real Banach space $\mathbf{W}$ which is dense in $\mathcal{Z}$ with respect to the topology of $\mathcal{Z}$ such that

$$(a) \mathcal{Z} \supset D(\mathcal{C}) \supset \mathbf{W}$$
$$(b) \mathcal{C} \in \mathbf{L}(\mathbf{W}, \mathbf{Y})$$
$$(c) T(t) \in \mathbf{L}(\mathcal{Z}, \mathbf{W}) \; for \; t > 0.$$

For our one-link flexible robot arm, $p = m = 1$. We can check that all the assumptions are satisfied for systems with noncolocated output measurement. These systems can be stabilized locally at any equilibrium point by applying this control scheme.

5 Summary and Conclusion

A typical one-link flexible robot arm is modeled as a nonlinear infinite dimensional dynamic system with gravity. For different output measurement, the stable and unstable zero-dynamics are appeared. For the system with stable zero-dynamics one can use nonlinear feedback to linearize and stablize the system, while for the system with unstable zero-dynamics, the nonlinear feedback control can not be used. To control the system, a very useful control scheme proposed in [19] is used here: control with periodic gain and sampled output.

For multi-link flexible robot, our conjecture is that except the complexity of the model, the profile of the spectrum of the system should be the same and one can apply the similar control laws as in this paper for stabilization of the system.

References

[1] A. De Luca, B. Siciliano, *Trajectory control of a non-linear one-link flexible arm*, Int. J. Control, 1989, Vol.50, No.5, 1699-1715.

[2] P. K. C. Wang and J. Wei, *Feedback Control of Vibration in a Moving Flexible Robot arm with Rotary and Prismatic Joints*, Proceedings of IEEE International Conference on Robotics and Automation , 1987.

[3] S. K. Biswas and R. D. Klafter, *Dynamic Modeling and Optimal Control of Flexible Robotic Manipulator*, Proceedings of the IEEE International Conference on Robotics and Automation, 1988, Vol.1.

[4] B. Gebler, *Feed-Forward Control Strategy for an Industrial Robot with Elastic Links and Joints*, Proceedings of the IEEE International Conference on Robotics and Automation, Vol.2, 1987.

[5] S. N. Singh and A. A. Schy, *Control of Elastic Robotic Systems by Nonlinear Inversion and Model Damping*, Sept. 1986, Vol.108, Transaction of the ASME .

[6] F. Pfeiffer and B. Gebler, *A Multistage-Approach to the Dynamics and Control of Elastic Robots*, IEEE International Conference on Robotics and Automation, April 1988, 24-29.

[7] A. De Luca and B. Siciliano, *Joint-Based Control of a Nonlinear Model of a Flexible Arm*, Proceedings, American Control Conference, Atlanta, June 1988.

[8] P. Chedmail and W. Khalil, *Nonlinear Decoupling of Flexible Robots*, Proceedings, ICAR, 1989.

[9] A. De Luca, P. Lucibello and F. Nicoló, *Automatic Symbolic Modeling and Nonlinear Control of Robots with Flexible Links*, IEEE Workshop on Robot Control, Oxford UK, April 1988.

[10] F. Khorrami, *Analysis of Multi-Link Flexible Manipulators via asymptotic Expansions*, Proceedings, 28th IEEE Conference on Decision and Control, 1989.

[11] Robert H. Cannon Jr., Eric Schmitz, *Initial experiments on the end-point control of a flexible one-link robot*, The International Journal of Robotics Research, Vol.3, No.3, Fall 1984.

[12] David Wang, M. Vidyasagar, *Transfer functions for a single flexible link*, The International Journal of Robotics Research, Vol.10, No.5, Oct. 1991.
[13] Jan Bontsema, *Dynamic stabilization of large flexible structures*, Ph.D. Dissertation, Dept. of mathematics, University of Groningen, Neitherland, June 1989.
[14] Toshio Fukuda, *Flexibility control of elastic robotic arms*, Journal of Robotic Systems, 2(1), 73-88, 1985.
[15] F. Bellezza, L. Lanari, G. Ulivi, *Exact modeling of the flexible slewing link*, Proceedings IEEE International Conference on Robotics and Automation, Cincinnati, Ohio, 734-739, 1990.
[16] A. De. Luca, P. Lucibello, G. Ulivi, *Inversion techniques for trajectory control of flexible robot arms*, Journal of Robotic Systems, 6(4), 325-344, 1989.
[17] X. Ding, T. J. Tarn, A. K. Bejczy, and C. Guo, *Nonlinear Feedback Control of Flexible Robot Arms with Infinite Dimensional Models*, to appear in Journal of Mathematical Systems, Estimation, and Control.
[18] Andrew D. Christian, Warren P. Seering, *Initial Experiments with a flexible robot*, Proceedings IEEE International Conference on Robotics and Automation, Cincinnati, Ohio, 722-727, 1990.
[19] T. J. Tarn, John R. Zavgren Jr., Xiaoming Zeng, *Stabilization of infinite-dimensional systems with periodic feedback gains and sampled output*, Automatica, Vol.24, No.1, 95-99, 1988.
[20] Xuru Ding, *An analytic study of general dynamic model and nonlinear feedback control of flexible robot arms*, Doctor's Dissertation, Dept. of Systems Science and Mathematics, Washington University, St. Louis, MO., USA, 1989.
[21] G. Chen, D. L. Russell, *A mathematical model for linear elastic systems with structural damping*, Quarterly of Applied Mathematics, 433-454, January 1982.
[22] D. L. Russell, *Mathematical models for the elastic beam with frequency-proportional damping*, Frontiers in Applied Mathematics(H.T. Banks, ed.) SIAM, to appear.
[23] Falun Huang, *On the mathematical model for linear elastic systems with analytic damping*, SIAM J. of Control and Optimization, Vol.26, No.3, may 1988.

Chapter 15
Covariance Based Control of Linear Hereditary Systems

James A. Reneke[*]

Abstract

Reproducing kernel Hilbert space methods are used to choose a parameterized feedback operator minimizing a functional of the terminal variance of the controlled system. The methods use the information contained in the data summarized in the discrete covariance function and a weak assumption on the form of the model. Computational results are presented for a process without a finite dimensional state space representation.

1 Introduction

The ultimate intelligent controller is an "embedded computer" or "chip" which sets the parameters of some feedback mechanism, based on inputs from system sensors, to improve system performance [6]. Ideally the algorithm would not be model based. Two trends motivate the development of such controllers.

Centralized supervision of large complex systems is difficult because of computing or communication limitations, large numbers of system components and sensors, and requirements for fast reaction times. A common response to these difficulties is to incorporate in the overall system design local sensors, data analysis and controller updates.

A second trend is the emerging philosophy of install and forget. Larger systems are assembled with increasing frequency from smart components; for instance, subsystems which have embedded controllers. Therefore the smart components must be able to respond to evolving systems or system environments.

A control design methodolgy adequate for the task would be data based and only weakly model dependent. The methodolgy would include a way to balance the robustness of the designed controller with the quality of the accummulated sensor data.

The approach presented here is based on a covariance description of the system. Given a zero mean vector stochastic process $Y_t, 0 \leq t$, we define the covariance function R by $R(s,t) = EY(s)Y(t)^*$. There is a one to one correspondence between processes and covariances [4]. We are assuming that the only information available to the control designer is contained in the process covariance function [see Figure 1].

In general, the covariance function R for a process $Y_t, 0 \leq t$, is nonnegative, i. e.,

$$\sum_{p,q=0}^{n} \langle c_p, R(s_p, s_q) c_q \rangle \geq 0$$

---

[*]Department of Mathematical Sciences, Clemson University, Clemson, SC 29634-1907, U.S.A. (Reneke@clemson.bitnet).

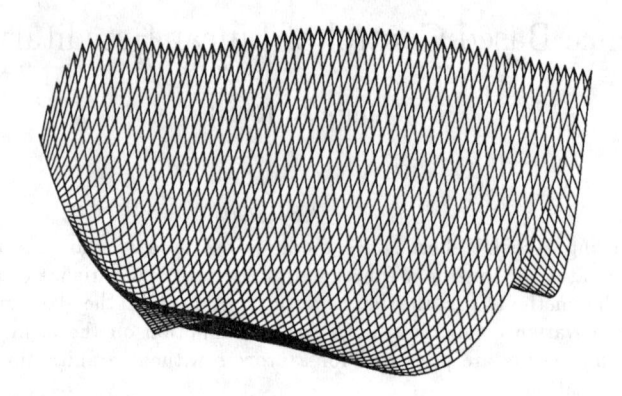

FIG. 1. *Views of a Typical Covariance Surface*

for each sequence $\{s_p\}_{p=0}^n$ in $[0,\infty)$ and sequence $\{c_p\}_{p=0}^n$ in R^d. A classic result [1] asserts the existence of a unique complete Hilbert space $\{G_R, Q_R\}$ of functions from $[0,\infty)$ into R^d with reproducing kernel R, i. e.,

1. $R(\cdot, t)c$ is in G_R for each t in $[0,\infty)$ and c in R^d and

2. $Q_R(f, R(\cdot, t)c) = \langle f(t), c \rangle$ for each f in G_R, t in $[0,\infty)$ and c in R^d.

Briefly, $\{G_R, Q_R\}$ is an RKH space with kernel R. While the geometry of finite dimensional inner product spaces carries over to Hilbert spaces, both the geometry and the linear algebra of finite dimensional inner product spaces carry over to reproducing kernel Hilbert spaces. We want to exploit this feature of the RKH space associated with the process $Y_t, 0 \leq t$, through the covariance function R.

This paper is intended as a roadmap tracing the application of RKH space methods to control through a single motivating infinite dimensional example. Questions remain about the utility of the approach. Are the data requirements realistic for practical applications? While the covariance determines the process, we are really interested in controlling the underlying system. Can we determine, based on observations, whether more sensors and actuators are required to control the system? For systems governed by PDE's the problem is to determine, based on observations, the placement in space of the sensors and actuators. In spite of the unanswered questions, exploring what can be done seems of interest, perhaps giving some urgency to these and similar questions.

2 Background

In Section 3 we will pose and solve a control problem based on a covariance description of the system. However, in order to develop and discuss robust methods we need a framework which includes a general class of systems [5].

There are two different time regimes possible for collecting data on a process. In the first, the system is started from a zero state and the covariance is estimated from an ensemble of sample paths on some fixed time interval. We can think of this as the transient problem.

In the second, the system has achieved steady state and the covariance is estimated from a single long sample path. We will be dealing with the first problem, i.e., improving system performance using feedback in the transient phase. Of course, the hope is that the second problem can be treated in a similar fashion.

Let R^d denote the space of d-tuples with the usual norm $|\cdot|$ and inner product $\langle \cdot, \cdot \rangle$. Let G denote the class of functions from $[0, \infty)$ into R^d, to which f belongs only in case $f(0) = 0$ and f is continuous. Let N_x, for each x in $[0, \infty)$, denote the pseudonorm defined on G by $N_x(f) = \sup_{z \leq x} |f(z)|$. For each positve number T define the projection P_T on G by

$$[P_T f](t) = \begin{cases} f(t) & -r \leq t \leq T \\ f(T) & T \leq t \end{cases}$$

for each f in G and $0 \leq t$. Note that $\{P_T G, N_T\}$ is a complete normed linear space.

Let $\mathcal{B}$ denote the set of linear operators on G to which B belongs only in case for each $T > 0$ there is a number b such that

$$|[Bf](t) - [Bf](s)| \leq b \int_s^t N_x(f) \, dx$$

for each f in G and $0 \leq s \leq t \leq T$. Let $\mathcal{A}$ denote the set of linear operators on G to which A belongs only in case $A - I$ is in $\mathcal{B}$, where I is the identity on G.

If B is in $\mathcal{B}$ then $I - B$ is a reversible operator from G onto G and $(I - B)^{-1}$ is in $\mathcal{A}$. If A is in $\mathcal{A}$ then A is a reversible operator from G onto G and $I - A^{-1}$ is in $\mathcal{B}$.

Example. Assume W is the standard Wiener process on $[0, \infty)$. Let

$$Y(t) = [AW](t) = \int_0^t \frac{1}{t - u + 1} \, dW(u)$$

for $t \geq 0$, then A is in $\mathcal{A}$. For future reference,

$$R(s, t) = \begin{cases} \frac{1}{t-s} \ln \left[\frac{(s+1)(t-s+1)}{t+1} \right] & 0 \leq s < t \\ \frac{s}{s+1} & s = t \end{cases}$$

One might think of AW as the source solution of a wave equation with a point noise source and a point sensor. An d-dimensional example might involve d point noise sources and d point sensors.

Our problem is to find a feedback control improving the performance of such a system in a setting where A is unknown and only a discrete version of R [see Figure 2] is available. Further, due to model uncertainty the feedback operator obtained must improve the performance of all nearby systems. The nearby systems would include infinite dimensional perturbations of the system for which the control is designed.

Let k denote the identity function defined on $[0, \infty)$. Let G_H denote the subspace of functions in G which are Hellinger integrable with respect to k, i. e., f is in G_H only in case there is a number M such that

$$\sum_{p=1}^n |f(s_p) - f(s_{p-1})|^2 / (k(s_p) - k(s_{p-1})) = \sum_s |df|^2 / dk \leq M$$

for each finite increasing sequence $\{s_p\}_0^n$ in $[0, \infty)$. The least such number M is denoted by $\int_0^\infty |df|^2 / dk$. The inner product for G_H is defined by

$$Q_H(f, g) = \int_0^\infty \langle df, dg \rangle / dk$$

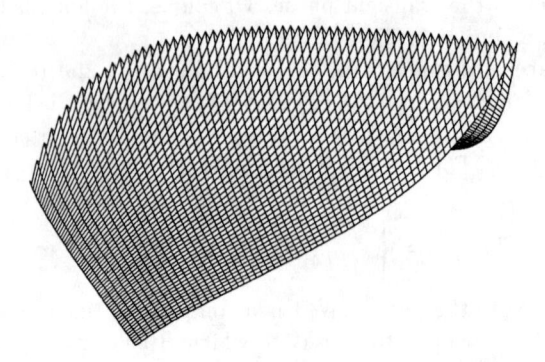

FIG. 2. *Exact covariance surface for the example.*

the limit through refinement of sums $\sum_s \langle df, dg \rangle / dk$.

The space $\{G_H, Q_H\}$ with inner product norm N_H is an RKH space with kernel K given by $K(s,t) = k(\min(s,t))$. Elements of $\mathcal{B}$ map G into G_H and elements of $\mathcal{A}$ map G_H onto G_H. From this point, we will only be concerned with restrictions of elements of $\mathcal{A}$ and $\mathcal{B}$ to G_H.

Assuming the process $Y_t, 0 \leq t$, with covariance function $R(s,t) = EY(s)Y(t)^*$ can be modelled by $Y = AW = W + BY$ we have the following two results.

THEOREM 2.1. *[5] For $0 \leq s \leq t$, $R(s,t) = [AA^*K(\cdot,t)](s)$, where A^* is the adjoint of A in $\{G_H, Q_H\}$.*

THEOREM 2.2. *[5] Let G_R denote the elements f of G_H such that $\int_0^\infty |d(I-B)f|^2/dk$ is finite and define $Q_R(f,g) = Q_H((I-B)f,(I-B)g)$ for each f and g in G_R. The space $\{G_R, Q_R\}$ is a complete Hilbert space with reproducing kernel R.*

If we introduce the matrix L defined by $L(s,t) = [A^*K(\cdot,t)](s) = ([AK(\cdot,s)](t))^T$ then $\langle R(s,t)c_1, c_2 \rangle = Q_H(L(\cdot,t)c_1, L(\cdot,s)c_2)$ for $0 \leq s,t$ and (c_1,c_2) in $R^d \times R^d$.

At this point, we will take everything to be real valued. This reduces the computational burden, but still leaves some interesting problems.

A class of polygonal functions, the K-polygonal functions, arises naturally in RKH spaces and can be used along with projection methods to develop finite dimensional approximations to the system operators [5]. Any function f on $[0,\infty)$ of the form $f(t) = \sum_{p=0}^n K(t,s_p)x_p$, where $s = \{s_p\}_0^n$ is an increasing sequence in $[0,\infty)$ and $\{x_p\}_0^n$ is a sequence of reals is called a K-polygonal function. The space of all K-polygonal functions determined by a fixed partition s of a finite interval $[0,T]$ is a closed linear subspace of G_H. We let Π_s denote the orthogonal projection of G_H onto this subspace.

THEOREM 2.3. *[5] The union of the finite dimensional subspaces $\Pi_s G_H$ is dense in $P_T G_H$ with respect to the inner product norm $N_H(\cdot) = Q_H(\cdot,\cdot)^{1/2}$.*

For f in G_H and increasing sequence $\{s_p\}_0^n$ in $[0,\infty)$ let $f_s = (f(s_0), f(s_1), \ldots, f(s_n))^T$. Similarly, let K_s denote the $(n+1) \times (n+1)$ matrix whose (p,q) element is given by $K_s(p,q) = K(s_p, s_q)$ for $0 \leq p,q \leq n$. If $\Pi_s f = \sum_{p=0}^n K(\cdot, s_p)x_p$ and $x = (x_0, x_1, \ldots, x_n)^T$

then $x = (K_s)^{-1} f_s$, where $(K_s)^{-1}$ denotes the pseudoinverse of K_s.

THEOREM 2.4. *[5] For each positive number T, f in G_H, C in $\mathcal{A} \bigcup \mathcal{B}$ and positve number c there is a partition s of $[0,T]$ such that if t refines s then $N_H(Cf - \Pi_t C \Pi_t f) < c$.*

Let LC_s denote the discrete matrix representation of C in $\mathcal{A} \bigcup \mathcal{B}$. Note that the first row and column of LC_s are both 0. If f is in G_H and $h = \Pi_s C \Pi_s f$ then $h_s = (LC_s)^T (K_s)^{-1} f_s$, where $(K_s)^{-1}$ denotes the pseudoinverse of K_s. We can avoid the complication of pseudoinverses by stripping off the first row and column of LC_s and K_s and compute with the $n \times n$ submatrices. We adopt this practice in what follows but now use LC_s and K_s to denote the $n \times n$ discretizations of L and K, respectively.

THEOREM 2.5. *Given B in $\mathcal{B}$, f in G_H, and positive numbers T and b, there is a partition s of $[0,T]$ such that if t refines s then*

$$(h_t - (I_t - (LB_t)^T (K_t)^{-1})^{-1} f_t)^T (K_t)^{-1} (h_t - (I_t - (LB_t)^T (K_t)^{-1})^{-1} f_t) \le b$$

where $h = (I - B)^{-1} f$.

LEMMA 2.6. *There is a partition $\{s_p\}_0^n$ of $[0,T]$ such that, for $p = 1, 2, \ldots, n$, $N_H((P_{s(p-1)} - P_{s(p)})B) < 1$.*

Proof. Let c be a number such that if $0 \le u \le v \le T$ and f is in G_H then

$$|[Bf](v) - [Bf](u)| \le c \int_u^v N_x(f) \, dx$$

Choose n so that $\frac{cT}{n} \sqrt{\frac{2q+1}{2}} < 1$ for $q = 1, 2, \cdots, n$. Let $s_q = \frac{qT}{n}$ for $q = 1, 2, \cdots, n$. If f is in G_H, $q = 1, 2, \cdots, n$ and t partitions $[s_{p-1}, s_p]$ then

$$\sum_p \frac{|[Bf](t_p) - [Bf](t_{p-1})|}{t_p - t_{p-1}} \le c^2 \sum_p N_{t(p)}(f)^2 (t_p - t_{p-1})$$
$$\le c^2 \sum_p N_H(P_{t(p)} f)^2 t_p (t_p - t_{p-1})$$
$$\le N_H(P_T f)^2 \frac{s(q)^2 - s(q-1)^2}{2}$$
$$= c^2 N_H(P_T f)^2 \frac{T^2}{2n^2} (2q+1)$$

Hence

$$N_H((P_{s(p)} - P_{s(p-1)}) Bf) \le \frac{c N_H(P_T f) T}{n} \sqrt{\frac{2q+1}{2}}$$
$$< N_H(f)$$

LEMMA 2.7. *For $p = 1, 2, \ldots, n$, let M_p denote the number set to which m belongs only in case there is an element g of G_H and a partition t of $[0,T]$ refining s (given by Lemma 2.6) such that*

1. $N_H(\Pi_t (I - B) \Pi_t g) \le 1$ and

2. $m = N_H(P_{s(p)} \Pi_t g)$.

Then M_p is bounded for each positive integer p.

Proof. Suppose $\sup M_1 = \infty$. If g is in G_H, t refines s and $N_H(\Pi_t (I - B) \Pi_t g) \le 1$ then

$$N_H(P_{s(1)} \Pi_t g) \le N_H(P_{s(1)} \Pi_t (I - B) \Pi_t g) + N_H(P_{s(1)} \Pi_t B \Pi_t g)$$
$$\le 1 + N_H(P_{s(1)} B) N_H(P_{s(1)} \Pi_t g)$$

and, when $N_H(P_{s(1)}\Pi_t g) \neq 0$

$$1 - \frac{1}{N_H(P_{s(1)}\Pi_t g)} \leq N_H(P_{s(1)}B)$$

Hence $1 \leq N_H(P_{s(1)}B)$, a contradiction, i.e., M_1 is bounded.

Suppose $p = 2, 3, \cdots, n$, $\sup M_p = \infty$ and M_{p-1} is bounded. If g is in G_H, t refines s and $N_H(\Pi_t(I - B)\Pi_t g) \leq 1$ then

$$\begin{aligned} N_H(P_{s(p)}\Pi_t g) &\leq N_H(P_{s(p)}\Pi_t(i - B)\Pi_t g) + N_H(P_{s(p)}\Pi_t B \Pi_t g) \\ &\leq 1 + N_H((P_{s(p)} - P_{s(p-1)})\Pi_t B \Pi_t g) \\ &\quad + N_H(P_{s(p-1)}\Pi_t g) \\ &\leq 1 + N_H((P_{s(p)} - P_{s(p-1)})\Pi_t B \Pi_t g) \\ &\quad + N_H(P_{s(p-1)}B) N_H(P_{s(p-1)}\Pi_t g) \end{aligned}$$

and if $N_H(P_{s(p)}\Pi_t g) \neq 0$ then

$$1 - \frac{1 + N_H(P_{s(p-1)}B) N_H(P_{s(p-1)}\Pi_t g)}{N_H(P_{s(p)}\Pi_t g)} \leq N_H((P_{s(p)} - P_{s(p-1)})B)$$

Hence $1 \leq N_H((P_{s(p)} - P_{s(p-1)})B)$, a contradiction. Therefore there is no such $p = 2, 3, \cdots, n$ and so $M - p$ is bounded for each positive integer p.

LEMMA 2.8. *There is a partition s of $[0, T]$ and a positive number c such that if t refines s then*

$$(K_t)^{-1}(K_t - LB_t)(K_t)^{-1}(K_t - (LB_t)^T)(K_t)^{-1} \geq c(K_t)^{-1}$$

or, equivalently,

$$(I_t - (LB_t)^T(K_t)^{-1})^T(I_t - (LB_t)^T(K_t)^{-1}) \geq c(K_t)^{-1}$$

Proof. There is a partition s of $[0, T]$ and a positive number b such that if g is in G_H, t refines s and $N_H(\Pi_t(I - B)\Pi_t g) \leq 1$ then $N_H(\Pi_t g) \leq b$. If g is in G_H and $N_H(\Pi_t(I - B)\Pi_t g) \neq 0$ then

$$N_H(\Pi_t(I - B)\Pi_t \frac{g}{N_H(\Pi_t(I - B)\Pi_t g)}) = 1$$

Hence

$$\frac{N_H(\Pi_t g)}{N_H(\Pi_t(I - B)\Pi_t g)} \leq b$$

or

$$\frac{1}{b} N_H(\Pi_t g) \leq N_H(\Pi_t(I - B)\Pi_t g)$$

Thus

$$\begin{aligned} &c(g_t)^T(K_t)^{-1} g_t \\ &\leq (g_t)^T(I_t - (LB_t)^T(K_t)^{-1})^T(K_t)^{-1}(I_t - (LB_t)^T(K_t)^{-1})(K_t)^{-1} g_t \end{aligned}$$

for all g_t, where $c = 1/b^2$, and so the conclusion follows.

Indication of proof for Theorem 2.5. Let s_1 and c be given by Lemma 2.8. If t refines s_1 then

$$\begin{aligned}
&((I_t - (LB_t)^T(K_t)^{-1})h_t - f_t)^T(K_t)^{-1}((I_t - (LB_t)^T(K_t)^{-1})h_t - f_t) \\
&= (h_t - (I_t - (LB_t)^T(K_t)^{-1}f_t)^T(I_t - (LB_t)^T(K_t)^T(K_t)^{-1} \cdot \\
&\quad (I_t - (LB_t)^T(K_t)(h_t - (I_t - (LB_t)^T(K_t)^{-1}f_t) \\
&\geq c(h_t - (I_t - (LB_t)^T(K_t)^{-1}f_t)^T(K_t)^{-1}(h_t - (I_t - (LB_t)^T(K_t)^{-1}f_t)
\end{aligned}$$

There is a partition s_2 of $[0,T]$ such that if t refines s_2 then $N_H(\Pi_t(I-B)\Pi_t h - \Pi_t f)^2 < cb$. Let s be a refinement of both s_1 and s_2. If t refines s then

$$\begin{aligned}
&(h_t - (I_t - (LB_t)^T(K_t)^{-1}f_t)^T(K_t)^{-1}(h_t - (I_t - (LB_t)^T(K_t)^{-1}f_t) \\
&\leq (1/c)((I_t - (LB_t)^T(K_t)^{-1})h_t - f_t)^T(K_t)^{-1}((I_t - (LB_t)^T(K_t)^{-1})h_t - f_t) \\
&< b
\end{aligned}$$

3 A Control Problem

Assume the uncontrolled system is of the form $Y = ACW$, A in $\mathcal{A}$ and C in $\mathcal{A} \bigcup \mathcal{B}$. The control problem we address is the following.

- Minimize:
$$J(\alpha) = EY(\infty)^2 + cE\{[B_\alpha Y](\infty)\}^2$$

- Subject to:
$$Y = AC(W + B_\alpha Y)$$

where the parameterized feedback B_α is of the form

$$[B_\alpha Y](t) = \alpha_1 \int_0^t (t-u)Y(u)\,du + \alpha_2 \int_0^t Y(u)\,du$$

Note that the system is more general than hypothesized in Theorem 2.2. However, the methods seem to work for a variety of examples. We will present here results for the example where the more restricted hypothesis holds.

We can write the controlled process $Y = AC(W + B_\alpha Y)$ as

$$Y = A_\alpha W = (I - ACB_\alpha)^{-1}ACW$$

We now proceed heuristically to rewrite the functional for the optimal control problem using the discrete versions of the system operators. If we denote the matrix representations of AC, B_α and A_α by LAC, LB_α and LA_α, respectively, then for each increasing sequence $\{t_p\}_0^n$ in $[0, \infty)$

$$(LA_\alpha)_t \sim (LAC)_t(K_t)^{-1}(I_t - (LB_\alpha)_t(K_t)^{-1}(LAC)_t(K_t)^{-1})^{-1}K_t$$

where I_t denotes the $n \times n$ identity matrix. Let R_t denote the discrete covariance of the uncontrolled process and R^U and K^U the upper Cholesky factors [2] of R_t and K_t, respectively. Then $(LAC)_t \sim (K^U)^T R^U$ and so

$$\begin{aligned}
(LA_\alpha)_t &\sim (LAC)_t(I_t - (K_t)^{-1}(LB_\alpha)_t(K_t)^{-1}(LAC)_t)^{-1} \\
&\sim (K^U)^T R^U (I_t - (K_t)^{-1}(LB_\alpha)_t(K^U)^{-1} R^U)^{-1}
\end{aligned}$$

c	α_1	α_2
1	0.0426	-0.1883
0.5	0.0847	-0.3803

TABLE 1

Produced by the Algorithm using the Exact R_t with $n = 64$

Let $C_\alpha = (K_t)^{-1}(LB_\alpha)_t(K^U)^{-1}$ and R_{tt} the discrete covariance of the controlled process. Then

$$\begin{aligned} R_{tt} &\sim ((LA_\alpha)_t)^T (K_t)^{-1}(LA_\alpha)_t \\ &= (R^U(I_t - C_\alpha R^U)^{-1})^T R^U(I_t - C_\alpha R^U)^{-1} \end{aligned}$$

i.e., $R^U(I_t - C_\alpha R^U)^{-1}$ is the upper Cholesky factor of R_{tt}. Furthermore, $R_{tt}(n,n) \sim |R^U(I_t - C_\alpha R^U)^{-1}(\cdot, n)|^2$.

Similarly, given the controlled process $Y = A_\alpha W = (I - ACB_\alpha)^{-1}ACW$ we have $B_\alpha Y = B_\alpha A_\alpha W$. The discrete matrix representation of $B_\alpha A_\alpha$ is $(LA_\alpha)_t (K_t)^{-1}(LB_\alpha)_t$. Hence

$$\begin{aligned} E\{[B_\alpha Y](T)\}^2 &\sim [((LA_\alpha)_t(K_t)^{-1}(LB_\alpha)_t)^T(K_t)^{-1}(LA_\alpha)_t(K_t)^{-1}(LB_\alpha)_t](n,n) \\ &= [((LB_\alpha)_t)^T(K_t)^{-1}R_{tt}(K_t)^{-1}(LB_\alpha)_t](n,n) \\ &\sim |R^U(I_t - C_\alpha R^U)^{-1}(K_t)^{-1}(LB_\alpha)_t(\cdot, n)|^2 \end{aligned}$$

The discrete optimization problem becomes
- Minimize: $J(\alpha) = x'x + cy'y$

- Subject to:
$$\begin{aligned} x &= R^U(I_t - C_\alpha R^U)^{-1}(\cdot, n) \\ y &= R^U(I_t - C_\alpha R^U)^{-1}(K_t)^{-1}(LB_\alpha)_t(\cdot, n) \end{aligned}$$

We can obtain x in two steps.
 1. z is the solution, for fixed α, of the triangular system $(I_t - C_\alpha R^U)z = e_n$ where e_n is the unit vector with nth component one and all others zero, and

 2. $x = R^U z$.

Similarly, for y.
 1. z is the solution, for fixed α, of the triangular system $(I_t - C_\alpha R^U)z = (K_t)^{-1}(LB_\alpha)_t(\cdot, n)$, and

 2. $y = R^U z$.

The point is that evaluating the discrete version of the functional only requires two solves, both of triangular systems. Thus the discrete optimization problem can be resolved with a general purpose optimization algorithm. The $Matlab\copyright$ command $fmins$ was used for the examples in this paper [see Table 1 and Figure 3].

In applications the covariance surface is not known exactly but must be estimated from data [see Figure 4]. This means the feedback produced by the algorithm is optimal for some system other than the true system. However, we still want the designed feedback control to improve the performance of the true system,i.e., we want the designed control to

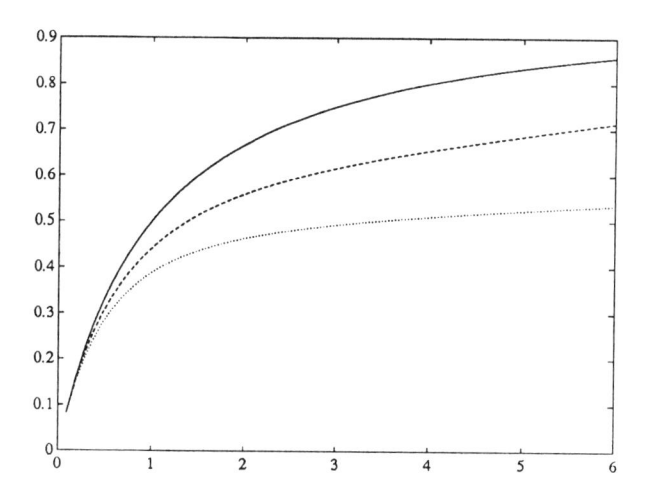

FIG. 3. *Uncontrolled System Variance (—) and Controlled with $c = 1$ (- - -) and $c = 0.5$ (· · ·)*

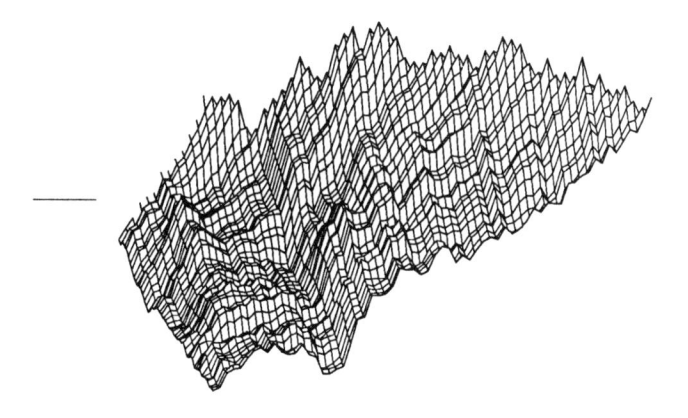

FIG. 4. *Estimated Covariance Surface Based on 125 Samples*

be robust. The problem is to match the robustness of the designed control with the quality of the estimate for the covariance surface.

A measure of robustness of the feedback control B_α is the largest number δ such that if $|R^U - \bar{R}^U|_2 \leq \delta$ then $|\bar{R}^U(I_t - C_\alpha \bar{R}^U)^{-1}| \leq |\bar{R}^U|$. The definition employs two different matrix norms

$$|X| = \sup_q (\sum_{p=1}^n |X_{pq}|^2)^{1/2}$$

and

$$|X|_2 = \sup_{u \neq 0} \frac{|Xu|_2}{|u|_2}$$

where $|u|_2 = \langle u, u \rangle^{1/2}$. If R_t is a discrete covariance with upper Cholesky factor R^U then $|R^U|^2 = R_t(n,n)$. Of course, $|\cdot|_2$ is an operator norm with the multiplicative property $|XY|_2 \leq |X|_2 |Y|_2$ and $|X| \leq |X|_2$.

THEOREM 3.1. *Assume* $|R^U| - |R^U(I_t - C_\alpha R^U)^{-1}| > 0$. *Let*

$$\begin{aligned}
a &= |R^U| - |R^U(I_t - C_\alpha R^U)^{-1}| \\
\eta &= |(I_t - C_\alpha R^U)^{-1}|_2 \\
b &= \eta(1 + \eta |R^U|_2 |C_\alpha|_2) \\
c &= \eta |C_\alpha|_2 \\
\Delta &= (ac + b + 1)^2 - 4ac
\end{aligned}$$

If $|\bar{R}^U - R^U| < \delta = \frac{ac+b+1-\sqrt{\Delta}}{2c}$ *then* $|\bar{R}^U(I_t - C_\alpha \bar{R}^U)^{-1}| \leq |\bar{R}^U|$.

Proof. We start by noting that

$$\begin{aligned}
\Delta &= (ac)^2 + b^2 + 1 + 2acb + 2ac + b - 4ac \\
&= (ac-1)^2 + b^2 + 2acb + b \\
&\geq 0
\end{aligned}$$

Also $\delta < 1/c$. If otherwise then

$$\begin{aligned}
\frac{ac+b+1-\sqrt{\Delta}}{2} &\geq 1 \\
(ac+b+1) - 2 &\geq \sqrt{\Delta} \\
(ac+b+1)^2 - 4(ac+b+1) + 4 &\geq (ac+b+1)^2 - 4ac \\
-4b &\geq 0
\end{aligned}$$

a contradiction. Hence

$$\begin{aligned}
|(I_t - C_\alpha \bar{R}^U)^{-1}|_2 &= |(I_t - (I_t - C_\alpha R^U)^{-1} C_\alpha (\bar{R}^U - R^U))^{-1} (I_t - C_\alpha R^U)^{-1}|_2 \\
&\leq \frac{\eta}{1 - |C_\alpha|_2 |\bar{R}^U - R^U|_2 \eta}
\end{aligned}$$

[3]. Since
$$|\bar{R}^U(I_t - C_\alpha \bar{R}^U)^{-1} - R^U(I_t - C_\alpha R^U)^{-1}|$$
$$\leq |\bar{R}^U - R^U|_2 |(I_t - C_\alpha \bar{R}^U)^{-1}|_2 + |R^U|_2 |(I_t - C_\alpha \bar{R}^U)^{-1} - (I_t - C_\alpha R^U)^{-1}|_2$$
$$\leq |\bar{R}^U - R^U|_2 |(I_t - C_\alpha \bar{R}^U)^{-1}|_2 + |R^U|_2 |C_\alpha|_2 |\bar{R}^U - R^U|_2 \eta |(I_t - C_\alpha \bar{R}^U)^{-1}|_2$$

c	α_1	α_2	δ
1	0.0364	-0.1555	0.4785
0.5	0.0694	-0.3048	0.4578

TABLE 2

125 Samples, $n = 8$: $|R_{exact} - R_{data}|_2 = 0.2279$

c	α_1	α_2	δ
1	0.0396	-0.1660	0.4790
0.5	0.0774	-0.3304	0.4580

TABLE 3

1000 Samples, $n = 8$: $|R_{exact} - R_{data}|_2 = 0.0670$

$$\leq |\bar{R}^U - R^U|_2 \{1 + |R^U|_2 |C_\alpha|_2 \eta\} \frac{\eta}{1 - |C_\alpha|_2 |\bar{R}^U - R^U|_2 \eta}$$

$$= |\bar{R}^U - R^U|_2 \frac{b}{1 - c|\bar{R}^U - R^U|_2}$$

we have

$$0 \leq \frac{a - (ac + b + 1)|\bar{R}^U - R^U|_2 + c|\bar{R}^U - R^U|_2^2}{1 - c|\bar{R}^U - R^U|_2}$$

$$= a - |\bar{R}^U - R^U|_2 - |\bar{R}^U - R^U|_2 \frac{b}{1 - c|\bar{R}^U - R^U|_2}$$

$$\leq |R^U| - |\bar{R}^U - R^U|_2 - |\bar{R}^U(I_t - C_\alpha \bar{R}^U)^{-1} - R^U(I_t - C_\alpha R^U)^{-1}|_2 -$$
$$|R^U(I_t - C_\alpha R^U)^{-1}|$$

$$\leq |\bar{R}^U| - |\bar{R}^U(I_t - C_\alpha \bar{R}^U)^{-1}|$$

We can begin to appreciate the balance required between the robustness of the designed control as measured by δ and the quality of the estimate of R_t by running the algorithm for two estimates based on different sample sizes [see Tables 2 and 3]. The differences in α values with Table 1 can be accounted for by the smaller n and the errors in the estimates.

4 Conclusions

We conclude by pointing out what has been accomplished and listing some questions that still have to be resolved.

1. Reproducing kernel Hilbert space methods have been applied to a class of optimal control problems for linear hereditary systems represented by system covariances.

2. With feasible feedbacks limited to a parameterized set of operators, the continuous time optimization problem has an RKH space discretization which is resolvable with general purpose optimization algorithms.

3. The RKH space setting which is associated with the covariance function representation of the system seems to provide a comprehensive set of tools for control design.

4. The relationship between control robustness and the quality of the estimates for the system covariance needs to be explored.

5. The measure of controller robustness suggested by the last theorem seems too conservative.

6. Everything needs to be extended to the steady state control problem.

References

[1] N. ARONSZAJN, *Theory of reproducing kernels*, Am.Math.Soc.Trans., 68 (1950), pp. 337–404.
[2] G. H. GOLUB AND C. F. VAN LOAN, *Matrix Computations*, The Johns Hopkins University Press, Baltimore, 1983.
[3] R. A. HORN AND C. R. JOHNSON, *Matrix Analysis*, Cambridge University Press, Cambridge, 1985.
[4] E. PARZEN, *An approach to time series analysis*, Ann.Math.Stat., 32 (1961), pp. 951–989.
[5] J. A. RENEKE, R. E. FENNELL, AND R. B. MINTON, *Structured Hereditary Systems*, Marcel Dekker, Inc., New York, 1987.
[6] R. SHOURESHI AND D. WORMLEY, *NSF/EPRI workshop on intelligent control systems*, tech. rep., National Science Foundation, 1990. Report of a workshop held in October, 1990 at Palo Alto, CA sponsored jointly by the National Science Foundation and the Electric Power Research Institute.